ADAPTIVE CONTROL OF SYSTEMS WITH ACTUATOR AND SENSOR NONLINEARITIES

Adaptive and Learning Systems for Signal Processing, Communications, and Control

Editor: Simon Haykin

Werbos / THE ROOTS OF BACKPROPAGATION: From Ordered Derivatives to Neural Networks and Political Forecasting

Krstić, Kanellakopoulos, and Kokotović / NONLINEAR AND ADAPTIVE CONTROL DESIGN

Nikias and Shao / SIGNAL PROCESSING WITH ALPHA-STABLE DISTRIBUTIONS AND APPLICATIONS

Diamantaras and Kung / PRINCIPAL COMPONENT NEURAL NETWORKS: THEORY AND APPLICATIONS

Tao and Kokotović / ADAPTIVE CONTROL OF SYSTEMS WITH ACTUATOR AND SENSOR NONLINEARITIES

ADAPTIVE CONTROL OF SYSTEMS WITH ACTUATOR AND SENSOR NONLINEARITIES

Gang Tao
University of Virginia

Petar V. Kokotović
University of California, Santa Barbara

A Wiley-Interscience Publication
JOHN WILEY & SONS, INC.
NEW YORK / CHICHESTER / BRISBANE / TORONTO / SINGAPORE

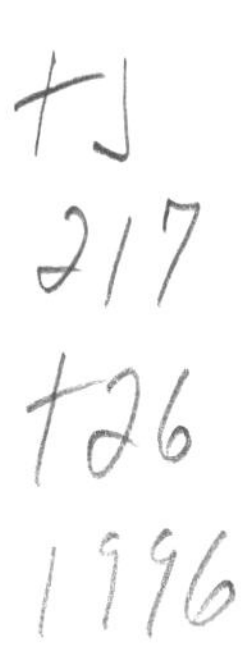

Library of Congress Cataloging in Publication Data

Tao, Gang.

Adaptive control of systems with actuator and sensor nonlinearities / Gang Tao, Petar V. Kokotović.

p. cm. — (Adaptive and learning systems for signal processing, communications, and control)

"A Wiley-Interscience Publication."

Includes bibliographical references and index.

ISBN 0-471-15654-X (cloth : alk. paper)

1. Adaptive control systems. 2. Nonlinear control theory.
I. Kokotović, Petar V. II. Title. III. Series.
TJ217.T36 1996
629.8′36—dc20 96-4666

Printed in the United States of America

10 9 8 7 6 5 4 3 2

To
Darrel, Doug, Jim, Vladimir
and other friends at Ford

Contents

Preface

Imperfections of system components, especially those of actuators and sensors, are among the factors that severely limit the performance of feedback control loops, the vital parts of industrial automation, consumer electronics, and defense and transportation systems. Most often, a critical imperfection is a nonlinearity which is poorly known, increases with wear and tear, and varies from component to component. Components without such imperfections are costly to manufacture, and their maintenance usually requires specialized personnel.

It is appealing to think of more intelligent approaches to increase the accuracy achievable with imperfect, but sturdy and inexpensive, components. Can the control system, after a period of learning or adaptation, recognize the imperfection and compensate for its harmful effects? With such adaptive controllers, the component specifications could be greatly relaxed, their cost reduced, and their reliability increased.

This book points to a direction in which this goal can be achieved for some of the most common component imperfections: *dead-zone, backlash*, and *hysteresis*. These "hard" nonlinearities are ubiquitous in a wide variety of components: mechanical, hydraulic, pneumatic, magnetic, piezoelectric, etc. They often serve as aggregate representations of more complex microscopic phenomena: friction, viscosity, elasticity, etc. While the "hard" nonlinearities have all but disappeared from the academic texts, they have become more common in engineering practice, because feedback controls have entered many new areas of applications. In particular, control systems have contributed to recent dramatic increases in fuel efficiency, drivability, and safety of passenger cars. Such successful applications show that it is more rational to improve performance with control algorithms than with more expensive mechanical components. The adaptive inverse methodology presented in this book is aimed in this direction.

The nonlinearities in this book are approximated by piecewise linear characteristics. A difficulty with such characteristics is that they have break-points, so that they are not differentiable. Existing adaptive control techniques are not applicable to such nonlinearities. However, a major advantage of the piecewise linear characteristics is that they admit linear parametrization with unknown break-point and slope parameters. This property is crucial for effective design and implementation of robust adaptive control, one of the main subjects of this book. The unifying theme of the book is its *adaptive inverse* approach. Not only are the nonlinear characteristics linear in their parameters, but so are their inverses, which, in the case of dead-zone and backlash, are discontinuous. While the inverses of the actuator nonlinearities are explicit, those of the sensors have a more complicated implicit form. The essence of the adaptive inverse approach is that, upon an adaptation transient, the inverse cancels the effects of the unknown nonlinear characteristic. In this way a significant improvement of accuracy and performance is achieved with inexpensive components. In other words, *the adaptation in the controller has "removed" the imperfection of the component.*

All the results in this book are new and have evolved from the recent journal papers of the authors. The style of presentation is aimed at an audience of practicing engineers and graduate students in electrical, mechanical, chemical, aeronautical, and computer engineering departments, as well as those pursuing interdisciplinary studies such as biomedical engineering. The assumed background is a standard course in control theory, while the required knowledge of model reference adaptive control is concisely presented in Appendix A.

Our interest in the problem of adaptive compensation of "hard" nonlinearities was ignited by Jim Winkelman and Doug Rhode, our colleagues at Ford Motor Company. Several years ago, they presented to us and Darrel Recker (then a Ph.D. student, now a researcher at Ford) a problem with a hydraulic valve dead-zone in an automotive suspension system. The dead-zone's purpose was to prevent the leakage and maintain the height when the car was parked and the engine was turned off. However, when the suspension was active, the effect of the dead-zone was harmful. In his Ph.D. thesis, Darrel Recker addressed the problem of using adaptation to remove the harmful effects of the dead-zone. His successful algorithms and experiments have encouraged us to pursue a broader investigation in this direction. We acknowledge with gratitude the pioneering contributions of Darrel Recker and his cooperation in this project. We also greatly benefited from the experience of Doug Rhode and Jim Winkelman. For our understanding of hydraulic components we are indebted to Vladimir Kokotović, also at Ford. For many years we have been inspired and helped by Petros Ioannou, University of Southern California, without whose vast knowledge of robust adaptive control a project like this would not have been possible. With their patience and understanding, our wives, Lanlin and Anna, generously contributed to the writing of this book.

Our research summarized in this book was not only initiated at, but also financially supported by, the Ford Motor Company. It was also supported by the National Science Foundation grant ECS-9203491 and RIA ECS-9307545 and by the Air Force Office of Scientific Research grant F-49620-92-J-0495.

GANG TAO
Charlottesville, Virginia

PETAR KOKOTOVIĆ
Santa Barbara, California

Symbols

$B(\cdot)$	Backlash
$BI(\cdot)$	Exact backlash inverse
$\widehat{BI}(\cdot)$	Backlash inverse estimate
$d_b(t)$	Backlash uncertainty
$d_d(t)$	Dead-zone uncertainty
$d_h(t)$	Hysteresis uncertainty
$d_N(t)$	Input or output nonlinearity uncertainty
$d_{N_i}(t)$	Input nonlinearity uncertainty
$d_{N_o}(t)$	Output nonlinearity uncertainty
$DZ(\cdot)$	Dead-zone
$DI(\cdot)$	Exact dead-zone inverse
$\widehat{DI}(\cdot)$	Dead-zone inverse estimate
$e(t)$	Tracking error $e(t) = y(t) - y_m(t)$
$G(D)$	Continuous-time or discrete-time linear plant
$G(s)$	Continuous-time linear plant
$G(z)$	Discrete-time linear plant
$H(\cdot)$	Hysteresis
$HI(\cdot)$	Exact hysteresis inverse
$\widehat{HI}(\cdot)$	Hysteresis inverse estimate
l^2, L^2	Signal spaces $l^2, L^2 : \{x(t) : \|x(\cdot)\|_2 < \infty\}$
l^∞, L^∞	Signal spaces $l^\infty, L^\infty : \{x(t) : \|x(\cdot)\|_\infty < \infty\}$
$N(\cdot)$	Input (actuator) or output (sensor) nonlinearity
$N_i(\cdot)$	Input (actuator) nonlinearity
$N_o(\cdot)$	Output (sensor) nonlinearity
$NI(\cdot)$	Exact input or output nonlinearity inverse
$NI_i(\cdot)$	Exact input nonlinearity inverse
$NI_o(\cdot)$	Exact output nonlinearity inverse
$\widehat{NI}(\cdot)$	Input or output nonlinearity inverse estimate
$\widehat{NI}_i(\cdot)$	Input nonlinearity inverse estimate
$\widehat{NI}_o(\cdot)$	Output nonlinearity inverse estimate
$r(t)$	Input to a reference model

$u(t)$	Input to $G(D)$, $G(s)$, or $G(z)$
$u_d(t)$	Input to $NI_i(\cdot)$ or $\widehat{NI}_i(\cdot)$
$v(t)$	Input to $N_i(\cdot)$
$y(t)$	Plant output
$y_m(t)$	Reference output
$W_m(D)$	Reference model transfer function
$z(t)$	Input to $N_o(\cdot)$
$\bar{z}(t)$, $\hat{z}(t)$	Outputs of $NI_o(\cdot)$, $\widehat{NI}_o(\cdot)$ with input $y(t)$
$z_m(t)$, $\hat{z}_m(t)$	Outputs of $NI_o(\cdot)$, $\widehat{NI}_o(\cdot)$ with input $y_m(t)$
$\epsilon(t)$	Estimation error for $G(D)$ unknown
$\epsilon_N(t)$	Estimation error for $G(D)$ known
$\chi[X]$	Indicator function of the event X
$\omega(t)$	Regressor vector of θ^* or θ
$\omega_1(t)$	Regressor vector of θ_1^* or θ_1
$\omega_2(t)$	Regressor vector of θ_2^* or θ_2
$\omega_b(t)$	Backlash inverse regressor vector
$\omega_d(t)$	Dead-zone inverse regressor vector
$\omega_h(t)$	Hysteresis inverse regressor vector
$\omega_N(t)$	Regressor vector of $\widehat{NI}(\cdot)$
$\omega_{N_i}(t)$	Regressor vector of $\widehat{NI}_i(\cdot)$
$\omega_{N_o}(t)$	Regressor vector of $\widehat{NI}_o(\cdot)$
θ^*, θ	Overall system parameter and its estimate
$\tilde{\theta}$	Parameter error $\tilde{\theta} = \theta - \theta^*$
θ_1^*, θ_3^*	Feedforward control parameters
θ_1, θ_3	Estimates of θ_1^*, θ_3^*
θ_2^*, θ_{20}^*	Feedback control parameters
θ_2, θ_{20}	Estimates of θ_2^*, θ_{20}^*
θ_4^*, θ_4	Kroneker product of $-\theta_1^*$ and $\theta_{N_i}^*$, and its estimate
θ_5^*, θ_5	Kroneker product of θ_2^* and $\theta_{N_o}^*$, and its estimate
θ_{50}^*, θ_{50}	Product of θ_{20}^* and $\theta_{N_o}^*$, and its estimate
θ_b^*, θ_b	Backlash parameter vector and its estimate
θ_d^*, θ_d	Dead-zone parameter vector and its estimate
θ_h^*, θ_h	Hysteresis parameter vector and its estimate
θ_N^*, θ_N	Parameter vector of $N(\cdot)$, and its estimate
$\tilde{\theta}_N$	Parameter vector of $\tilde{\theta}_N = \theta_N - \theta_N^*$
$\theta_{N_i}^*$, θ_{N_i}	Parameter vector of $N_i(\cdot)$, and its estimate
$\theta_{N_o}^*$, θ_{N_o}	Parameter vector of $N_o(\cdot)$, and its estimate
$\\|x\\|_2$	l^2 vector norm of $x \in R^n : \\|x\\|_2 = \sqrt{x_1^2 + \cdots + x_n^2}$
$\\|x\\|_\infty$	l^∞ vector norm of $x \in R^n : \\|x\\|_\infty = \max_{1 \le i \le n} \mid x_i \mid$
$\\|x(\cdot)\\|_2$	l^2 signal norm of $x(t) : \\|x(\cdot)\\|_2 = \sqrt{\sum_{t=0}^{\infty} \\|x(t)\\|_2}$
	or L^2 signal norm of $x(t) : \\|x(\cdot)\\|_2 = \sqrt{\int_0^\infty \\|x(t)\\|_2 dt}$
$\\|x(\cdot)\\|_\infty$	l^∞ or L^∞ signal norm of $x(t) : \\|x(\cdot)\\|_\infty = \sup_{t \ge 0} \\|x(t)\\|_\infty$

Chapter 1

Introduction

The immense growth of real-time computing power has led to new solutions of long-standing engineering problems. One of them is the goal of achieving high accuracy and performance with imprecise, but sturdy and inexpensive, components. Applicability of a noisy sensor can be dramatically broadened by adaptive filtering and other forms of signal conditioning. Actuator performance can similarly be improved by redesigning the control signal.

Actuator and sensor nonlinearities are among the key factors limiting both static and dynamic performance of feedback control systems. Harmful effects of backlash in gears are well known. Backlash prevents accurate positioning and may lead to chattering and limit-cycle-type instabilities. This increases wear and tear of the gears, which, in turn, increases backlash. An escape from this difficulty are various designs of anti-backlash gears. However, their cost is high and they introduce extra weight and friction.

An example of backlash naturally introduces the topic of this book: Can real-time computations be employed to remove the harmful effects of backlash in a nonmechanical fashion, without cumbersome and expensive anti-backlash gears?

Backlash is the simplest form of hysteresis. Examples of more complex hysteresis are magnetic and piezoelectric phenomena in solenoid actuated valves and micro-motion scanners. As a rule, materials with low hysteresis are costly. Can inexpensive magnetic and piezoelectric materials be used, but with their hysteresis effects removed by real-time computations? An analogous question can be raised for actuators and sensors with "dead-zones"—that is, with insensitivity to small magnitude signals.

In this book we answer the above questions for piecewise linear models of unknown dead-zone, backlash, and hysteresis. While our results can be extended to piecewise linear approximations of more complex nonlinearities, our dead-zone, backlash, and hysteresis characteristics are representative for a large majority of components used in industry, defense, bioengineering, transportation, scientific instrumentation, etc. The literature on this subject is rich: [1], [5], [12], [31], [52], [66], [67], [83], [125], [129], [130], [133], [134].

1.1 Adaptive Inverse Control

Nonlinearities in actuators and sensors are usually poorly known, increase with wear and tear, and, in mass production, change from component to component. Our piecewise linear dead-zone, backlash, and hysteresis characteristics allow us to express their uncertainties in terms of unknown slopes of linear segments and their break-points. For all three characteristics we have achieved linear parametrization—that is, linear dependence on unknown parameters.

The main idea of our approach is to cancel the harmful effects of actuator and sensor nonlinearities by implementing their inverses inside the controller. Our first concern is that the inverses, possibly discontinuous, indeed exist. Our second concern is that they also be linearly parametrized as linear functions of the unknown parameters.

Having passed these two hurdles, we look for an adaptive implementation of the linearly parametrized inverses, continuously adjusted by adaptive update laws. If they would converge to the true inverses of the unknown nonlinearities, the ideal goal of canceling the nonlinear effects would be achieved: The control loops would perform as if their inexpensive actuators and sensors were perfect.

To see if our adaptive inverses can be implemented using the existing adaptive control theory, let us take a brief look at the adaptive control literature. The fundamental issues of adaptive control—controller structures, parametrization, error models, parameter update laws, and stability—have been extensively studied for linear plants. Systematic design procedures have been developed for model reference adaptive control [2], [3], [8], [19], [23], [28], [30], [39], [61], [63], [70], [71], [75], [80]–[82], [85], [96], [99], [104], [110], [114], self-tuning regulators [3], [19], pole placement control [20], [27], [30], [39], [56], multivariable adaptive control [7], [15], [16], [17], [21], [22], [29], [30], [48], [98], [100], [103], [109], [128], adaptive control of partially known systems [3], [4], [10], [13], [80], and nonlinear adaptive controllers for linear systems [58], [59]. Other references of interest are [6], [11], [14], [18], [26], [62], [64], [69], [72]–[76], [88], [127], [131], [135]–[137].

Robustness of adaptive control schemes with respect to modeling errors such as unmodeled dynamics, parameter variations, and external disturbances has been a problem of continuing interest [25], [32], [34], [36], [41], [51], [84],

[94], [95], [101], [102], [111], [112], [126]. The robustness issue is important because an adaptive control scheme designed under some ideal modeling assumptions may become unstable in the presence of modeling errors. Nonrobust behavior of an "idealized" adaptive controller can be eliminated by modifying the adaptive update law. Among the robust adaptive update laws which guarantee boundedness are dead-zone modifications [54], [68], [86], σ-modification [35], [37], switching σ-modification [33], [42], [43], parameter projection modifications [53], [55], [77], ϵ-modification [79], normalization techniques [42], [43], [54], [68], [90], and some of their extensions to multivariable adaptive control [107], [108], [113]. Unifications of these robust designs can be found in [3], [30], [39], [80], [96]. The key features of robust adaptive laws are the uniform boundedness of parameter estimates and the smallness in a mean sense of the estimation error [38]. Applied together, they are sufficient for the closed-loop signal boundedness in the presence of modeling errors.

Adaptive control of nonlinear systems has recently made significant advances. Adaptive controllers have been developed for special classes of nonlinear systems such as *pure-feedback systems* and *feedback linearizable systems* [9], [44]–[47], [49], [57], [59], [60], [65], [78], [89], [97].

The above adaptive nonlinear results employ integrator backstepping which requires that the nonlinearities be continuously differentiable a sufficient number of times. These results cannot be used to design adaptive inverses of the piecewise linear characteristics which are not differentiable. For this reason, the development of adaptive inverses for unknown dead-zone, backlash, and hysteresis has followed an independent path.

The study of adaptive control for systems with unknown dead-zones at the input was initiated in [91], where an adaptive scheme was proposed with full state measurement. The more realistic situations when only a single output is available for measurement were addressed using a direct control method in [115], [116], [119], [120], [123] for actuator nonlinearities and in [117], [121], [122] for sensor nonlinearities. In [118], [124], unified adaptive inverse control designs for systems with unknown actuator or sensor nonsmooth nonlinearities were presented. In [92], [93], an indirect adaptive inverse method was used for control systems with unknown actuator dead-zones. In [105], reduced-order adaptive inverse controllers were developed using some partial knowledge of systems. In [106], adaptive control schemes were designed for systems with both actuator and sensor nonsmooth nonlinearities.

In this book we unify and expand our adaptive inverse methodology introduced in [105], [106], and [115]–[124].

1.2 Book Outline

The development of the adaptive inverse control approach begins in Chapter 2, where we present examples and parametrized models for the dead-zone, backlash, and hysteresis nonlinearities. Each of these nonlinearities is described by a multiregion equation characterized by a set of parameters and a set of indicator functions. These descriptions are given for both continuous-time and discrete-time operations. We consider three types of plants to be controlled: plants with input (actuator) nonlinearities, plants with output (sensor) nonlinearities, and plants with both input and output nonlinearities.

Chapter 3 develops inverses for the input (actuator) nonlinearities: dead-zone, backlash, and hysteresis. The output signal of the inverse is applied to the actuator input. The inverse is also characterized by a set of parameters and a set of indicator functions, and can be designed in either continuous or discrete time. While the output of a continuous-time backlash or hysteresis inverse depends on the derivative of its input signal, this is not the case with their discrete-time counterparts which are therefore easier to implement. If implemented with the true parameters the inverses cancel the effects of the nonlinearities. If implemented with parameter estimates they result in a control error which can be expressed in two parts, one of which is parametrizable. The unparametrizable part due to the nonsmoothness of the nonlinearities is treated as an unknown disturbance. It is crucial that this disturbance is always bounded. The control error expression is instrumental in the development of an adaptive law which updates the estimates of the unknown parameters.

Continuous-time and discrete-time control schemes with fixed inverses of input nonlinearities are designed in Chapter 4. Fixed inverses are implemented with fixed estimates of the unknown parameters. When the remaining linear part of the plant is known, the corresponding part of the controller is linear. If not, this part of the controller is adaptive. In either case this part of the controller is designed as if the nonlinearity is absent—that is, as if its inverse is exact. Even when the inverse is implemented with inaccurate parameter estimates and the nonlinear effects are not cancelled, the resulting controller still ensures closed-loop signal boundedness. As shown by simulations, when the parameter error is small, such a controller appreciably improves the tracking performance. However, when the parameter error in the fixed inverse is large, the tracking error between the plant output and the reference output may be unacceptably large for some applications. This motivates our adaptive inverse designs presented in the remaining chapters of the book.

In Chapter 5, three illustrative examples are used to introduce the adaptive inverse control approach: a continuous-time adaptive dead-zone inverse, a continuous-time adaptive backlash inverse, and a discrete-time adaptive backlash inverse. In each of the examples the control scheme is made up of an adaptive inverse and a linear controller which, in the absence of the input (actuator) nonlinearity, would achieve asymptotic output tracking. The presence of the dead-zone or backlash nonlinearity makes it impossible for the linear controller alone to achieve output tracking. However, an adaptive inverse with the same linear controller ensures asymptotic output tracking. The backlash examples also explain how the derivative of the signal from the linear controller structure is used to implement a continuous-time adaptive backlash inverse and why such a signal is not needed for discrete-time implementation.

Continuous-time adaptive inverse control for plants with unknown input nonlinearities is developed in Chapter 6. The output of the nonlinear part is not accessible for either control or measurement. Two adaptive control problems are investigated: when only the nonlinear part of the plant is unknown, and when the whole plant is unknown. The adaptive inverse controller consists of a linear controller structure with an adaptive dead-zone, backlash, or hysteresis inverse. When only the nonlinear part is unknown, the adaptive inverse controller contains an adaptive inverse and a fixed linear structure. When the whole plant is unknown, the adaptive inverse controller has a new adaptive linear structure which results in a linearly parametrized closed-loop system suitable for the development of an adaptive law. For both problems, the linear controller structure generates the input signal to the adaptive inverse whose output is then applied to the plant with an unknown input nonlinearity. The adaptive inverse parameters are updated by adaptive laws with modifications for robustness with respect to a bounded "disturbance"—an unparametrizable error due to the nonsmoothness of the dead-zone, backlash, and hysteresis nonlinearities. In addition, parameter projection is employed to ensure that the parameter estimates stay in a prespecified region. Extensive simulation results show significant improvements of the system tracking performance.

Chapter 7 presents discrete-time adaptive inverse control designs—the counterparts of those in Chapter 6. The designs are simpler because discrete-time backlash and hysteresis inverses do not employ the derivative of their input signals. Proofs of the closed-loop signal boundedness are given for dead-zone, backlash, and hysteresis cases. Simulations also show significant system performance improvements.

Sensor nonlinearities—that is, the nonlinearities at the plant output—are introduced in Chapter 8 where fixed discrete-time inverse control schemes are designed. A continuous-time design is practically infeasible and only the discrete-time designs are presented. The difficulty with the output (sensor) dead-zone or backlash is that they make the sensor input unobservable from its output. Despite this difficulty, when both the linear part and the nonlinear part are known, an inverse control scheme is developed for achieving the output tracking of a given reference signal. When the nonlinear part is unknown, a fixed inverse, combined with a fixed or adaptive linear controller structure, ensures the closed-loop signal boundedness in the presence of small estimation errors in the slopes of the nonlinear characteristics.

In Chapter 9 an adaptive inverse controller for a plant with an unknown sensor nonlinearity is designed which consists of a linear part and two adaptive inverses: one to invert the plant output and the other to invert a given reference output. Two designs are presented: one for a known linear part and the other for an unknown linear part. There are two choices for a plant output inverse for the first design: one with an explicit adaptive inverse and the other with an implicit adaptive inverse. The second design employs an implicit adaptive inverse. Such adaptive inverse structures ensure linearly parametrized error models based on which robust adaptive laws are designed to update the parameter estimates. In simulations the designed adaptive output inverse controllers also lead to significant improvements of the tracking performance.

Chapter 10 addresses some design issues in adaptive control of partially known plants. Modified model reference adaptive control schemes are designed for linear plants whose stable zero dynamics or pole dynamics are partially known. Adaptive inverse controllers for partially known plants significantly reduce the adaptive system order.

Chapter 11 presents discrete-time inverse control schemes for plants with both input (actuator) and output (sensor) nonlinearities. First, an output matching controller is developed which ensures output tracking of a given reference signal when the nonlinearities are known. Such a controller uses an inverse for the input nonlinearity and two inverses for the output nonlinearity. Then, two adaptive inverse control schemes are presented for plants with a known linear part and unknown nonlinearities. Finally, an adaptive inverse control scheme is proposed for plants with unknown linear and nonlinear parts. Each of these adaptive inverse controllers results in a linearly parametrized estimation error which serves as a basis for an adaptive update law design.

Appendix A presents a unified model reference adaptive control theory for the reader to review. Appendix B contains the proof of signal boundedness of the closed-loop control systems with continuous-time adaptive inverses for actuator nonlinearities, while Appendix C has its discrete-time counterpart. The signal boundedness proof for sensor nonlinearities is in Appendix D.

Chapter 2

Dead-Zone, Backlash, and Hysteresis

Dead-zone, backlash, and hysteresis are often called "common" or "typical" nonlinearities because they are ubiquitous in mechanical, hydraulic, magnetic, and other types of system components. In most cases they are treated as imperfections of component characteristics. Dead-zone is a static "memoryless" nonlinearity which describes the component's insensitivity to small signals. In addition to this type of insensitivity, backlash and hysteresis also include delays and are, in fact, dynamic.

There are also applications in which nonlinear characteristics are intentionally introduced, such as in heating-cooling systems, where a dead-zone is needed to prevent simultaneous heating and cooling. In some hydraulic valves an intentional dead-zone prevents flow of fluid when the valve is inactive.

Whether intentional or not, dead-zone, backlash, and hysteresis always have undesirable effects on feedback loop dynamics and control system performance. This will be illustrated by a few examples in this chapter. Through these examples we will also motivate our models of dead-zone, backlash, and hysteresis. It should be clear from the outset that these models are only simplified descriptions of complex physical phenomena. This is particularly true for hysteresis phenomena which have been described by models ranging from a single hysteresis loop to those including integrodifferential operators.

In our pragmatic approach we have chosen simple *piecewise linear models* with a sufficient number of adjustable parameters which provide significant flexibility in matching real situations. This flexibility will be exploited for our *adaptive inverse control* of systems with dead-zone, backlash, and hysteresis nonlinearities.

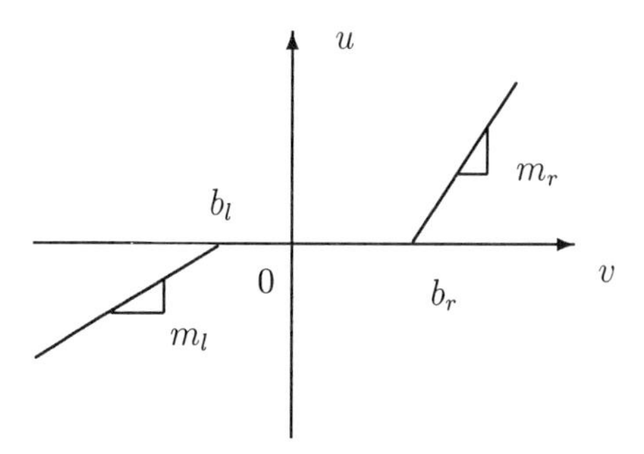

Figure 2.1: Dead-zone.

2.1 Dead-Zone

Dead-zone is a static input-output relationship which for a range of input values gives no output. Once the output appears, the slope between the input and the output is constant. A graphical representation of the dead-zone is shown in Figure 2.1, where v is the input and u is the output. In general, neither the *break-points* $b_r \geq 0$, $b_l \leq 0$ nor the *slopes* $m_r > 0$, $m_l > 0$ are equal. There is no loss of generality in assuming that the zero input point is inside the dead-zone because this can always be achieved with a redefinition of the input v.

The simple dead-zone model appears in numerous studies of a wide variety of phenomena, not limited to man-made systems. We briefly describe four typical examples, starting with a bioengineering application.

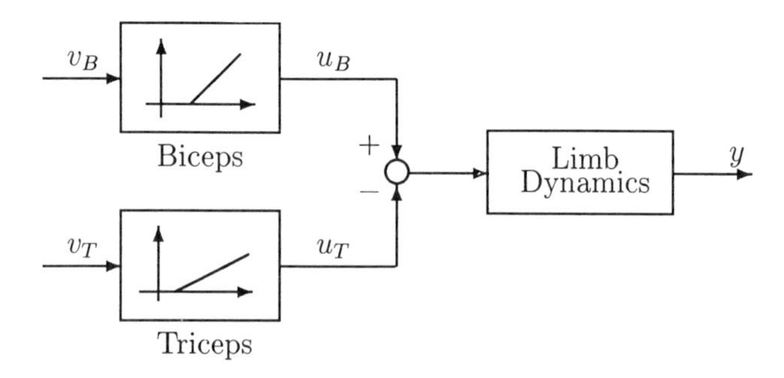

Figure 2.2: Dead-zones in the upper-limb model of [1].

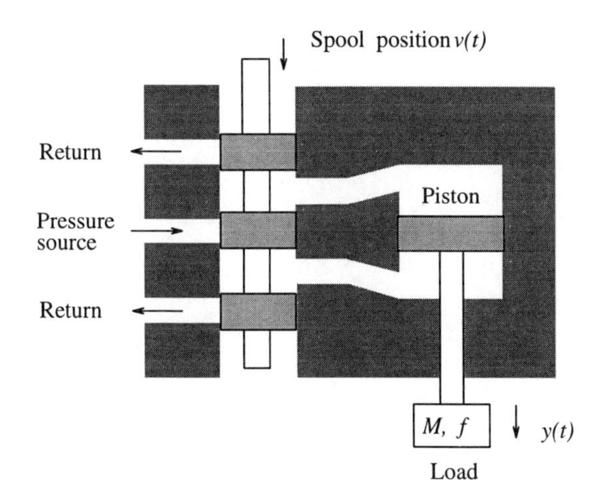

Figure 2.3: Dead-zone in a servo-valve.

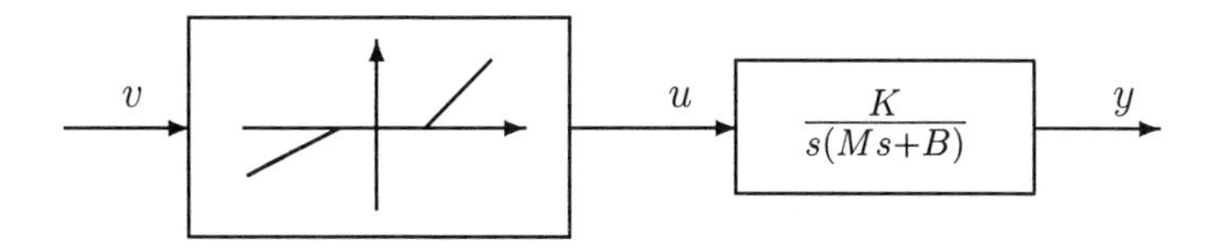

Figure 2.4: Block diagram of the servo-valve in Figure 2.3.

2.1.1 Upper-Limb Model

In functional neuromuscular stimulation a controlled electrical stimulus v is applied to the intact nerve in an attempt to replace upper motor neuron control which may be lost through cerebral stroke, brain injury, tumor, or spinal cessation. In [1] this approach has been applied to the stimulation of the upper limb, concentrating on elbow flexion/extension. Two dead-zone models are employed to represent the biceps and triceps nonlinear "gains" appearing at the input of the limb dynamics block in Figure 2.2. A similar model was employed in [5], [31] to adaptively control the knee joint of the paraplegics.

2.1.2 Servo-Valve

A common example from industrial applications is the servo-valve in Figure 2.3. Its spool occludes the orifice with some overlap so that for a range of

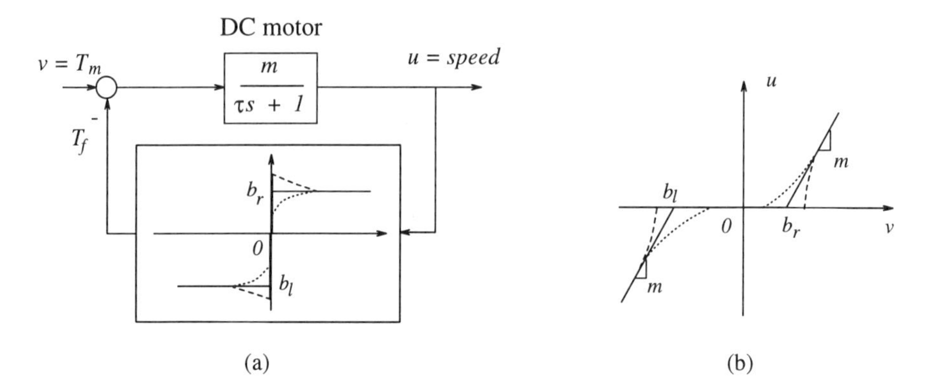

Figure 2.5: Dead-zone caused by friction.

spool positions v there is no fluid flow u. This overlap prevents leakage losses which increase with wear and tear. Considering the spool position as the input v, and the load position y as the output, the hydraulic system in Figure 2.3 is represented in Figure 2.4 as a dead-zone block. It is located at the input of the linear dynamics with transfer function $G(s) = \frac{K}{Ms^2+B}$, where $K = \frac{Ak_x}{k_p}$, $B = f + \frac{A^2}{k_p}$, $k_x = \frac{\partial g}{\partial x}$, $k_p = \frac{\partial g}{\partial P}$, $g = g(x, P) =$ flow, $A =$ area of piston, $P =$ pressure, and $f =$ viscous friction.

2.1.3 DC Motor with Friction

A dead-zone effect is often caused by friction. In such applications the simple dead-zone model serves as an aggregate static approximation of more complex microscopic dynamic phenomena. Perhaps the most common example is a DC motor with Coulomb friction, represented in Figure 2.5(a). Considering motor torque T_m as the input, the transfer function in the forward path is a first-order lag with motor time constant τ. When this time constant is negligible, the low-frequency approximation of the feedback loop is given by the dead-zone in Figure 2.5(b). This approximation can be rigorously justified as a singular perturbation [50]. It is important to observe that the friction torque characteristic is responsible for the break-points b_r and b_l, while the feedforward gain m determines the slope.

2.1.4 Solenoid-Controlled Servo-Valve

Thus far we have considered dead-zone elements at the inputs of linear dynamic blocks. There are also applications when the dead-zone element is at the

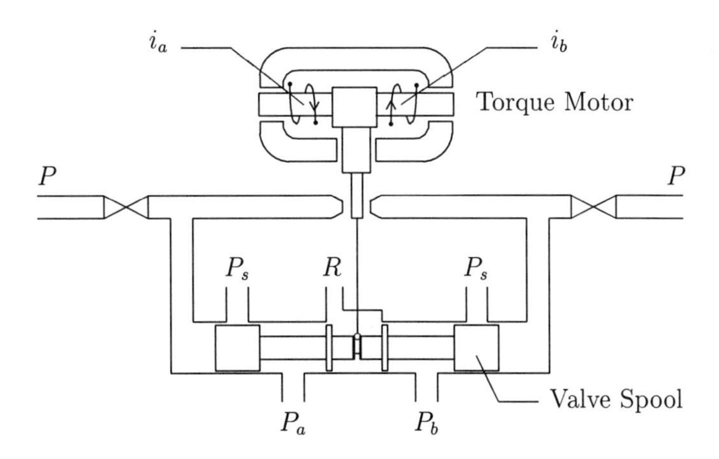

Figure 2.6: Dead-zone due to the overlap at the output of a two-stage servo-valve. (Courtesy of Darrel Recker.)

output. A two-stage servo-valve with position feedback shown in Figure 2.6 can serve as an example, considering (1) the difference of the electric currents i_a and i_b in the two coils of the torque motor solenoid as the input and (2) the difference of the fluid flows P_a and P_b as the output. As in Figure 2.4, the dead-zone characteristic is due to the spool overlap. The crucial difference is that now we consider the valve itself as the system under investigation. For this reason, the dead-zone now appears at the output of the linear block representing the dynamics of the torque motor, the nozzle-flapper stage, and the spool. Typically the linear block is a fourth-order transfer function [67].

2.1.5 Input and Output Dead-Zones

The two common types of dead-zone systems illustrated above are presented in Figures 2.7 and 2.8. The notation $G(D)$ means that we are operating in either continuous time, in which case D is the Laplace transform variable, or in discrete time, in which case D is the z-transform variable. The analytical expression of the dead-zone characteristic is

$$u(t) = DZ(v(t)) = \begin{cases} m_r(v(t) - b_r) & \text{if } v(t) \geq b_r \\ 0 & \text{if } b_l < v(t) < b_r \\ m_l(v(t) - b_l) & \text{if } v(t) \leq b_l. \end{cases} \tag{2.1}$$

This characteristic and notation will be used throughout the book, with t being either continuous or discrete time.

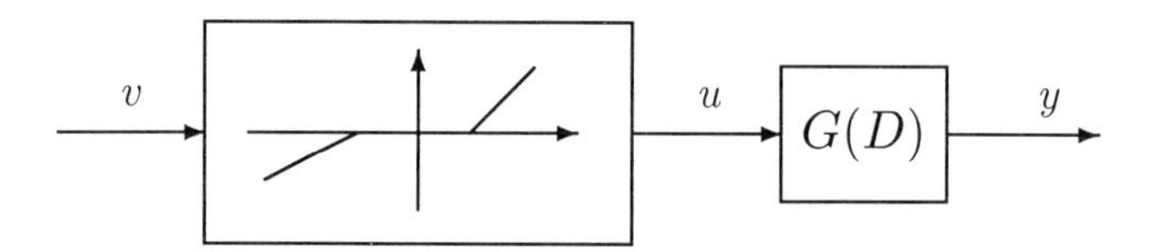

Figure 2.7: Plant with an input dead-zone.

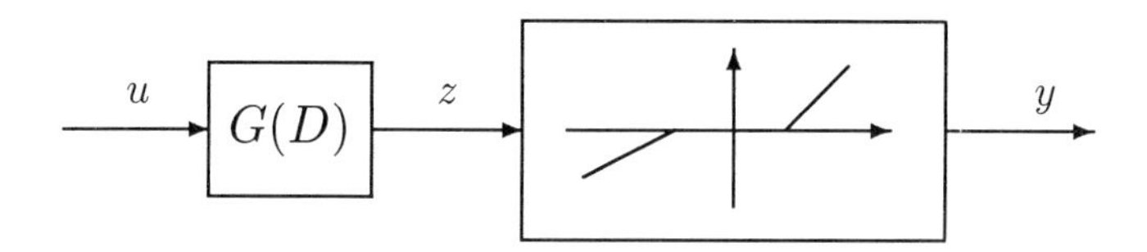

Figure 2.8: Plant with an output dead-zone.

2.2 Backlash

Backlash is a phenomenon which has haunted the constructors of control systems for more than 50 years: from the servomechanisms in the 1940s to the modern high precision robotic manipulators. The omnipresent concern for backlash is visible, for example, in the 1958 *Control Engineer's Handbook*, which also describes anti-backlash gear boxes. Catalogs of servo-lenses for active vision experiments are more current illustration. The price of backlash-compensated lenses is much higher than that of those with backlash. Typically the concept of backlash is associated with gear trains and similar mechanical couplings. Sometimes backlash can be used to approximate description of the delays in drives with elastic cables or in long pipes.

2.2.1 Backlash in Mechanical Connections

In contrast to the memoryless dead-zone, backlash has an element of memory and is, in a certain sense, dynamic. A widely accepted characteristic of backlash [130] is shown in Figure 2.9 where v is the input, u is the output, and $c_r > 0$ is the right "crossing," while $c_l < 0$ is the left "crossing."

This characteristic looks deceptively simple and, to appreciate its com-

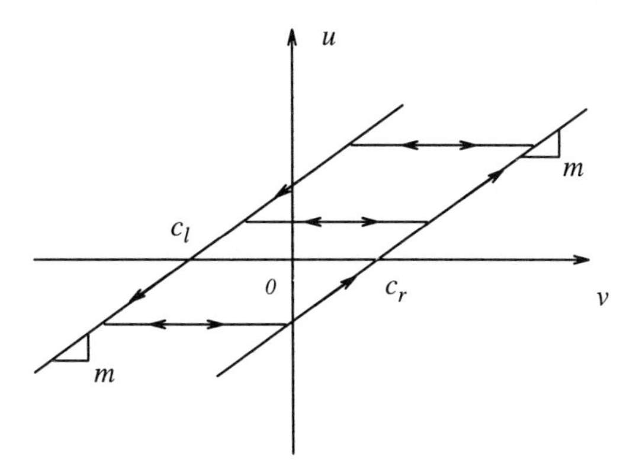

Figure 2.9: Backlash.

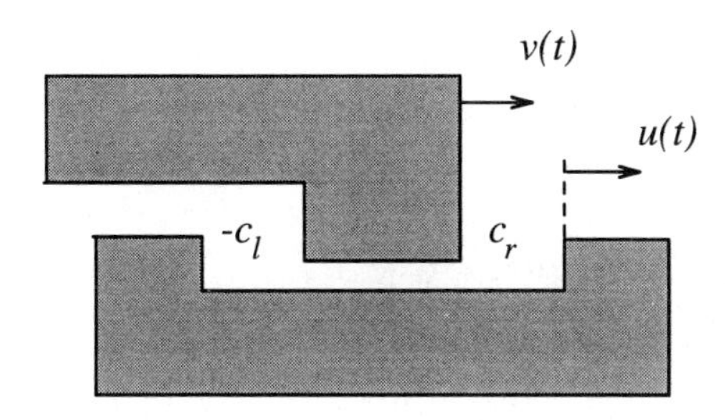

Figure 2.10: Schematic representation of backlash.

plexity, we will explain it step by step with the help of Figure 2.10, where an L-shaped object is driving a U-shaped object with the contact gap $c_r - c_l$. The input v is the position of the L-shaped object and the output u is the position of the U-shaped object. Both objects are inertialess and only their positions are of interest.

Let the positions in Figure 2.10 be $v = 0$ and $u = c_r$ and suppose that v begins to move to the right. When v reaches $v = c_r = u$ the contact is established and u follows v as described by the upward slope of the characteristic. If at some point v stops and begins to move to the left, u will remain motionless and "remember" its position. The motion during that period is represented by the horizontal transition to the left. It is easy to see from Figure 2.10 that the length of the horizontal segment is $c_r - c_l$, recalling that $c_l < 0$. At the end of this segment the contact is established with the left side of the U-shaped object. Then u begins to move to the left jointly with v—that is, along the downward slope on the characteristic. If at some point v again stops and then moves to the right, u will stop and wait until v traverses the whole segment $c_r - c_l$. The motion is again along a horizontal segment, this time to the right. Of course, v can change its direction before it traverses the segment $c_r - c_l$ and the next contact may be to the left. Or v can stop before it reaches a new contact.

2.2.2 Backlash Description

Let us summarize this narrative by concluding that the backlash characteristic $u(t) = B(v(t)) = B(m, c_r, c_l; v(t))$ is described by two parallel straight lines connected with horizontal line segments. The upward side is active when both $v(t)$ and $u(t)$ increase:

$$u(t) = m(v(t) - c_r),\ \dot{v}(t) > 0,\ \dot{u}(t) > 0.$$

The downward side is active when both $v(t)$ and $u(t)$ decrease:

$$u(t) = m(v(t) - c_l),\ \dot{v}(t) < 0,\ \dot{u}(t) < 0,$$

where $m > 0$, $c_l < c_r$ are constant parameters. The motion on any inner segment is characterized by $\dot{u}(t) = 0$.

A compact description of backlash $B(\cdot)$ is

$$\dot{u}(t) = \begin{cases} m\dot{v}(t) & \text{if } \dot{v}(t) > 0 \text{ and } u(t) = m(v(t) - c_r), \text{ or} \\ & \text{if } \dot{v}(t) < 0 \text{ and } u(t) = m(v(t) - c_l) \\ 0 & \text{otherwise.} \end{cases} \tag{2.2}$$

With the help of Figure 2.9 it is also easy to visualize a *discrete-time* version of the backlash model. All we need to imagine is that the positions $v(t)$ and $u(t)$ are observed at time instants $t = 0, 1, 2, \ldots$. Instead of using $\dot{v}(t)$ to characterize the direction of v, we compare $v(t)$ with v_l and v_r given by

$$v_l = \frac{u(t-1)}{m} + c_l,\ v_r = \frac{u(t-1)}{m} + c_r.$$

These values are the v-axis projections of the intersections of the two parallel lines of slope m with the horizontal inner segment containing $u(t-1)$. Hence, the discrete-time version of the backlash model is

$$u(t) = B(v(t)) = \begin{cases} m(v(t) - c_l) & \text{if } v(t) \leq v_l \\ m(v(t) - c_r) & \text{if } v(t) \geq v_r \\ u(t-1) & \text{if } v_l < v(t) < v_r. \end{cases} \tag{2.3}$$

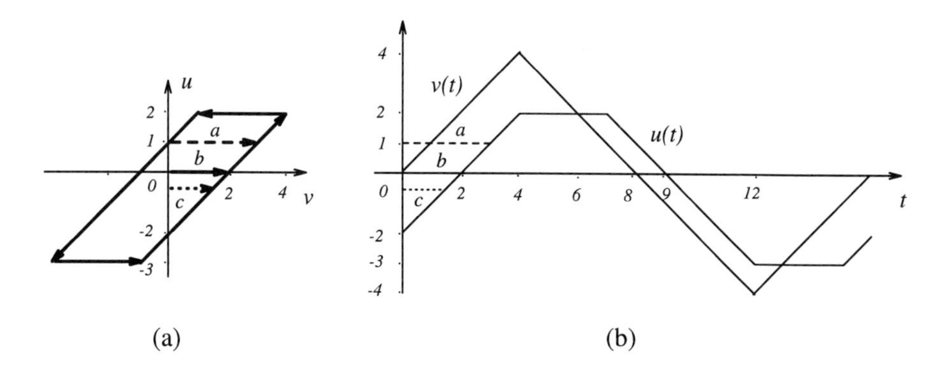

Figure 2.11: Backlash response to a sawtooth input.

2.2.3 Periodic Response

A further insight into the nature of backlash can be gained from the waveform of the output $u(t)$ when the input $v(t)$ is the sawtooth signal in Figure 2.11. For this illustration the backlash parameters are $m = 1$, $c_r = 2$, $c_l = -1$; that is, the two sides of the backlash characteristic are the straight lines $u(t) = v(t)+1$, $u(t) = v(t) - 2$. For each of the three initial conditions $u(0) = 1$, $u(0) = 0$, $u(0) = -0.5$, the output $u(t)$ is constant over an initial period:

$$u(t) = u(0) = 1, \quad \text{for } t \in [0, 3]$$

$$u(t) = u(0) = 0, \quad \text{for } t \in [0, 2]$$

$$u(t) = u(0) = -0.5, \quad \text{for } t \in [0, 1.5].$$

In Figure 2.11 these "transients" of $u(t)$ are marked by a, b, c. After such an initial "transient," $u(t)$ settles in its periodic steady state. It is important to observe that there are initializations with which the periodic steady state is reached without "transients," as is the case with $u(0) = -2$.

The periodic steady state of $u(t)$ reveals the two fundamental features of backlash. First, it introduces a phase delay. Second, it causes a loss of information by "chopping" the peaks of $v(t)$. These effects of backlash have been studied by the describing function method [83], [129].

2.2.4 Input and Output Backlash

We will consider systems in which a backlash element $B(\cdot)$ is either at the input or at the output, as shown in Figure 2.12 and Figure 2.13.

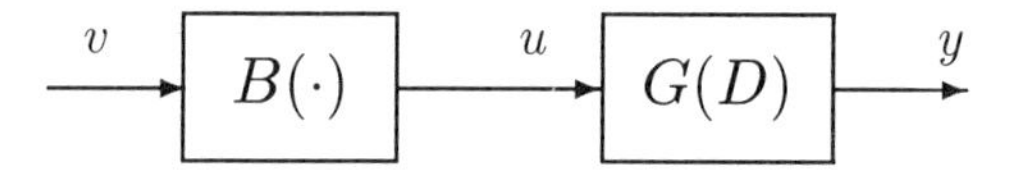

Figure 2.12: Plant with input backlash.

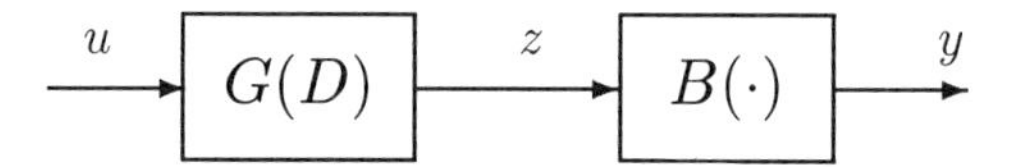

Figure 2.13: Plant with output backlash.

An *input backlash* example is shown in Figure 2.14, where the backlash is in the valve control mechanism and $G(s) = \frac{k}{s}$ is the transfer function relating the liquid level h with the difference between the controlled inflow u and the uncontrolled outflow d.

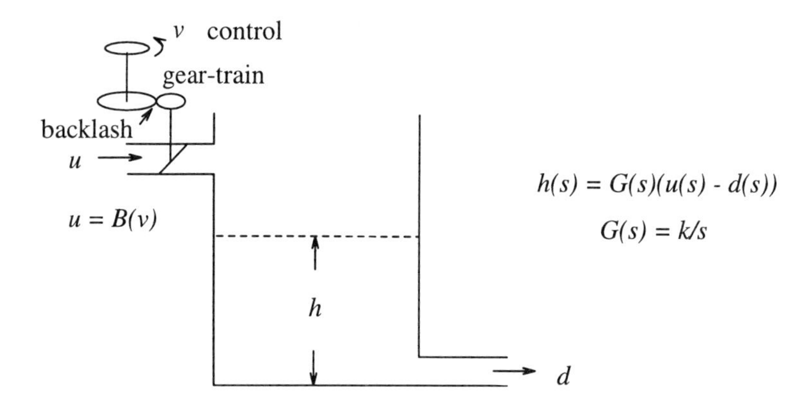

Figure 2.14: Backlash in the valve control mechanism of a liquid tank.

An example of the *output backlash* is a simple servo for positioning of a low inertia object (mirror, potentiometer, etc.), as shown in Figure 2.15. In this case, $G(D)$ is the transfer function of the amplifier/motor unit. Effects of gear-train backlash in such classical servomechanisms have been extensively studied by phase plane and describing function methods [130].

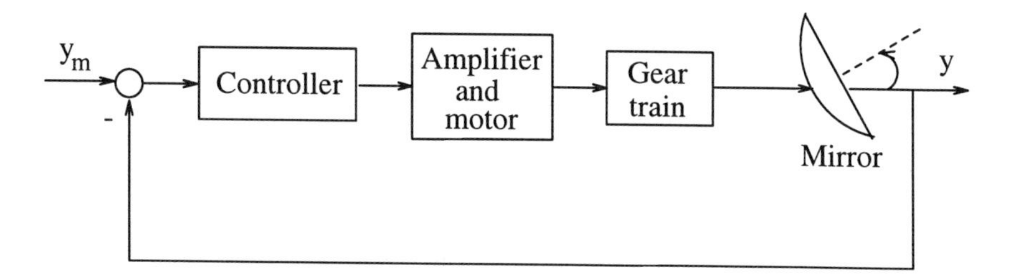

Figure 2.15: Output backlash in a positioning system.

2.3 Hysteresis

Hysteresis phenomena are even more numerous and diverse than those modeled by dead-zone and backlash characteristics. While ferromagnetic hysteresis is the best known type of hysteresis, similar characteristics are common in plastic, piezoelectric, and other materials. Extremely sophisticated mathematical models of hysteresis have appeared in the literature [12], [52], [66], [134]. There are also many detailed empirical models for specific components such as solenoids, piezoelectric actuators, and automobile tires (e.g., see [87], [133]).

2.3.1 Hysteresis Description

Although it is unrealistic to expect that a single hysteresis model can serve a vast variety of applications, the adaptive control approach helps to make a simple model more universal. Our piecewise linear model will be of this type. Its main hysteresis loop and two minor loops are shown in Figure 2.16.

This characteristic is considerably more complex than backlash, which has only three parameters: c_r, c_l, and m. The hysteresis characteristic in Figure 2.16 can be tuned by as many as eight parameters: four slopes m_l, m_r, m_b, m_t and four crossing parameters c_l, c_r, c_b, c_t, where the subscripts l, r, b, and t, respectively, indicate "left," "right," "bottom," and "top" sides of the hysteresis loop. The difference between the slopes m_b and m_t allows for the appearance of local loops.

A narrative description more involved than the backlash explanation can

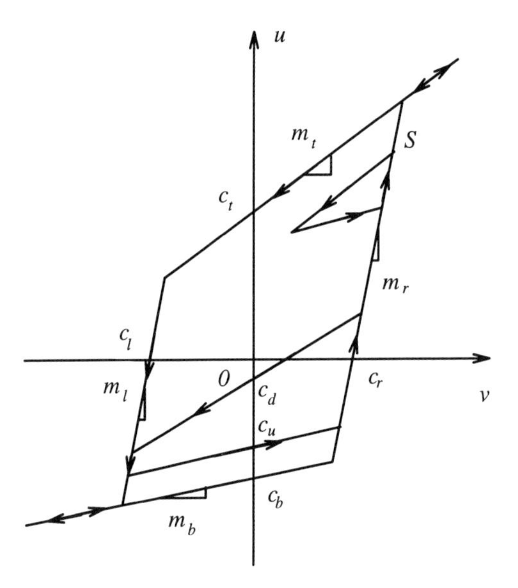

Figure 2.16: Hysteresis.

be given for the hysteresis characteristic. We will only give its summary by stating that the input $v(t)$ and output are confined to two half-lines, two line segments, and the quadrilateral formed by those half-lines and segments. The half-lines are described by

$$u(t) = m_t v(t) + c_t,\ v(t) > v_1 = \frac{c_t + m_l c_l}{m_l - m_t} \tag{2.4}$$

$$u(t) = m_b v(t) + c_b,\ v(t) < v_2 = \frac{c_b + m_r c_r}{m_r - m_b} \tag{2.5}$$

and the line segments are

$$u(t) = m_r(v(t) - c_r),\ v_2 \leq v(t) < v_3 = \frac{c_t + m_r c_r}{m_r - m_t} \tag{2.6}$$

$$u(t) = m_l(v(t) - c_l),\ \frac{c_b + m_l c_l}{m_l - m_b} = v_4 < v(t) \leq v_1, \tag{2.7}$$

where v_1, v_2, v_3, v_4 are the values of $v(t)$ at the upper-left, lower-right, upper-right, and lower-left corners of the quadrilateral.

Along the segments the time derivatives of $u(t)$, $v(t)$ are of constant sign, namely,

$$\dot{u}(t) > 0,\ \dot{v}(t) > 0 \text{ for } u(t) = m_r(v(t) - c_r) \tag{2.8}$$

$$\dot{u}(t) < 0,\ \dot{v}(t) < 0 \text{ for } u(t) = m_l(v(t) - c_l). \tag{2.9}$$

The hysteresis phenomena occur inside the loop formed by the half-lines (2.4) - (2.5) and the segments (2.6) - (2.7). Inside the hysteresis loop, the relationship between $u(t)$ and $v(t)$ is

$$u(t) = \begin{cases} m_t v(t) + c_d & \text{if } \dot{v}(t) < 0 \\ m_b v(t) + c_u & \text{if } \dot{v}(t) > 0, \end{cases} \tag{2.10}$$

where c_d and c_u are the intersections with the u-coordinate axis in Figure 2.16. Their values are restricted to the intervals $c_d \in (c_t, c_1)$ and $c_u \in (c_2, c_b)$ and whose ends c_1 and c_2 are determined from

$$c_1 = \begin{cases} (m_b - m_t)\frac{c_b + m_l c_l}{m_l - m_b} + c_b & \text{if } m_t < m_b \\ (m_b - m_t)\frac{c_b + m_r c_r}{m_r - m_b} + c_b & \text{if } m_t > m_b \\ c_b & \text{if } m_t = m_b \end{cases}$$

$$c_2 = \begin{cases} (m_t - m_b)\frac{c_t + m_r c_r}{m_r - m_t} + c_t, & \text{if } m_t > m_b \\ (m_t - m_b)\frac{c_t + m_l c_l}{m_l - m_t} + c_t & \text{if } m_t < m_b \\ c_t & \text{if } m_t = m_b. \end{cases}$$

It should be observed that c_d and c_u depend on the point where $\dot{v}(t)$ changes the sign and on the past trajectories of $(v(t), u(t))$.

The relationship (2.10) holds also for parts of the half-lines: When $m_t > m_b$, on the half-line (2.4) with $v(t) < v_3$, the signal motion is

$$u(t) = m_t v(t) + c_t \text{ for } \dot{v}(t) < 0$$

and when $m_t < m_b$, on the half-line (2.5) with $v_4 < v(t)$, the signal motion is

$$u(t) = m_b v(t) + c_b \text{ for } \dot{v}(t) > 0.$$

On other parts of these two half-lines the signs of $\dot{u}(t)$ and $\dot{v}(t)$ are not restricted:

$$u(t) = m_t v(t) + c_t,\ v(t) \geq v_3$$

$$u(t) = m_b v(t) + c_b,\ v(t) \leq v_4$$

$$u(t) = m_t v(t) + c_t,\ v_1 < v(t) < v_3 \text{ when } m_t < m_b$$

$$u(t) = m_b v(t) + c_b,\ v_4 < v(t) < v_2 \text{ when } m_t > m_b.$$

The hysteresis $u(t) = H(v(t))$ representing the motion of $u(t)$ and $v(t)$ on the half-lines (2.4) - (2.5) and the segments (2.6) - (2.7) and inside the hysteresis loop is fully described by

$$\dot{u}(t) = \begin{cases} m_t\dot{v}(t) & \text{if } v(t) \geq v_3 \text{ and } u(t) = m_t v(t) + c_t, \text{ or} \\ & \text{if } v_4 < v(t) < v_3,\ \dot{v}(t) < 0, \\ & u(t) = m_t v(t) + c_d, \\ & u(t) \neq m_l(v(t) - c_l) \text{ and} \\ & u(t) \neq m_b v(t) + c_b, \text{ or} \\ & \text{if } m_t < m_b,\ v_4 < v(t) < v_3, \\ & u(t) = m_b v(t) + c_b \text{ and } \dot{v}(t) < 0, \text{ or} \\ & \text{if } m_t < m_b,\ v_4 < v(t) < v_3, \\ & \dot{v}(t) > 0 \text{ and } u(t) = m_t v(t) + c_t \\ m_b\dot{v}(t) & \text{if } v(t) \leq v_4 \text{ and } u(t) = m_b v(t) + c_b, \text{ or} \\ & \text{if } v_4 < v(t) < v_3,\ \dot{v}(t) > 0, \\ & u(t) = m_b v(t) + c_u, \\ & u(t) \neq m_r(v(t) - c_r) \text{ and} \\ & u(t) \neq m_t v(t) + c_t, \text{ or} \\ & \text{if } m_t > m_b,\ v_4 < v(t) < v_3, \\ & \dot{v}(t) > 0 \text{ and } u(t) = m_t v(t) + c_t, \text{ or} \\ & \text{if } m_t > m_b,\ v_4 < v(t) < v_3, \\ & \dot{v}(t) < 0 \text{ and } u(t) = m_b v(t) + c_b \\ m_r\dot{v}(t) & \text{if } v_4 < v(t) < v_3,\ \dot{v}(t) > 0 \text{ and} \\ & u(t) = m_r(v(t) - c_r) \\ m_l\dot{v}(t) & \text{if } v_4 < v(t) < v_3,\ \dot{v}(t) < 0 \text{ and} \\ & u(t) = m_l(v(t) - c_l) \\ 0 & \text{if } \dot{v}(t) = 0. \end{cases} \tag{2.11}$$

These expressions indicate that hysteresis is a complex nonlinear dynamic system defined by piecewise linear relationships between the input $v(t)$, output $u(t)$, and their time derivatives $\dot{v}(t)$, $\dot{u}(t)$.

For a *discrete-time* form of the hysteresis model we proceed as in the backlash case. We replace the time derivatives by comparing $v(t)$ with v_d, v_u such that $v_4 \leq v_d \leq v_u \leq v_3$:

$$v_d = \frac{m_l c_l + c_d}{m_l - m_t}, \quad v_u = \frac{m_r c_r + c_u}{m_r - m_b}$$

$$c_d = u(t-1) - m_t v(t-1), \quad c_u = u(t-1) - m_b v(t-1).$$

Then the discrete-time hysteresis characteristic $u(t) = H(v(t))$ is

$$u(t) = \begin{cases} u(t-1) & \text{if } v(t) = v(t-1) \\ m_t v(t) + c_t & \text{if } v(t) \geq v_3, \text{ or} \\ & \text{if } m_t < m_b, \\ & u(t-1) = m_t v(t-1) + c_t \text{ and} \\ & v(t-1) < v(t) < v_3 \\ m_b v(t) + c_b & \text{if } v(t) \leq v_4, \text{ or} \\ & \text{if } m_t > m_b, \\ & u(t-1) = m_b v(t-1) + c_b \text{ and} \\ & v_4 < v(t) < v(t-1) \\ m_t v(t) + c_d & \text{if } v_d < v(t) < v(t-1) \\ m_b v(t) + c_u & \text{if } v(t-1) < v(t) < v_u \\ m_l (v(t) - c_l) & \text{if } v_d \geq v(t) \geq v_4 \\ m_r (v(t) - c_r) & \text{if } v_u \leq v(t) \leq v_3. \end{cases} \tag{2.12}$$

We will consider systems in which such hysteresis elements are at either the *input* or the *output* of a linear block.

2.3.2 Magnetic Suspension with Hysteresis

Typical examples of control systems with input hysteresis are magnetic suspensions and bearings. An oversimplified schematic representation of a magnetic suspension systems is shown in Figure 2.17. The position of an iron ball is detected by a light source L and a photocell P and compared with a desired reference r. The error signal $y - r$ is fed to the controller which generates the control signal – the electromagnet current I.

The magnetic force acting upon the iron ball is a nonlinear function of the ball position y and the magnetic flux ϕ. For small displacements the nonlinearity in y can be linearized. Then the remaining nonlinearity is the ferromagnetic hysteresis characteristic $\phi(I)$. To cast this system in the form of a hysteresis block at the input of linear dynamics, we consider the current I as the signal $v(t)$ and the magnetic force F acting upon the iron ball as the signal $u(t)$. To hold the ball at some desired position $y = r$ the required force is u_s. The amount of current v needed to generate this force depends on the operating point on the hysteresis characteristic. Suppose that the operating point is the upper corner S of the minor loop in Figure 2.16. If from that point the current $v(t)$ is first reduced and then increased to the same value, the motion is along a minor loop.

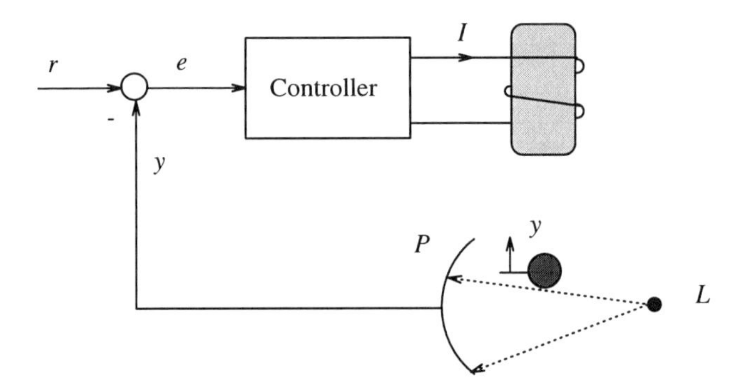

Figure 2.17: A magnetic suspension with solenoid hysteresis.

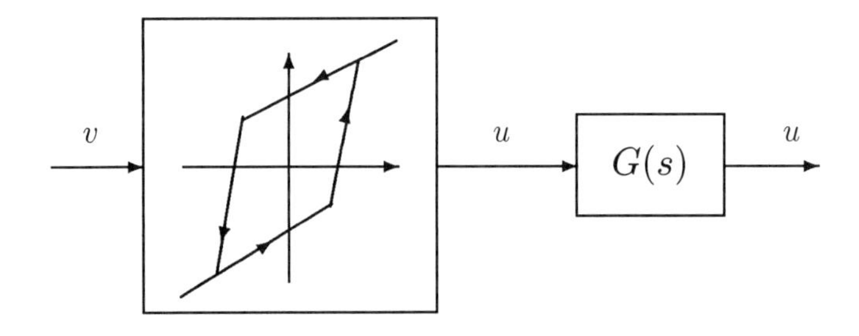

Figure 2.18: Plant with input hysteresis.

The input-output model of the magnetic suspension system from the current I to the ball position y can be represented by the block diagram in Figure 2.18 where under certain simplifying assumptions the transfer function $G(s)$ has two poles: $G(s) = \frac{K}{(s-p_1)(s-p_2)}$. One of the poles, say p_1, is necessarily unstable, $p_1 > 0$, because the magnetic force decays with the distance and the gravitation force is constant.

2.3.3 Hysteresis in Piezoelectric Positioning

Among the "smart materials" used for micromotion control and high-accuracy positioning, the most common are piezoelectric ceramic materials. Under the action of an electric field they expand, and this expansion can be employed for micropositioning. However, the expansion is not a linear function of the strength of the applied electric field—that is, of the input voltage. As presented in manufacturer's catalog [87], the dependence of the expansion on the voltage clearly exhibits a hysteresis phenomenon. The maximum width of the

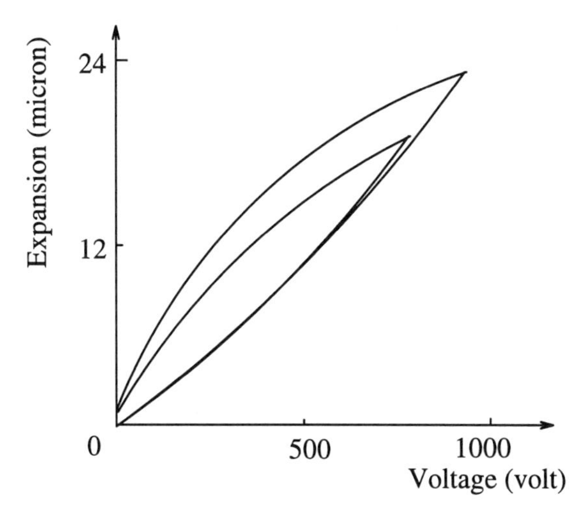

Figure 2.19: Characteristics of a piezoelectric actuator at two peak voltages.

hysteresis can be as large as 15% of the path covered [87]; see Figure 2.19. Such large hysteresis effects make it impossible to achieve high positioning accuracy without additional feedback compensation.

2.3.4 Hysteresis in Brakes

Disk brakes are becoming important actuators in advance automotive control systems which improve safety, drivability, and the overall performance of passenger cars and trucks. A common air disk (ADB-1560) for trucks has an input-output force characteristic similar to that in Figure 2.20. In a feedback control loop, this large hysteresis would limit the achievable dynamic performance. Therefore, for high dynamic performance the effect of hysteresis must be compensated.

2.4 Combined Nonlinearities

Thus far we have presented physical examples of dead-zone, backlash, and hysteresis appearing individually. More complex types of nonsmooth nonlinearities, which are also common in applications, can often be interpreted as combinations of these three elementary nonlinear characteristics. Our piecewise linear approximation, with the slopes and the break-points as the adjustable parameters, is flexible enough to practically handle such combined nonlinearities. As an example, we show in Figure 2.21 a combination of a

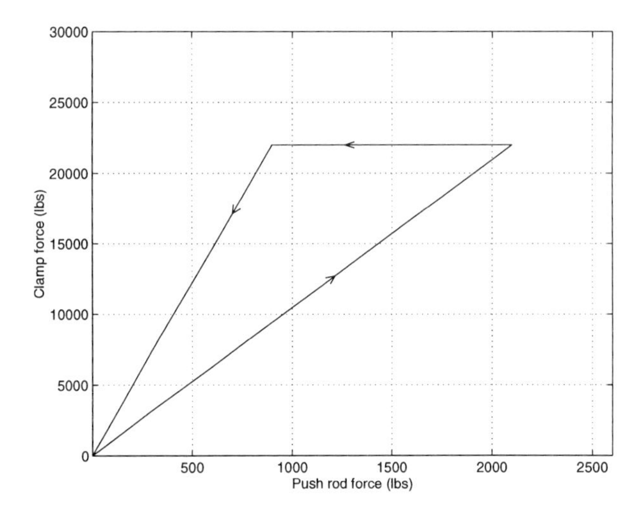

Figure 2.20: Piecewise linear approximation of an air disk brake hysteresis. (Courtesy of John Cheng, Rockwell.)

dead-zone with two hysteresis loops.

The characteristic in Figure 2.21 is a piecewise linear approximation of a characteristic obtained experimentally by Darrel Recker [92] for an electrohydraulic actuator of a gas spring. The gas spring system is schematically shown in Figure 2.22, where the inflow q_f of the hydraulic fluid is controlled by the fill valve, while the outflow q_v is controlled by the vent valve. The net flow q controls the presure of the gas spring. This system has been investigated as a part of modern automotive suspensions.

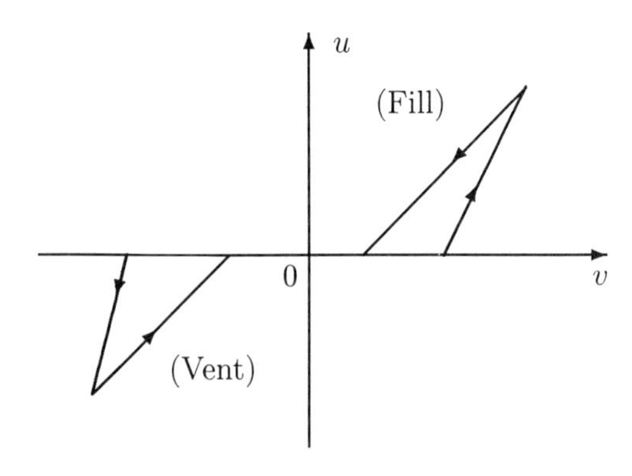

Figure 2.21: A combined nonlinearity: dead-zone with two hysteresis loops.

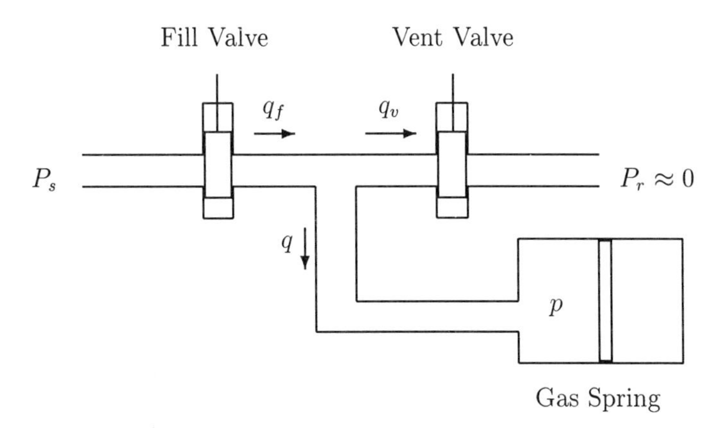

Figure 2.22: An electrohydraulically controlled gas spring. (Courtesy of Darrel Recker.)

The experimental fill and vent valve characteristics obtained by Recker in several overlapping runs are shown in Figure 2.23. They show the valve flows (in $\mathrm{cm}^3/\mathrm{sec}$) as functions of the solenoid voltage (in fractions of the pulse-width modulation (pwm) duty cycle).

2.5 Performance Deterioration

It is easy to see that the presence of a dead-zone, backlash, or hysteresis adversely affects the static accuracy of feedback control systems. With a linear controller the static accuracy is limited by the width of the dead-zone, backlash, or hysteresis. Attempts to improve it by increasing the gain of the feedback loop lead to sustained oscillations which may cause rapid wear of gear trains, valves, and other components. Similar difficulty is encountered with integral action (PI control). Backlash and hysteresis are also harmful for the dynamic performance because of their inherent phase lag.

The loss of performance due to backlash will now be illustrated by a single integral plant in the feedback loop with a PI controller shown in Figure 2.24. A physical example of this plant is the liquid tank in Figure 2.14. To be specific, let the input backlash be described by the upward line $u(t) = v(t) - 1$ and the downward line $u(t) = v(t) + 1$.

It can be verified that with a proportional controller $v = -\alpha e$, $\beta = 0$, the error $e = y - r$ cannot be reduced to zero except in the trivial case when $y(0) = r$ and $u(0) = 0$. The static error is either $\frac{1}{\alpha}$ or $\frac{-1}{\alpha}$. In this very

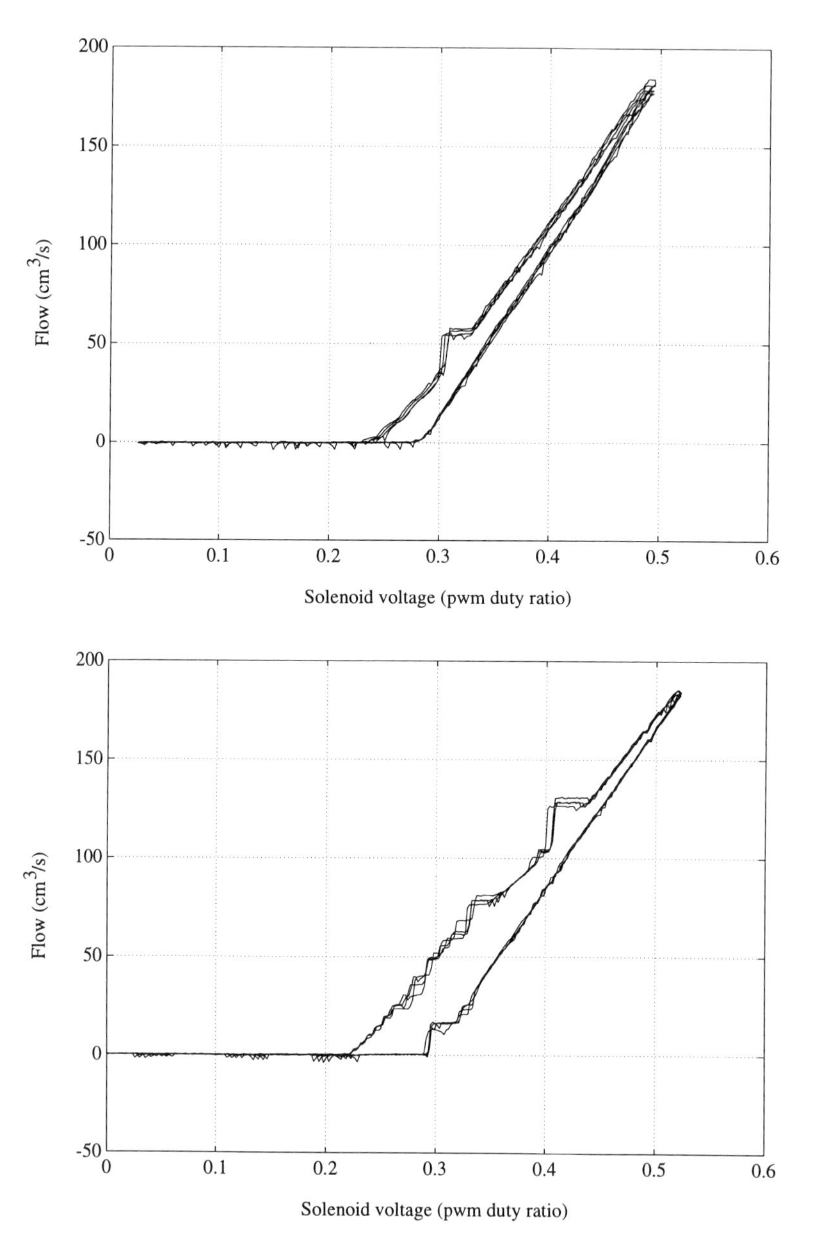

Figure 2.23: The characteristics of the "fill" (above) and "vent" (below) valves. (Courtesy of Darrel Recker.)

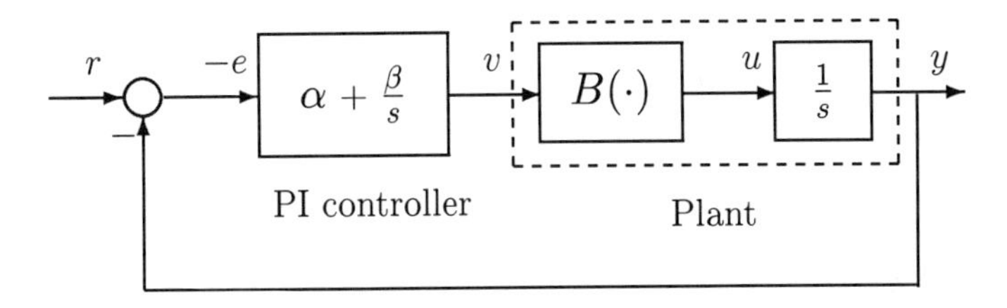

Figure 2.24: One-integrator plant with input backlash and by a PI controller.

simple case this error can be reduced by increasing the gain α, but with more complex plants a big increase in α is undesirable. Another way of achieving static accuracy is the introduction of the integration term with $\beta > 0$. For the sake of illustration we select $\alpha = \beta = 1$. In the absence of backlash the characteristic equation with this choice of gain would be $s^2 + s + 1 = 0$, which means that the complex pair of poles would have a damping factor 0.5. What happens in the presence of backlash is shown in Figure 2.25(a) - (c) for the case when $r = 10$, $y(0) = 5$, so that the initial error is $e(0) = -5$. The controller integral action is initialized at zero so that $v(0) = -\alpha e(0) = 5$.

The trajectory on the (e, v)-plane, as shown in Figure 2.25(a), settles into a limit cycle with $e(t)$ oscillating in the interior of backlash between 1 and -1. This can also be seen in the (u, v)-plane in Figure 2.25(b). The unsatisfactory time response is shown in Figure 2.25(c).

Although in this example the size of the limit cycle is overdramatized for clarity, it is true in general that linear controllers cannot cope with dead-zone, backlash, and hysteresis. Some form of nonlinear compensation is needed to improve the performance of feedback loops with such common nonlinearities.

2.6 Plant Models

From the examples in this chapter we deduce that the following three classes of nonlinear plants are common in applications: plants with input nonlinearities, plants with output nonlinearities, and plants with both input and output nonlinearities. They are shown in Figures 2.26(a) - (c), respectively, where $N(\cdot)$, $N_i(\cdot)$, and $N_o(\cdot)$ represent a dead-zone, backlash, or hysteresis, and $G(D)$ is the transfer function of a linear dynamics.

For such plants we will develop an adaptive inverse to cancel the effects of the nonlinearities so that a linear controller structure can achieve improved output tracking performance.

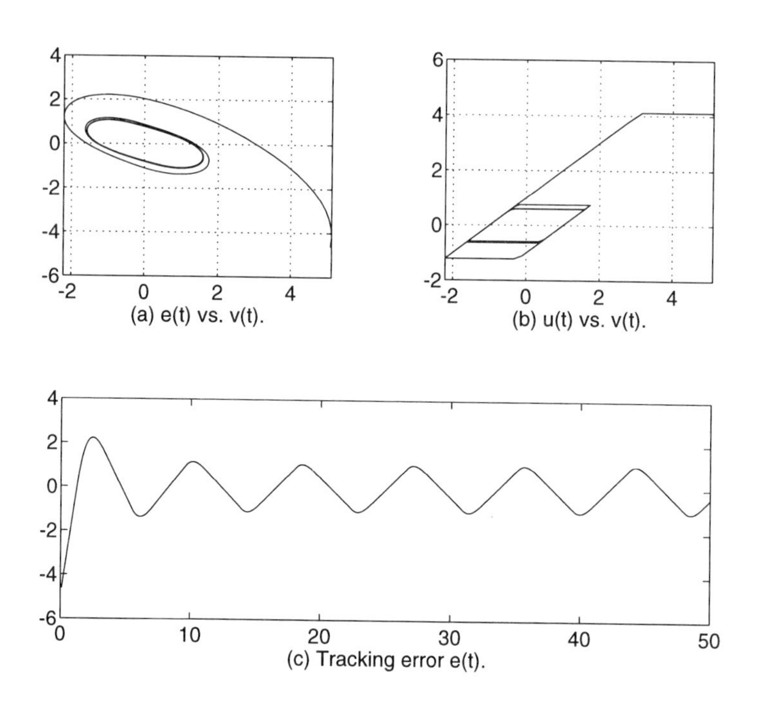

Figure 2.25: Sustained oscillation (limit cycle) in the feedback control system with backlash shown in Figure 2.24.

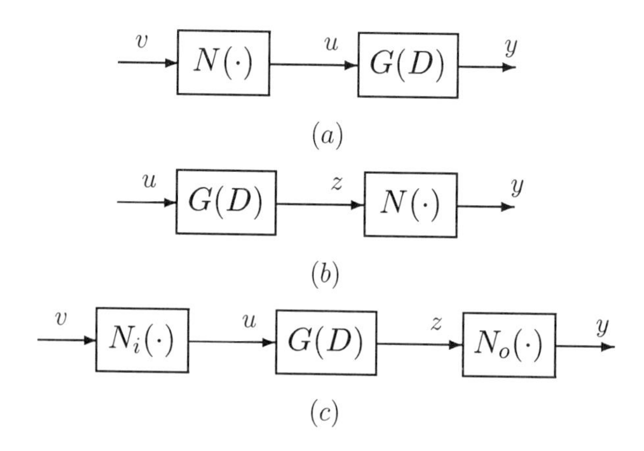

Figure 2.26: Plant models: (a) with input nonlinearities, (b) with output nonlinearities, and (c) with input and output nonlinearities.

Chapter 3

Inverse Models

In the presence of dead-zone, backlash, or hysteresis it is not possible to achieve high-precision tracking using only linear controllers. Our approach is to add to linear controllers specially designed nonlinear compensators which include fixed or adaptive inverses of dead-zone, backlash, or hysteresis.

We begin with the case of a known input dead-zone, whose effect we cancel with a discontinuous dead-zone inverse at the controller output. The situation is less clear with input backlash and hysteresis. Do these nonsmooth nonlinearities have inverses and, if so, are they implementable? This question has first been raised in [115], [120]. Following these references, we will show that for the backlash and hysteresis models given in Chapter 2, the answer is affirmative and leads to an effective compensation of backlash and hysteresis. Compensation of the output nonlinearity is more difficult and requires a reparametrization and redesign of the linear controller as well, which will be addressed in Chapter 8 and Chapter 9.

We let $u = N(v)$ represent either the dead-zone $DZ(\cdot)$, backlash $B(\cdot)$, or hysteresis $H(\cdot)$. Then the compensator for the input nonlinearity $N(\cdot)$ is $v = NI(u_d)$, where $NI(\cdot)$ is the desired inverse and u_d is the control input which would achieve the control objective in the absence of $N(\cdot)$. When the inverse is exact, then $u = N(NI(u_d)) = u_d$ achieves the control objective as if $N(\cdot)$ were absent. It should be pointed out that $NI(\cdot)$ is the *right inverse* of $N(\cdot)$. Neither $DZ(\cdot)$ nor $B(\cdot)$ has a left inverse because for these nonlinearities different input signals may produce the same output.

In reality, $NI(\cdot)$ is seldom exactly known and only an estimate $\widehat{NI}(\cdot)$ of the inverse $NI(\cdot)$ can be implemented. Our goal is to adaptively update the inverse estimate to make the inverse $\widehat{NI}(\cdot)$ adaptive.

Our compensation scheme in Figure 3.1 is described by

$$v(t) = \widehat{NI}(u_d(t)) \tag{3.1}$$

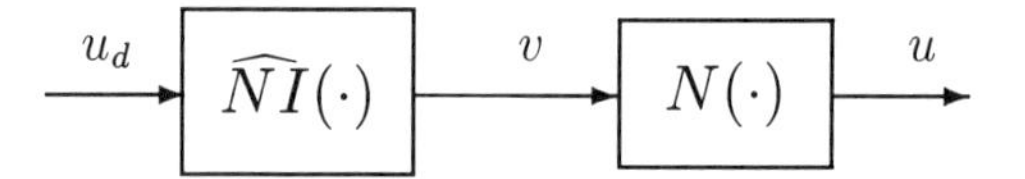

Figure 3.1: Compensation of the input nonlinearity $N(\cdot)$.

where the inverse estimate $\widehat{NI}(\cdot)$ can be either fixed or adaptive. We will derive the right dead-zone inverse $\widehat{DI}(\cdot)$, backlash inverse $\widehat{BI}(\cdot)$, and hysteresis inverse $\widehat{HI}(\cdot)$ as $\widehat{NI}(\cdot)$ in (3.1) for $DZ(\cdot)$, $B(\cdot)$, and $H(\cdot)$. For each of the three inverses we reparametrize the slope, break-point, or crossing parameters to obtain a linear form needed for adaptive control.

3.1 Dead-Zone Inverse

Let us first develop a parametrized dead-zone inverse whose parameters can be either fixed or adaptively updated. When implemented with true parameters the dead-zone inverse cancels the dead-zone effect exactly, and when implemented with parameter estimates it results a control error which is linearly parametrized by the parameter error plus a bounded disturbance.

3.1.1 Inverse Characteristic

For a linear parametrization of a dead-zone inverse, we use the estimates $\widehat{m_r b_r}$, $\widehat{m_r}$, $\widehat{m_l b_l}$, $\widehat{m_l}$ of $m_r b_r$, m_r, $m_l b_l$, m_l, respectively. The estimates of b_r, b_l are then obtained from

$$\widehat{b_r} = \frac{\widehat{m_r b_r}}{\widehat{m_r}}, \ \widehat{b_l} = \frac{\widehat{m_l b_l}}{\widehat{m_l}}.$$

The inverse model for the dead-zone characteristic (2.1) is depicted in Figure 3.2 (a) and described by

$$v(t) = \widehat{DI}(u_d(t)) = \begin{cases} \frac{u_d(t)+\widehat{m_r b_r}}{\widehat{m_r}} & \text{if } u_d(t) > 0 \\ 0 & \text{if } u_d(t) = 0 \\ \frac{u_d(t)+\widehat{m_l b_l}}{\widehat{m_l}} & \text{if } u_d(t) < 0. \end{cases} \tag{3.2}$$

This dead-zone inverse expression will be used for both the *continuous-time* and *discrete-time* designs.

As a "soft" alternative to the "hard" dead-zone inverse (3.2) as shown in Figure 3.2(a), we can use any smooth characteristic in the shaded area in

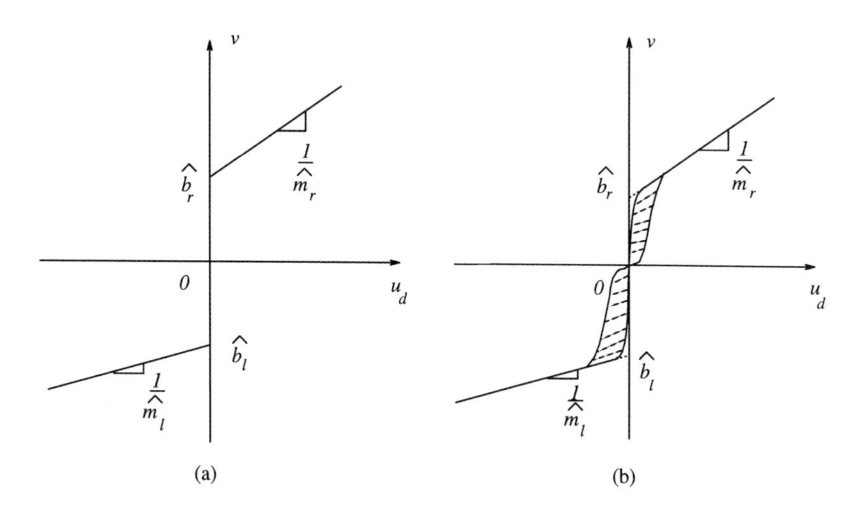

Figure 3.2: Dead-zone inverse: (a) hard, (b) soft.

Figure 3.2 (b) with bounded $\widehat{b_r}$, $\widehat{b_l}$. The soft inverse avoids the discontinuity at $u_d = 0$, but it also introduces an error in the cancellation of dead-zone effects.

To ensure that the dead-zone inverse (3.2) is implementable—that is, to ensure that $\widehat{m_r} > m_0$, $\widehat{m_l} > m_0$ for some $m_0 > 0$ and $\widehat{b_r} \geq 0$, $\widehat{b_l} \leq 0$—we will use parameter projection based on the assumption that $b_l \leq 0 \leq b_r$, and $m_r \geq m_0$, $m_l \geq m_0$, for some known $m_0 > 0$.

Denote the exact dead-zone inverse as

$$DI(\cdot) = \widehat{DI}(\cdot)|_{\widehat{b_l}=b_l, \widehat{m_l}=m_l, \widehat{b_r}=b_r, \widehat{m_r}=m_r}.$$

Then the following property holds for the dead-zone inverse (3.2).

Lemma 3.1 *(Dead-Zone Inverse) When implemented with the true parameters $m_r b_r$, m_r, $m_l b_l$, m_l, the dead-zone inverse (3.2) cancels the dead-zone effect; that is,*

$$v(t) = DI(u_d(t)) \Rightarrow u(t) = u_d(t), \ \forall t \geq 0.$$

3.1.2 Parametrization

For more compact notation, we introduce the indicator function $\chi[X]$ of the event X:

$$\chi[X] = \begin{cases} 1 & \text{if } X \text{ is true} \\ 0 & \text{otherwise.} \end{cases}$$

The indicator function employing estimates will be denoted by $\widehat{\chi}$. In this notation we rewrite the dead-zone inverse (3.2) as a single expression linear in parameter estimates:

$$u_d(t) = \widehat{m_r}\widehat{\chi_r}(t)v(t) - \widehat{m_r b_r}\widehat{\chi_r}(t) + \widehat{m_l}\widehat{\chi_l}(t)v(t) - \widehat{m_l b_l}\widehat{\chi_l}(t), \tag{3.3}$$

where

$$\widehat{\chi_r}(t) = \chi[v(t) > 0]$$
$$\widehat{\chi_l}(t) = \chi[v(t) < 0].$$

With the parameter vectors θ_d^*, θ_d, and "regressor" $\omega_d(t)$:

$$\theta_d^* = (m_r, m_r b_r, m_l, m_l b_l)^T \tag{3.4}$$

$$\theta_d = (\widehat{m_r}, \widehat{m_r b_r}, \widehat{m_l}, \widehat{m_l b_l})^T \tag{3.5}$$

$$\omega_d(t) = (-\widehat{\chi_r}(t)v(t), \widehat{\chi_r}(t), -\widehat{\chi_l}(t)v(t), \widehat{\chi_l}(t)^T \tag{3.6}$$

the dead-zone inverse expression (3.3) acquires a compact form:

$$u_d(t) = -\theta_d^T \omega_d(t). \tag{3.7}$$

It is important to see that $\widehat{DI}(\cdot)$, implemented with inaccurate parameter estimates, results in a *control error*:

$$u(t) - u_d(t) = (\theta_d - \theta_d^*)^T \omega_d(t) + d_d(t). \tag{3.8}$$

In this control error expression, the unparametrized part

$$\begin{aligned} d_d(t) = &-m_r \chi[0 < v(t) < b_r](v(t) - b_r) \\ &-m_l \chi[b_l < v(t) < 0](v(t) - b_l) \end{aligned} \tag{3.9}$$

is an important quantity which in most compensation schemes acts as an unknown "disturbance". The following result, which can be easily verified, characterizes this disturbance.

Proposition 3.1 *The unparametrizable part of the dead-zone inverse $d_d(t)$ has the following properties:*

(i) $d_d(t)$ is bounded for all $t \geq 0$.

(ii) $d_d(t) = 0$ whenever $\widehat{b_r} \geq b_r$ and $\widehat{b_l} \leq b_l$; that is, $d_d(t)$ disappears if the break-points b_r and b_l are overestimated.

(iii) $d_d(t) = 0$ whenever $v(t) \geq b_r$ or $v(t) \leq b_l$—that is, when $u(t)$ and $v(t)$ are outside the dead-zone.

(iv) $d_d(t)$ is also bounded if a soft dead-zone inverse is used.

It follows from this proposition that $d_d(t) = 0$ whenever $\theta_d = \theta_d^*$. As will be seen in the subsequent chapters, these properties are favorable for adaptive dead-zone inverse designs.

3.2 Backlash Inverse

In this section we will provide an affirmative answer to the following question: Does the backlash have an inverse? We will show that there is a backlash inverse characteristic $BI(\cdot)$ which can ensure that

$$u(t) = B(BI(u_d(t))) = u_d(t);$$

that is, the cancellation of backlash is achieved. Since the backlash characteristic is dynamic and more complex than the dead-zone, as shown by (2.2) or (2.3), its inverse characteristic is also dynamic and more complex than the dead-zone inverse.

3.2.1 Inverse Characteristic

We have seen in Chapter 2 that one of the damaging effects of backlash on system performance is the delay corresponding to the time needed to traverse an inner segment of $B(\cdot)$. The ideal backlash inverse $BI(\cdot)$ will make the traverse of this segment instantaneous and thus cancel this undesirable backlash effect. Another undesirable effect of backlash is the information loss occurring on an inner segment when the output $u(t)$ remains constant while the input $v(t)$ continues to change, as illustrated by "chopped" peaks of $u(t)$ in Figure 2.11. These two undesirable effects are eliminated with the backlash inverse $v(t) = BI(u_d(t))$:

$$\dot{v}(t) = \begin{cases} \frac{1}{m}\dot{u}_d(t) & \text{if } \dot{u}_d(t) > 0,\ v(t) = \frac{u_d(t)}{m} + c_r, \text{ or} \\ & \text{if } \dot{u}_d(t) < 0,\ v(t) = \frac{u_d(t)}{m} + c_l \\ 0 & \text{if } \dot{u}_d(t) = 0 \\ g(t,t) & \text{if } \dot{u}_d(t) > 0,\ v(t) = \frac{u_d(t)}{m} + c_l \\ -g(t,t) & \text{if } \dot{u}_d(t) < 0,\ v(t) = \frac{u_d(t)}{m} + c_r. \end{cases} \tag{3.10}$$

In this definition, the inverse of a horizontal segment of the backlash characteristic is a vertical jump defined as the time integral of the impulse function

$$g(\tau, t) = \delta(\tau - t)(c_r - c_l), \tag{3.11}$$

where $\delta(t)$ is the Dirac δ-function. Thus an upward jump in the backlash inverse is

$$v(t^+) = v(t^-) + \int_{t^-}^{t^+} g(\tau, t)d\tau = \frac{u_d(t^-)}{m} + c_r. \tag{3.12}$$

The effect of this jump in $BI(\cdot)$ will be to eliminate the delay caused by a segment in $B(\cdot)$. In a similar manner the use of (3.10) restores the information that would have been lost in (2.2). We show this by proving that the characteristic $BI(\cdot)$ defined in (3.10) is the *right inverse* of the backlash $B(\cdot)$ defined by (2.2).

Lemma 3.2 *(Backlash Inverse) The characteristic $BI(\cdot)$ defined by (3.10) is the right inverse of the characteristic $B(\cdot)$ defined by (2.2) in the sense that*

$$u_d(t_0) = B(BI(u_d(t_0))) \Rightarrow B(BI(u_d(t))) = u_d(t), \ \forall t \geq t_0. \tag{3.13}$$

Proof: Suppose that $\dot{u}_d(t) > 0$ for $t \in [t_0, t_1]$ and some $t_1 > t_0$. First, if $v(t_0) = \frac{u_d(t_0)}{m} + c_r$ and $u(t_0) = m(v(t_0) - c_r)$, then it follows from (3.10), (2.2) that $\dot{u}(t) = m\dot{v}(t) = m\frac{\dot{u}_d(t)}{m} = \dot{u}_d(t)$ for $t \in [t_0, t_1]$ with $u(t_0) = u_d(t_0)$. Hence $B(BI(u_d(t))) = u_d(t)$ for any $t \in [t_0, t_1]$. Second, if $v(t_0) = \frac{u_d(t_0)}{m} + c_l$ and $u(t_0) = m(v(t_0) - c_l)$, then, according to (3.10), $v(t)$ will have a jump at $t = t_0$ so that $v(t) = \frac{u_d(t)}{m} + c_r$ for $t = t_0^+$. The jump in $v(t)$ makes $u(t)$ traverse an inner segment so that $u(t_0^+) = m(v(t_0^+) - c_r)$, which reduces to the first case above. When $\dot{u}_d(t) < 0$ for $t \in [t_0, t_1]$, a similar analysis shows that $B(BI(u_d(t))) = u_d(t)$ for any $t \in [t_0, t_1]$. If $\dot{u}_d(t) = 0$ for $t \in [t_0, t_1]$, then $B(BI(u_d(t))) = u_d(t)$ holds for any $t \in [t_0, t_1]$.

If $\dot{u}_d(t)$ changes the sign at $t = t_1$, then according to (3.10), $v(t)$ will have a jump at $t = t_1$ so that $v(t) = \frac{u_d(t)}{m} + c_r$ for $\dot{u}_d(t_1^+) > 0$, and $v(t) = \frac{u_d(t)}{m} + c_l$ for $\dot{u}_d(t_1^+) < 0$. The jump in $v(t)$ makes $u(t)$ traverse an inner segment so that $u(t_1^+) = m(v(t_1^+) - c_r)$ for $\dot{u}_d(t_1^+) > 0$, and $u(t_1^+) = m(v(t_1^+) - c_l)$ for $\dot{u}_d(t_1^+) < 0$. Then we can repeat the above procedure to show that $B(BI(u_d(t))) = u_d(t)$ for any $t \geq t_1$. ∇

The mapping (3.10) - (3.12) may not define a backlash inverse only if the signal $u_d(t)$ is such that $v(t)$ and $u(t)$ never leave an inner segment. This situation can happen only if $v(0)$, $u(0)$ are initially on an inner segment and $\dot{u}_d(t) = 0$ for $t \geq 0$ or if $\dot{u}_d(t)$ does not change sign but the total increment of $\frac{u_d(t)}{m}$ is insufficient for $v(t)$, $u(t)$ to leave the segment.

As $u_d(t)$ is a signal at our disposal, the above situation can be remedied. If $u_d(t)$ does not reach t_0 defined in (3.13), then $BI(\cdot)$ can be initialized as follows:

$$v(t_0^+) = \begin{cases} \frac{u_d(t_0)}{m} + c_r & \text{if } v(t_0) = \frac{u_d(t_0)}{m} + c_l \\ \frac{u_d(t_0)}{m} + c_l & \text{if } v(t_0) = \frac{u_d(t_0)}{m} + c_r. \end{cases}$$

This will always result in $u_d(t_0^+) = B(BI(u_d(t_0^+)))$ so that (3.13) will hold for all $t > t_0$.

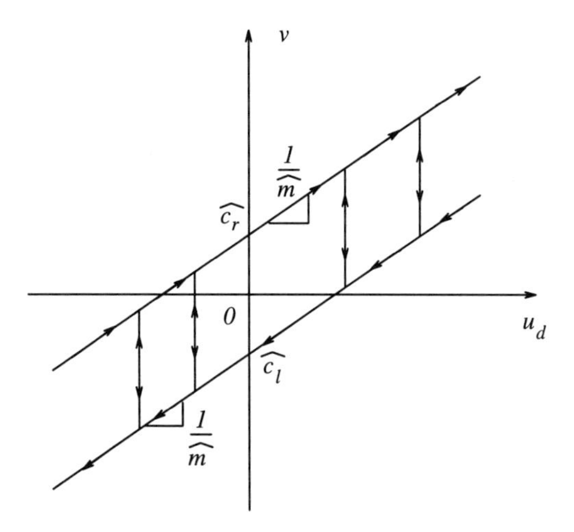

Figure 3.3: Backlash inverse.

When the backlash parameters m, c_r, c_l are unknown, we will use their estimates $\widehat{m}$, $\widehat{c}_l$, $\widehat{c}_r$ to design a backlash inverse estimate: $v(t) = \widehat{BI}(u_d(t))$, characterized by

$$\dot{v}(t) = \begin{cases} \frac{1}{\widehat{m}}\dot{u}_d(t) & \text{if } \dot{u}_d(t) > 0 \text{ and } v(t) = \frac{u_d(t)}{\widehat{m}} + \widehat{c}_r, \text{ or} \\ & \text{if } \dot{u}_d(t) < 0 \text{ and } v(t) = \frac{u_d(t)}{\widehat{m}} + \widehat{c}_l \\ 0 & \text{if } \dot{u}_d(t) = 0 \\ \widehat{g}t, t) & \text{if } \dot{u}_d(t) > 0 \text{ and } v(t) = \frac{u_d(t)}{\widehat{m}} + \widehat{c}_l \\ -\widehat{g}t, t) & \text{if } \dot{u}_d(t) < 0 \text{ and } v(t) = \frac{u_d(t)}{\widehat{m}} + \widehat{c}_r, \end{cases} \tag{3.14}$$

where

$$\widehat{g}(\tau, t) = \delta(\tau - t)(\widehat{c}_r - \widehat{c}_l)$$

is the estimate of $g(\tau, t)$ defined in (3.11). Graphically, the backlash inverse (3.14) with parameter estimates is depicted in Figure 3.3 by two straight lines and vertical jumps between the lines, where the downward side is

$$v(t) = \frac{u_d(t)}{\widehat{m}} + \widehat{c}_l, \ \dot{u}_d(t) < 0 \tag{3.15}$$

and the upward side is

$$v(t) = \frac{u_d(t)}{\widehat{m}} + \widehat{c}_r, \ \dot{u}_d(t) > 0. \tag{3.16}$$

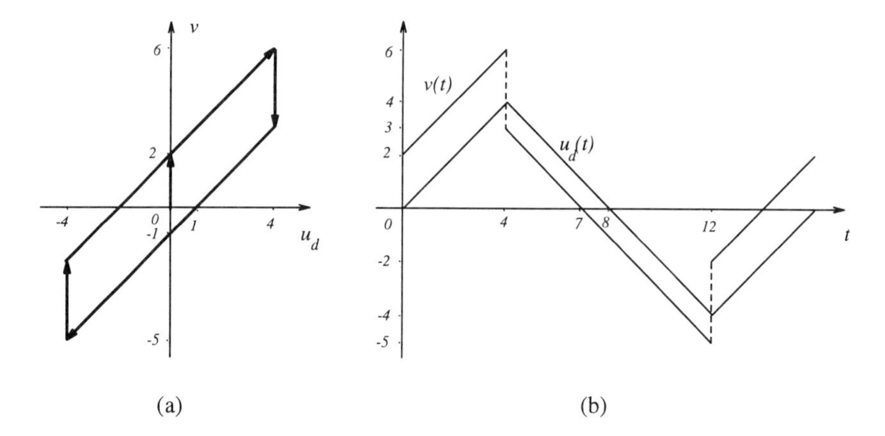

Figure 3.4: Backlash inverse response to a sawtooth input $u_d(t)$.

Vertical jumps of $v(t)$ occur whenever $\dot{u}_d(t)$ changes its sign. It is the knowledge of $sign(\dot{u}_d(t))$ rather than the exact value of $\dot{u}_d(t)$ that is used to determine whether and when there is a jump of $v(t)$. To see this, suppose that the motion is on the upward side of $\widehat{BI}(\cdot)$; that is, $v(t_1) = \frac{u_d(t_1)}{\widehat{m}} + \widehat{c}_r$ for some $t_1 \geq 0$. If $\dot{u}_d(t_1) < 0$, then a vertical jump of $v(t)$ occurs so that $v(t_1^+) = \frac{u_d(t_1^+)}{\widehat{m}} + \widehat{c}_l$. If $\dot{u}_d(t) \geq 0$ for $t \in [t_1, t_2]$, then $v(t)$ and $u_d(t)$ will stay on the upward side so that $v(t) = \frac{u_d(t)}{\widehat{m}} + \widehat{c}_r$. Therefore it is sufficient to know when $\dot{u}_d(t)$ changes sign in order to implement a jump for $v(t)$.

3.2.2 Example: Backlash Compensation

Let us now return to the sawtooth response in Figure 2.11, where backlash caused a phase delay and "chopped" the peaks. To compensate for these distortions, we implement the exact backlash inverse $BI(\cdot)$ shown in Figure 3.4(a), for the backlash characteristic $B(\cdot)$ in Figure 2.11(a). The output of the backlash inverse is applied to the backlash input: $v(t) = BI(u_d(t))$, which is shown in Figure 3.4(b), where the motion starts with $u_d(0) = v(0) = 0$. At $t = 0$, $v(t)$ jumps to $v(0^+) = 2$ on the upward side. For $t \in (0, 4)$, $\dot{u}_d(t) = 1$ does not change so the motion stays on the upward side (3.16). For $t \in (4, 12)$, $\dot{u}_d(t) = -1$, the downward side (3.15) is active. At $t = 4$, the sign of $\dot{u}_d(t)$ changes, causing a jump of $v(t)$ to the downward side. Because of the subsequent sign change of $\dot{u}_d(t)$, one more vertical jump of $v(t)$ occurs at $t = 12$. The magnitude of the jump of $v(t)$ is equal to the length of the inner segment of $B(\cdot)$, which in this case is 3.

The compensation of backlash is accomplished, because it can be easily verified that if $v(t)$ from Figure 3.4(b) is applied to the backlash characteristic shown in Figure 2.11(a) then the backlash output $u(t)$ is identical to $u_d(t)$ as desired: $u(t) = B(BI(u_d(t))) = u_d(t)$.

3.2.3 Soft Inverses

To avoid the presence of vertical jumps in the backlash inverse, we can employ continuous approximations of $\widehat{BI}(\cdot)$, such as

$$v(t) = \frac{1}{\widehat{m}} u_d(t) + \bar{\widehat{c}}(t), \tag{3.17}$$

where, for a continuous $\dot{u}_d(t)$, $\bar{\widehat{c}}(t)$ is defined as

$$\bar{\widehat{c}}(t) = \begin{cases} \widehat{c}_r & \text{if } 0 \leq \dot{u}_d(t) \text{ and } v(t) = \frac{u_d(t)}{\widehat{m}} + \widehat{c}_r \\ \widehat{c}_l & \text{if } \dot{u}_d(t) \leq 0 \text{ and } v(t) = \frac{u_d(t)}{\widehat{m}} + \widehat{c}_l \\ f_r(\dot{u}_d(t)) & \text{if } 0 < \dot{u}_d(t) \leq \beta_r \text{ and } \bar{\widehat{c}}(t_-) < \widehat{c}_r \\ f_l(\dot{u}_d(t)) & \text{if } -\beta_l \leq \dot{u}_d(t) < 0 \text{ and } \bar{\widehat{c}}(t_-) > \widehat{c}_l \end{cases} \tag{3.18}$$

with $\beta_r > 0$, $\beta_l > 0$ being constants, and $f_r(\cdot)$, $f_l(\cdot)$ being continuous and monotonically increasing functions such that

$$f_r(0) = \widehat{c}_l,\ f_l(0) = \widehat{c}_r,\ f_r(\beta_r) = \widehat{c}_r,\ f_l(-\beta_l) = \widehat{c}_l.$$

As a comparison, we express the backlash inverse $\widehat{BI}(\cdot)$ shown in Figure 3.3 and defined by (3.15) - (3.16) as

$$v(t) = \widehat{BI}(u_d(t)) = \frac{1}{\widehat{m}} u_d(t) + \widehat{c}(t),$$

where

$$\widehat{c}(t) = \begin{cases} \widehat{c}_r & \text{if } v(t) = \frac{u_d(t)}{\widehat{m}} + \widehat{c}_r \\ \widehat{c}_l & \text{if } v(t) = \frac{u_d(t)}{\widehat{m}} + \widehat{c}_l. \end{cases}$$

Another choice of $\bar{\widehat{c}}(t)$ is

$$\dot{\bar{\widehat{c}}}(t) = -\alpha \bar{\widehat{c}}(t) + \alpha \widehat{c}(t),\ \alpha > 0,\ \bar{\widehat{c}}(0) \in [\widehat{c}_l, \widehat{c}_r]. \tag{3.19}$$

These definitions of $\bar{\widehat{c}}(t)$ have the following properties:

(i) $|\widehat{c}(t) - \bar{\widehat{c}}(t)| \leq max\{\widehat{c}_r, |\widehat{c}_l|\}$ for $\bar{\widehat{c}}(t)$ in (3.18), and $|\widehat{c}(t) - \bar{\widehat{c}}(t)| \leq \widehat{c}_r - \widehat{c}_l$ for $\bar{\widehat{c}}(t)$ in (3.19).

(ii) $\bar{\widehat{c}}(t) \to \widehat{c}(t)$, as $\beta_r \to 0$, $\beta_l \to 0$, or $\alpha \to \infty$, for $\dot{u}_d \neq 0$.

(iii) For β_r, β_l, or α given, the signal $\bar{\hat{c}}(t)$ in (3.18) becomes closer to $\hat{c}(t)$ if $\dot{u}_d(t)$ has a larger magnitude, while $\bar{\hat{c}}(t)$ in (3.19) becomes closer to $\hat{c}(t)$ if $\dot{u}_d(\tau)$ has not changed its sign over the interval $[t-T, t]$ for a larger $T > 0$.

Property (i) implies that an approximate backlash inverse introduces a bounded error in $v(t)$, while properties (ii) and (iii) characterize qualitatively the accuracy of approximation.

Another way to approximate the backlash inverse $BI(\cdot)$ defined by (3.10) is to replace the vertical jumps between its upward and downward lines by continuous curves with bounded slopes. For example, a vertical transition can be replaced with a line segment which links two sides of $\widehat{BI}(\cdot)$ and has a finite positive slope.

3.2.4 Discrete-Time Representation

For discrete-time applications the backlash model (2.2) is not appropriate because of the discontinuity of signals. In Chapter 2, we have obtained the following discrete-time backlash model (see (2.3)):

$$u(t) = B(v(t)) = \begin{cases} m(v(t) - c_l) & \text{if } v(t) \leq v_l \\ m(v(t) - c_r) & \text{if } v(t) \geq v_r \\ u(t-1) & \text{if } v_l < v(t) < v_r, \end{cases} \tag{3.20}$$

where

$$v_l = \frac{u(t-1)}{m} + c_l$$

$$v_r = \frac{u(t-1)}{m} + c_r.$$

For this backlash model, the exact backlash inverse is

$$v(t) = BI(u_d(t)) = \begin{cases} v(t-1) & \text{if } u_d(t) = u_d(t-1) \\ \frac{u_d(t)}{m} + c_l & \text{if } u_d(t) < u_d(t-1) \\ \frac{u_d(t)}{m} + c_r & \text{if } u_d(t) > u_d(t-1). \end{cases} \tag{3.21}$$

The discrete-time version of Lemma 3.1 states that the characteristic $BI(\cdot)$ defined by (3.21) is the right inverse of the characteristic $B(\cdot)$ defined by (3.20) such that

$$u_d(t_0) = B(BI(u_d(t_0))) \Rightarrow B(BI(u_d(t))) = u_d(t),\ \forall t \geq t_0.$$

An initialization of $BI(\cdot)$:

$$v(t_0^+) = \begin{cases} \frac{u_d(t_0)}{m} + c_r & \text{if } v(t_0) = \frac{u_d(t_0)}{m} + c_l \\ \frac{u_d(t_0)}{m} + c_l & \text{if } v(t_0) = \frac{u_d(t_0)}{m} + c_r \end{cases}$$

results in $u_d(t_{0^+}) = B(BI(u_d(t_{0^+})))$, as $v(t)$ jumps from $v(t_0)$ to $v(t_0^+)$.

Based on the parameter estimates $\widehat{m}$, $\widehat{c}_l$, $\widehat{c}_r$, the *discrete-time* backlash inverse estimate $v(t) = \widehat{BI}(u_d(t))$ is

$$v(t) = \begin{cases} v(t-1) & \text{if } u_d(t) = u_d(t-1) \\ \frac{u_d(t)}{\widehat{m}} + \widehat{c}_l & \text{if } u_d(t) < u_d(t-1) \\ \frac{u_d(t)}{\widehat{m}} + \widehat{c}_r & \text{if } u_d(t) > u_d(t-1). \end{cases} \tag{3.22}$$

Similar to the dead-zone inverse, to ensure that the estimated backlash inverse is implementable, we assume that $c_l \leq 0 \leq c_r$ and $m \geq m_0$ for some known $m_0 > 0$ and choose the estimates $\widehat{m}$, $\widehat{c}_r$, $\widehat{c}_l$ such that $\widehat{m} \geq m_0$, $\widehat{c}_r \geq 0$, $\widehat{c}_l \leq 0$.

Unlike its continuous-time counterpart the discrete-time backlash inverse does not require the knowledge of $\dot{u}_d(t)$ for implementation. This makes discrete-time adaptive inverse controllers more practical than continuos-time adaptive controllers.

3.2.5 Parametrization

Our next task is to express the backlash inverse control error $u(t) - u_d(t)$ in terms of a parametrizable part and an unparametrizable but bounded part. To give a compact description for the backlash inverse estimate (3.14) or (3.22), we introduce the backlash inverse indicator functions

$$\widehat{\chi_r}(t) = \chi[v(t) = \frac{u_d(t)}{\widehat{m}} + \widehat{c}_r] \tag{3.23}$$

$$\widehat{\chi_l}(t) = \chi[v(t) = \frac{u_d(t)}{\widehat{m}} + \widehat{c}_l]. \tag{3.24}$$

Since one of these mutually exclusive events must take place, we have

$$\widehat{\chi_r}(t) + \widehat{\chi_l}(t) = 1 \tag{3.25}$$

$$\widehat{\chi_l}^2(t) = \widehat{\chi_l}(t),\ \widehat{\chi_r}^2(t) = \widehat{\chi_r}(t),\ \widehat{\chi_l}(t)\widehat{\chi_r}(t) = 0. \tag{3.26}$$

Using (3.15), (3.16), (3.23) - (3.26), we express $v(t)$ as

$$\begin{aligned} v(t) &= (\widehat{\chi_r}(t) + \widehat{\chi_l}(t))v(t) \\ &= \frac{\widehat{\chi_r}(t)}{\widehat{m}}(u_d(t) + \widehat{m}\widehat{c}_r) + \frac{\widehat{\chi_l}(t)}{\widehat{m}}(u_d(t) + \widehat{m}\widehat{c}_l). \end{aligned} \tag{3.27}$$

We also introduce the backlash indicator functions

$$\chi_r(t) = \chi[\dot{u}(t) > 0],\ \chi_l(t) = \chi[\dot{u}(t) < 0],\ \chi_s(t) = \chi[\dot{u}(t) = 0].$$

These functions satisfy the following obvious relationships:

$$\chi_r(t) + \chi_l(t) + \chi_s(t) = 1$$

$$\chi_l^2(t) = \chi_l(t),\ \chi_r^2(t) = \chi_r(t),\ \chi_s^2(t) = \chi_s(t)$$
$$\chi_l(t)\chi_r(t) = 0,\ \chi_l(t)\chi_s(t) = 0,\ \chi_s(t)\chi_r(t) = 0.$$

With the help of these relationships, we arrive at the compact expression for the output $u(t)$ of the backlash $B(\cdot)$:

$$\begin{aligned} u(t) &= (\chi_r(t) + \chi_l(t) + \chi_s(t))u(t) \\ &= \chi_r(t)m(v(t) - c_r) + \chi_l(t)m(v(t) - c_l) + \chi_s(t)u_s, \end{aligned} \tag{3.28}$$

where u_s is a generic constant corresponding to the value of $u(t)$ at any active inner segment characterized by

$$\frac{u_s}{m} + c_l \leq v(t) \leq \frac{u_s}{m} + c_r.$$

Multiplying both sides of (3.27) by $\widehat{\chi_l}(t)$ and using (3.26), we obtain

$$\widehat{\chi_l}(t)u_d(t) = \widehat{\chi_l}(t)(\widehat{m}v(t) - \widehat{m}\widehat{c_l}). \tag{3.29}$$

Similarly we have

$$\widehat{\chi_r}(t)u_d(t) = \widehat{\chi_r}(t)(\widehat{m}v(t) - \widehat{m}\widehat{c_r}). \tag{3.30}$$

Using (3.25), (3.29), and (3.30) we express $u_d(t)$ as follows:

$$\begin{aligned} u_d(t) &= (\widehat{\chi_l}(t) + \widehat{\chi_r}(t))u_d(t) \\ &= \widehat{\chi_l}(t)(\widehat{m}v(t) - \widehat{m}\widehat{c_l}) + \widehat{\chi_r}(t)(\widehat{m}v(t) - \widehat{m}\widehat{c_r}). \end{aligned} \tag{3.31}$$

From (3.25), (3.28), (3.31), we have the following relationship between $u(t)$ and $u_d(t)$:

$$\begin{aligned} u(t) = u_d(t) &+ \widehat{\chi_r}(t)(m(v(t) - c_r) - \widehat{m}v(t) + \widehat{m}\widehat{c_r}) \\ &+ \widehat{\chi_l}(t)(m(v(t) - c_l) - \widehat{m}v(t) + \widehat{m}\widehat{c_l}) + d_b(t). \end{aligned} \tag{3.32}$$

Once again we have encountered the important quantity

$$\begin{aligned} d_b(t) = (\chi_r(t) &- \widehat{\chi_r}(t))(m(v(t) - c_r)) \\ &+ (\chi_l(t) - \widehat{\chi_l}(t))(m(v(t) - c_l)) + \chi_s(t)u_s, \end{aligned} \tag{3.33}$$

which represents the unparametrizable part of the error $u(t) - u_d(t)$.

From (3.33), we see that $d_b(t)$ is reduced to zero if

$$\chi_r(t) - \widehat{\chi_r}(t) = \chi_l(t) - \widehat{\chi_l}(t) = \chi_s(t) = 0.$$

This condition is satisfied if $\widehat{m} = m$, $\widehat{c_l} = c_l$, and $\widehat{c_r} = c_r$, because, after initialization, the motion of $v(t)$, $u(t)$ will not be on any one of the inner segments, and $u(t)$, $v(t)$ are on the upward (or downward) side of $B(\cdot)$ if and only if $u_d(t)$, $v(t)$ are on the upward (or downward) side of $BI(\cdot)$.

When the parameter estimation errors are present, then $d_b(t) \neq 0$. However, as we show next, $d_b(t)$ is bounded for any $t \geq 0$.

Proposition 3.2 *The unparametrizable part $d_b(t)$ of the backlash inverse control error $u(t)$ - $u_d(t)$ is bounded for any $t \geq 0$. Furthermore, it reduces to zero for $t \geq t_0$ when the backlash parameter estimates are equal to their true values—that is, $\widehat{m} = m$, $\widehat{c_r} = c_r$, $\widehat{c_l} = c_l$—and the backlash inverse is initialized at t_0: $B(BI(u_d(t_0))) = u_d(t_0)$.*

Proof: There are three different cases to be examined:

(i) If $\chi_l(t) = 1$, $\chi_r(t) = \chi_s(t) = 0$, then

$$d_b(t) = \begin{cases} 0 & \text{if } \widehat{\chi_l}(t) = 1,\ \widehat{\chi_r}(t) = 0 \\ m(c_r - c_l) & \text{if } \widehat{\chi_l}(t) = 0,\ \widehat{\chi_r}(t) = 1. \end{cases}$$

(ii) If $\chi_r(t) = 1$, $\chi_l(t) = \chi_s(t) = 0$, then

$$d_b(t) = \begin{cases} 0 & \text{if } \widehat{\chi_l}(t) = 0,\ \widehat{\chi_r}(t) = 1 \\ m(c_l - c_r) & \text{if } \widehat{\chi_l}(t) = 1,\ \widehat{\chi_r}(t) = 0. \end{cases}$$

(iii) If $\chi_s(t) = 1$, $\chi_l(t) = \chi_r(t) = 0$, then

$$d_b(t) = \begin{cases} m(c_l - c_{ls}) & \text{if } \widehat{\chi_l}(t) = 1,\ \widehat{\chi_r}(t) = 0 \\ m(c_r - c_{rs}) & \text{if } \widehat{\chi_l}(t) = 0,\ \widehat{\chi_r}(t) = 1, \end{cases}$$

where c_{ls} or c_{rs} depends on the motion of $v(t)$ and $u(t)$ on the inner segment when $\chi_s(t) = 1$, whose values are in the interval (c_l, c_r). From these expressions, it is clear that $d_b(t)$ is bounded.

The second part follows directly from Lemma 2.2 and (3.32). ∇

To arrive at a compact expression for (3.32), we redefine the parameter estimates $\widehat{mc_r} = \widehat{m}\widehat{c_r}$, $\widehat{mc_l} = \widehat{m}\widehat{c_l}$ and introduce the parameter vectors θ_b^*, θ_b and the regressor vector $\omega_b(t)$:

$$\theta_b^* = (mc_r, m, mc_l)^T \tag{3.34}$$

$$\theta_b = (\widehat{mc_r}, \widehat{m}, \widehat{mc_l})^T \tag{3.35}$$

$$\omega_b(t) = (\widehat{\chi_r}(t), -v(t), \widehat{\chi_l}(t))^T. \tag{3.36}$$

Using (3.31), (3.35), and (3.36), we express the backlash inverse (3.10) as

$$u_b(t) = -\theta_b^T \omega_b(t). \tag{3.37}$$

Finally, using (3.32), (3.34) - (3.36), we obtain the parametrized expression for the backlash inverse error $u(t) - u_d(t)$ in terms of the parameter error $\theta_b(t) - \theta_b^*$ with a bounded unparametrized part $d_b(t)$:

$$u(t) - u_d(t) = (\theta_b - \theta_b^*)^T \omega_b(t) + d_b(t). \tag{3.38}$$

Equations (3.37) and (3.38) also hold for the discrete-time case. These parametrized expressions will be used for our adaptive inverse designs in subsequent chapters.

3.3 Hysteresis Inverse

Now we show that the hysteresis characteristic (2.11) or (2.12) also has a parametrized inverse which, when implemented with true parameters, can cancel the hysteresis effect and, when implemented with parameter estimates, also leads to a linearly parametrized control error $u(t) - u_d(t)$, similar to the dead-zone and backlash cases.

3.3.1 Inverse Characteristic

We first show that the inverse of the hysteresis model (2.11) is the hysteresis-like characteristic $v(t) = \widehat{HI}(u_d(t))$ in Figure 3.5. In this characteristic the motion is confined to two half-lines, two line segments, and the quadrilateral formed by these half-lines and segments.

With some estimates $\widehat{m_t}$, $\widehat{c_t}$, $\widehat{m_b}$, $\widehat{c_b}$, $\widehat{m_r}$, $\widehat{c_r}$, $\widehat{m_l}$, $\widehat{c_l}$ of the unknown hysteresis parameters m_t, c_t, m_b, c_b, m_r, c_r, m_l, c_l, the half-lines are

$$v(t) = \frac{1}{\widehat{m_t}}(u_d(t) - \widehat{c_t}), \ u_d(t) > u_1 = \frac{\widehat{m_l}(\widehat{m_t}\widehat{c_l} + \widehat{c_t})}{\widehat{m_l} - \widehat{m_t}} \tag{3.39}$$

$$v(t) = \frac{1}{\widehat{m_b}}(u_d(t) - \widehat{c_b}), \ u_d(t) < u_2 = \frac{\widehat{m_r}(\widehat{m_b}\widehat{c_r} + \widehat{c_b})}{\widehat{m_r} - \widehat{m_b}} \tag{3.40}$$

and the line segments are

$$v(t) = \frac{1}{\widehat{m_r}} u_d(t) + \widehat{c_r}, \ u_2 \le u_d(t) < u_3 = \frac{\widehat{m_r}(\widehat{m_t}\widehat{c_r} + \widehat{c_t})}{\widehat{m_r} - \widehat{m_t}} \tag{3.41}$$

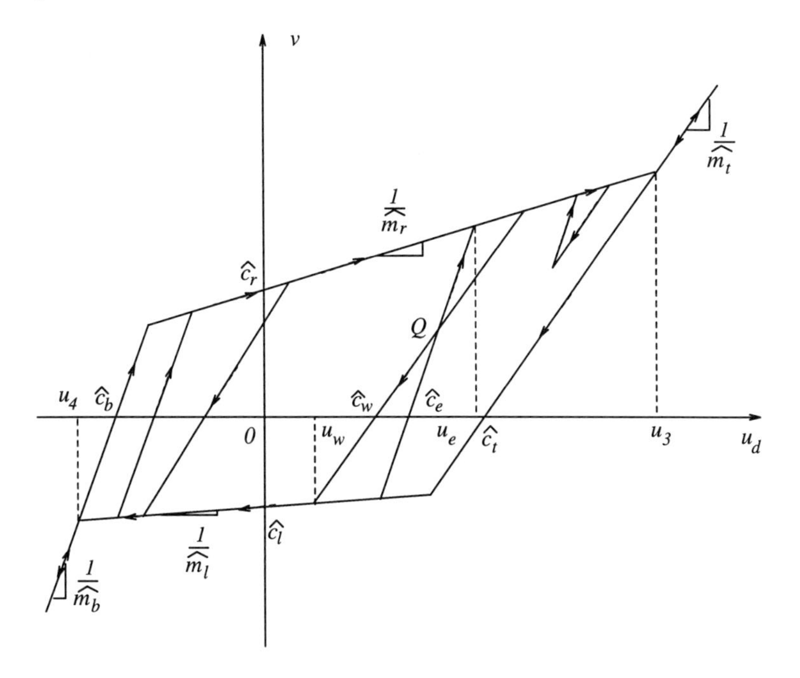

Figure 3.5: Hysteresis inverse.

$$v(t) = \frac{1}{\widehat{m_l}} u_d(t) + \widehat{c_l}, \ \frac{\widehat{m_l}(\widehat{m_b}\widehat{c_l} + \widehat{c_b})}{\widehat{m_l} - \widehat{m_b}} = u_4 < u_d(t) \leq u_1, \tag{3.42}$$

where u_1, u_2, u_3, and u_4 are the values of $u_d(t)$ at the lower-right, upper-left, upper-right, and lower-left corners of the quadrilateral formed by the half-lines (3.39) - (3.40) and segments (3.41) - (3.42).

Along the segments the time derivatives of $v(t)$, $u_d(t)$ are of constant sign, namely,

$$\dot{u}_d(t) > 0, \ \dot{v}(t) > 0 \text{ if } v(t) = \frac{1}{\widehat{m_r}} u_d(t) + \widehat{c_r}$$

$$\dot{u}_d(t) < 0, \ \dot{v}(t) < 0 \text{ if } v(t) = \frac{1}{\widehat{m_l}} u_d(t) + \widehat{c_l}.$$

Inside the loop formed by the half-lines (3.39) - (3.40) and the segments (3.41) - (3.42), $v(t)$ is

$$v(t) = \begin{cases} \frac{1}{\widehat{m_t}}(u_d(t) - \widehat{c_w}) & \text{if } \dot{u}_d(t) < 0 \\ \frac{1}{\widehat{m_b}}(u_d(t) - \widehat{c_e}) & \text{if } \dot{u}_d(t) > 0, \end{cases} \tag{3.43}$$

where $\widehat{c_w} \in (\widehat{c_t}, \widehat{c_1})$, $\widehat{c_e} \in (\widehat{c_2}, \widehat{c_b})$ are quantities which depend on the point

$(u_d(t), v(t))$ where $\dot{u}_d(t)$ changes the sign, with

$$\widehat{c}_1 = \begin{cases} (\widehat{m_b} - \widehat{m_t})\frac{\widehat{c}_b + \widehat{m_l}\widehat{c}_l}{\widehat{m_l} - \widehat{m_b}} + \widehat{c}_b & \text{if } \widehat{m_t} < \widehat{m_b} \\ (\widehat{m_b} - \widehat{m_t})\frac{\widehat{c}_b + \widehat{m_r}\widehat{c}_r}{\widehat{m_r} - \widehat{m_b}} + \widehat{c}_b & \text{if } \widehat{m_t} > \widehat{m_b} \\ \widehat{c}_b & \text{if } \widehat{m_t} = \widehat{m_b} \end{cases}$$

$$\widehat{c}_2 = \begin{cases} (\widehat{m_t} - \widehat{m_b})\frac{\widehat{c}_t + \widehat{m_r}\widehat{c}_r}{\widehat{m_r} - \widehat{m_t}} + \widehat{c}_t & \text{if } \widehat{m_t} > \widehat{m_b} \\ (\widehat{m_t} - \widehat{m_b})\frac{\widehat{c}_t + \widehat{m_l}\widehat{c}_l}{\widehat{m_l} - \widehat{m_t}} + \widehat{c}_t & \text{if } \widehat{m_t} < \widehat{m_b} \\ \widehat{c}_t & \text{if } \widehat{m_t} = \widehat{m_b}. \end{cases}$$

The relationship (3.43) also holds for a part of each half-line:

(i) When $\widehat{m_t} > \widehat{m_b}$, on the common part of the half-line (3.39) and the hysteresis inverse quadrilateral, the signal motion is

$$v(t) = \frac{u_d(t) - \widehat{c}_t}{\widehat{m_t}}, \ \dot{u}_d(t) < 0.$$

(ii) When $\widehat{m_t} < \widehat{m_b}$, on the common part of the half-line (3.40) and the hysteresis inverse quadrilateral, the signal motion is

$$v(t) = \frac{u_d(t) - \widehat{c}_b}{\widehat{m_b}}, \ \dot{u}_d(t) > 0.$$

Furthermore, $\dot{u}_d(t)$ and $\dot{v}(t)$ are allowed to have both positive and negative signs on other parts of the half-lines (3.39) - (3.40):

$$v(t) = \frac{1}{\widehat{m_t}}(u_d(t) - \widehat{c}_t), \ u_d(t) \geq u_3$$

$$v(t) = \frac{1}{\widehat{m_b}}(u_d(t) - \widehat{c}_b), \ u_d(t) \leq u_4$$

$$v(t) = \frac{1}{\widehat{m_t}}(u_d(t) - \widehat{c}_t), \ u_1 < u_d(t) < u_3, \quad \text{when } \widehat{m_t} < \widehat{m_b}$$

$$v(t) = \frac{1}{\widehat{m_b}}(u_d(t) - \widehat{c}_b), \ u_4 < u_d(t) < u_2, \quad \text{when } \widehat{m_t} > \widehat{m_b}.$$

Figure 3.5 also indicates the inverse characteristics for the two minor hysteresis loops shown in Figure 2.16. With the exact estimates of the true hysteresis parameters, these two inverse loops cancel the effects of these hysteresis loops, after an initialization of the hysteresis inverse, as shown in Section 3.3.3.

The motion of $u_d(t)$ and $v(t)$ along the half-lines (3.39) - (3.40), along the segments (3.41) - (3.42), and around inside loops is completely described by

$$\dot{v}(t) = \begin{cases} \frac{1}{\widehat{m_t}}\dot{u}_d(t) & \text{if } u_d(t) \geq u_3 \text{ and} \\ & v(t) = \frac{1}{\widehat{m_t}}(u_d(t) - \widehat{c_t}), \text{ or} \\ & \text{if } u_4 < u_d(t) < u_3,\ \dot{u}_d(t) < 0, \\ & v(t) = \frac{1}{\widehat{m_t}}(u_d(t) - \widehat{c_w}), \\ & v(t) \neq \frac{1}{\widehat{m_l}}u_d(t) + \widehat{c_l} \text{ and} \\ & v(t) \neq \frac{1}{\widehat{m_b}}(u_d(t) - \widehat{c_b}), \text{ or} \\ & \text{if } \widehat{m_t} < \widehat{m_b},\ u_4 < u_d(t) < u_3, \\ & \dot{u}_d(t) < 0 \text{ and } v(t) = \frac{1}{\widehat{m_b}}(u_d(t) - \widehat{c_b}), \text{ or} \\ & \text{if } \widehat{m_t} < \widehat{m_b},\ u_4 < u_d(t) < u_3, \\ & \dot{u}_d(t) > 0 \text{ and } v(t) = \frac{1}{\widehat{m_t}}(u_d(t) - \widehat{c_t}), \text{ or} \\ \frac{1}{\widehat{m_b}}\dot{u}_d(t) & \text{if } u_d(t) \leq u_4 \text{ and} \\ & v(t) = \frac{1}{\widehat{m_b}}(u_d(t) - \widehat{c_b}), \text{ or} \\ & \text{if } u_4 < u_d(t) < u_3,\ \dot{u}_d(t) > 0, \\ & v(t) = \frac{1}{\widehat{m_b}}(u_d(t) - \widehat{c_e}), \\ & v(t) \neq \frac{1}{\widehat{m_r}}u_d(t) + \widehat{c_r} \text{ and} \\ & v(t) \neq \frac{1}{\widehat{m_t}}(u_d(t) - \widehat{c_t}), \text{ or} \\ & \text{if } \widehat{m_t} > \widehat{m_b},\ u_4 < u_d(t) < u_3, \\ & \dot{u}_d(t) > 0 \text{ and } v(t) = \frac{1}{\widehat{m_t}}(u_d(t) - \widehat{c_t}), \text{ or} \\ & \text{if } \widehat{m_t} > \widehat{m_b},\ u_4 < u_d(t) < u_3, \\ & \dot{u}_d(t) < 0 \text{ and } v(t) = \frac{1}{\widehat{m_b}}(u_d(t) - \widehat{c_b}) \\ \frac{1}{\widehat{m_r}}\dot{u}_d(t) & \text{if } u_4 < u_d(t) < u_3,\ \dot{u}_d(t) > 0 \text{ and} \\ & v(t) = \frac{1}{\widehat{m_r}}u_d(t) + \widehat{c_r} \\ \frac{1}{\widehat{m_l}}\dot{u}_d(t) & \text{if } u_4 < u_d(t) < u_3,\ \dot{u}_d(t) < 0 \text{ and} \\ & v(t) = \frac{1}{\widehat{m_l}}u_d(t) + \widehat{c_l} \\ 0 & \text{if } \dot{u}_d(t) = 0. \end{cases} \tag{3.44}$$

Similar to the continuous-time backlash inverse, the continuous-time hysteresis inverse (3.44) needs the knowledge of the sign of $\dot{u}_d(t)$ for its implementation. As will be shown next, a discrete-time hysteresis inverse can eliminate this difficulty.

3.3.2 Discrete-Time Representation

For discrete-time applications, we need a discrete-time hysteresis inverse for the discrete-time hysteresis model (2.12). Since the characteristic (2.12) is dynamic, its inverse is also dynamic and the hysteresis inverse output $v(t) = HI(u_d(t))$ is determined not only by $u_d(t)$ but also by $u_d(t-1)$ and $v(t-1)$.

Let us look at Figure 3.5, where the point Q represents the state of the (u_d, v)-trajectory at $t-1$:

$$Q = (u_d(t-1), v(t-1)).$$

Passing this point Q, there are two segments: One has slope $\frac{1}{\widehat{m_b}}$, which intersects the u_d-axis at $\widehat{c_e}$ and intersects the segment $v(t) = \frac{u_d(t)}{\widehat{m_r}} + \widehat{c_r}$ at a point whose projection on the u_d-axis is u_e; the other has slope $\frac{1}{\widehat{m_t}}$, which intersects the u_d-axis at $\widehat{c_w}$ and intersects the segment $v(t) = \frac{u_d(t)}{\widehat{m_l}} + \widehat{c_r}$ at a point whose projection on the u_d-axis is u_w. More precisely, these quantities satisfy $u_4 \leq u_w \leq u_e \leq u_3$ and are given as

$$\begin{aligned}
\widehat{c_e} &= u_d(t-1) - \widehat{m_b} v(t-1) \\
\widehat{c_w} &= u_d(t-1) - \widehat{m_t} v(t-1) \\
u_e &= \frac{\widehat{m_r}(\widehat{m_b}\widehat{c_r} + \widehat{c_e})}{\widehat{m_r} - \widehat{m_b}} \\
u_w &= \frac{\widehat{m_l}(\widehat{m_t}\widehat{c_l} + \widehat{c_w})}{\widehat{m_l} - \widehat{m_t}}.
\end{aligned}$$

Then $v(t)$, the output of the hysteresis inverse, is determined as

$$v(t) = \begin{cases}
v(t-1) & \text{if } u_d(t) = u_d(t-1) \\
\frac{1}{\widehat{m_t}}(u_d(t) - \widehat{c_t}) & \text{if } u_d(t) \geq u_3, \text{ or} \\
 & \text{if } \frac{1}{\widehat{m_t}} > \frac{1}{\widehat{m_b}}, \\
 & u_d(t-1) < u_d(t) < u_3, \text{ and} \\
 & v(t-1) = \frac{1}{\widehat{m_t}}(u_d(t-1) + \widehat{c_t}) \\
\frac{1}{\widehat{m_b}}(u_d(t) - \widehat{c_b}) & \text{if } u_d(t) \leq u_4, \text{ or} \\
 & \text{if } \frac{1}{\widehat{m_t}} < \frac{1}{\widehat{m_b}}, \\
 & u_4 < u_d(t) < u_d(t-1), \text{ and} \\
 & v(t-1) = \frac{1}{\widehat{m_b}}(u_d(t-1) + \widehat{c_b}) \\
\frac{1}{\widehat{m_r}} u_d(t) + \widehat{c_r} & \text{if } u_3 \geq u_d(t) \geq u_e \\
\frac{1}{\widehat{m_l}} u_d(t) + \widehat{c_l} & \text{if } u_4 \leq u_d(t) \leq u_w \\
\frac{1}{\widehat{m_t}}(u_d(t) - \widehat{c_w}) & \text{if } u_w < u_d(t) < u_d(t-1) \\
\frac{1}{\widehat{m_b}}(u_d(t) - \widehat{c_e}) & \text{if } u_d(t-1) < u_d(t) < u_e.
\end{cases} \tag{3.45}$$

The *discrete-time* hysteresis inverse (3.45) does not require the knowledge of $\dot{u}_d(t)$ for implementation, making a discrete-time adaptive inverse control design more practical.

3.3.3 Inverse Lemma

When the hysteresis inverse is implemented with the true parameters m_t, c_t, m_b, c_b, m_r, c_r, m_l, c_l, it has the following property for both the continuous-time and discrete-time cases.

Lemma 3.3 *(Hysteresis Inverse) The characteristic $HI(\cdot)$ defined by (3.44) is the right inverse of the hysteresis characteristic $H(\cdot)$ defined by (2.11), in the sense that*

$$H(HI(u_d(t_0))) = u_d(t_0) \Rightarrow H(HI(u_d(t))) = u_d(t),\ \forall t \geq t_0.$$

This lemma can be proved in a similar way to that for Lemma 3.2 which establishes the right invertibility of the backlash characteristic.

The mapping (3.44) or (3.45) may fail to define a hysteresis inverse only if $u_d(t)$ is such that $v(t)$ and $u(t)$ never leave the hysteresis loop. Such a situation happens when $u_d(t)$ is inside the hysteresis inverse loop and $u(t) \neq u_d(t)$ initially. Then if the motion of $u_d(t)$ is such that $u(t)$ never leaves the hysteresis loop, it is impossible to correct the error between this $u_d(t)$ and $u(t)$. However, as $u_d(t)$ is a signal at our disposal, an initialization of the hysteresis inverse by an appropriate choice of $u_d(t_0)$ can always force $v(t)$ and $u(t)$ to leave the interior of the hysteresis loop at t_0 so that $u(t_0) = u_d(t_0)$. Then Lemma 3.3 applies and $u(t) = u_d(t)$ for all $t > t_0$.

To implement the hysteresis inverse (3.44) or (3.45), we need to prevent the hysteresis slope estimates from being zero, especially in the adaptive case when these estimates are obtained from an adaptive update law. To this end, we assume that positive constants m_{t1}, m_{t2}, m_{b1}, m_{b2}, m_{r0}, m_{l0} and constants c_{b1}, c_{bt}, c_{t2}, c_{l1}, c_{lr}, c_{r2} are known such that $m_{t1} \leq m_t \leq m_{t2}$, $m_{b1} \leq m_b \leq m_{b2}$, $\max\{m_{t2}, m_{b2}\} < m_{r0} \leq m_r$, $\max\{m_{t2}, m_{b2}\} < m_{l0} \leq m_l$, and $c_{b1} \leq c_b \leq c_{bt} \leq c_t \leq c_{t2}$, $c_{l1} \leq c_l \leq c_{lr} \leq c_r \leq c_{r2}$. The estimates $\widehat{m_t}$, $\widehat{c_t}$, $\widehat{m_b}$, $\widehat{c_b}$, $\widehat{m_r}$, $\widehat{c_r}$, $\widehat{m_l}$, $\widehat{c_l}$ are chosen to satisfy the same condition.

3.3.4 Parametrization

To proceed for a parametrization of the hysteresis inverse, we define the hysteresis inverse indicator functions

$$\widehat{\chi_t}(t) = \chi[v(t) = \tfrac{u_d(t)-\widehat{c_t}}{\widehat{m_t}}] \tag{3.46}$$

$$\widehat{\chi_b}(t) = \chi[v(t) = \tfrac{u_d(t)-\widehat{c_b}}{\widehat{m_b}}] \tag{3.47}$$

$$\widehat{\chi_r}(t) = \chi[v(t) = \tfrac{1}{\widehat{m_r}} u_d(t) + \widehat{c_r}] \tag{3.48}$$

$$\widehat{\chi_l}(t) = \chi[v(t) = \tfrac{1}{\widehat{m_l}}u_d(t) + \widehat{c_l}] \quad (3.49)$$

$$\widehat{\chi_w}(t) = \chi[v(t) = \tfrac{u_d(t)-\widehat{c_w}}{\widehat{m_t}}] \quad (3.50)$$

$$\widehat{\chi_e}(t) = \chi[v(t) = \tfrac{u_d(t)-\widehat{c_e}}{\widehat{m_b}}] \quad (3.51)$$

and the hysteresis indicator functions

$$\chi_t(t) = \chi[u(t) = m_t v(t) + c_t] \quad (3.52)$$

$$\chi_b(t) = \chi[u(t) = m_b v(t) + c_b] \quad (3.53)$$

$$\chi_r(t) = \chi[u(t) = m_r(v(t) - c_r)] \quad (3.54)$$

$$\chi_l(t) = \chi[u(t) = m_l(v(t) - c_l)] \quad (3.55)$$

$$\chi_d(t) = \chi[u(t) = m_t v(t) + c_d] \quad (3.56)$$

$$\chi_u(t) = \chi[u(t) = m_b v(t) + c_u]. \quad (3.57)$$

In defining these indicator functions, we do not repeatedly count any intersection of the half-lines and the segments; for example, at the top right corner of $\widehat{HI}(\cdot)$ we define $\widehat{\chi_t}(t) = 1$ and $\widehat{\chi_r}(t) = 0$ and when it goes down passing the $\widehat{c_b}(t)$ point we define $\widehat{\chi_b}(t) = 1$ and $\widehat{\chi_w}(t) = 0$. The same statement applies to the hysteresis indicator functions. With this constraint, only one of these indicator functions is nonzero at any given time t, namely,

$$\widehat{\chi_t}(t) + \widehat{\chi_b}(t) + \widehat{\chi_r}(t) + \widehat{\chi_l}(t) + \widehat{\chi_w}(t) + \widehat{\chi_e}(t) = 1 \quad (3.58)$$

$$\widehat{\chi_k}^2(t) = \widehat{\chi_k}(t), \ \widehat{\chi_i}(t)\widehat{\chi_j}(t) = 0 \quad (3.59)$$

for $i \neq j, \ i, j, k \in \{t, b, r, l, w, e\}$.

Similarly, for the hysteresis indicator functions, we have

$$\chi_t(t) + \chi_b(t) + \chi_r(t) + \chi_l(t) + \chi_d(t) + \chi_u(t) = 1$$

$$\chi_k^2(t) = \chi_k(t), \ \chi_i(t)\chi_j(t) = 0$$

for $i \neq j, \ i, j, k \in \{t, b, r, l, d, u\}$.

Using (2.4) - (2.10), (3.52) - (3.59), we express $u(t)$ as

$$\begin{aligned}
u(t) &= \chi_t(t)(m_t v(t) + c_t) + \chi_b(t)(m_b v(t) + c_b) \\
&\quad + \chi_r(t)(m_r(v(t) - c_r)) + \chi_l(t)(m_l(v(t) - c_l)) \\
&\quad + \chi_d(t)(m_t v(t) + c_d(t)) + \chi_u(t)(m_b v(t) + c_u(t)) \\
&= u_d(t) + \widehat{\chi_t}(t)(m_t v(t) + c_t - \widehat{\chi_t}(t)u_d(t)) \\
&\quad + \widehat{\chi_b}(t)(m_b v(t) + c_b - \widehat{\chi_b}(t)u_d(t))
\end{aligned}$$

$$+\widehat{\chi_r}(t)(m_r(v(t)-c_r)-\widehat{\chi_r}(t)u_d(t))$$
$$+\widehat{\chi_l}(t)(m_l(v(t)-c_l)-\widehat{\chi_l}(t)u_d(t))$$
$$+\widehat{\chi_w}(t)(m_tv(t)+c_d(t)-\widehat{\chi_w}(t)u_d(t))$$
$$+\widehat{\chi_e}(t)(m_bv(t)+c_u(t)-\widehat{\chi_e}(t)u_d(t))+d_1(t), \quad (3.60)$$

where

$$\begin{aligned} d_1(t) = &(\chi_t(t)-\widehat{\chi_t}(t))(m_tv(t)+c_t) \\ &+(\chi_b(t)-\widehat{\chi_b}(t))(m_bv(t)+c_b) \\ &+(\chi_r(t)-\widehat{\chi_r}(t))(m_r(v(t)-c_r)) \\ &+(\chi_l(t)-\widehat{\chi_l}(t))(m_l(v(t)-c_l)) \\ &+(\chi_d(t)-\widehat{\chi_w}(t))(m_tv(t)+c_d(t)) \\ &+(\chi_u(t)-\widehat{\chi_e}(t))(m_bv(t)+c_u(t)). \end{aligned} \quad (3.61)$$

Since the estimates of the hysteresis parameters are bounded, the hysteresis inverse loop is bounded. If $v(t)$ is large such that the state $(u_d(t), v(t))$ is outside the hysteresis inverse loop so that the state $(v(t), u(t))$ is outside the hysteresis loop, then all $\chi(t)$'s and $\widehat{\chi}(t)$'s are zero except for $\chi_t(t)$ (or $\chi_b(t)$) and $\widehat{\chi_t}(t)$ (or $\widehat{\chi_b}(t)$) so that $d_1(t) = 0$ from (3.61). This implies that $d_1(t)$ is bounded whenever the estimates of the hysteresis parameters are bounded, which is crucial for our adaptive hysteresis inverse designs. In fact, to have a bounded hysteresis inverse loop, it is sufficient to ensure bounded parameter estimates $\widehat{c_t}$, $\widehat{c_b}$, $\widehat{c_l}$, and $\widehat{c_r}$. Therefore, the unparametrizable part $d_1(t)$ is bounded if the estimates $\widehat{c_t}$, $\widehat{c_b}$, $\widehat{c_l}$, and $\widehat{c_r}$ are bounded.

Using (3.39) - (3.42), (3.43), and (3.46) - (3.51), we express $v(t)$ as

$$\begin{aligned} v(t) = &\frac{\widehat{\chi_t}(t)}{\widehat{m_t}}(u_d(t)-\widehat{c_t}) + \frac{\widehat{\chi_b}(t)}{\widehat{m_b}}(u_d(t)-\widehat{c_b}) \\ &+\frac{\widehat{\chi_r}(t)}{\widehat{m_r}}(u_d(t)+\widehat{m_r}\widehat{c_r}) + \frac{\widehat{\chi_l}(t)}{\widehat{m_l}}(u_d(t)+\widehat{m_l}\widehat{c_l}) \\ &+\frac{\widehat{\chi_w}(t)}{\widehat{m_t}}(u_d(t)-\widehat{c_w}) + \frac{\widehat{\chi_e}(t)}{\widehat{m_b}}(u_d(t)-\widehat{c_e}). \end{aligned} \quad (3.62)$$

In view of (3.59), from (3.62), we obtain

$$\widehat{\chi_t}(t)u_d(t) = \widehat{\chi_t}(t)(\widehat{m_t}v(t)+\widehat{c_t}) \quad (3.63)$$

$$\widehat{\chi_b}(t)u_d(t) = \widehat{\chi_b}(t)(\widehat{m_b}v(t)+\widehat{c_b}) \quad (3.64)$$

$$\widehat{\chi_r}(t)u_d(t) = \widehat{\chi_r}(t)(\widehat{m_r}v(t)-\widehat{m_r}(t)\widehat{c_r}) \quad (3.65)$$

$$\widehat{\chi_l}(t)u_d(t) = \widehat{\chi_l}(t)(\widehat{m_l}v(t)-\widehat{m_l}(t)\widehat{c_l}) \quad (3.66)$$

$$\widehat{\chi_w}(t)u_d(t) = \widehat{\chi_w}(t)(\widehat{m_t}v(t) + \widehat{c_w}) \tag{3.67}$$

$$\widehat{\chi_e}(t)u_d(t) = \widehat{\chi_e}(t)(\widehat{m_b}v(t) + \widehat{c_e}). \tag{3.68}$$

Substituting (3.63) - (3.68) in (3.60), denoting $\widehat{m_r c_r} = \widehat{m_r}\widehat{c_r}$ and $\widehat{m_l c_l} = \widehat{m_l}\widehat{c_l}$ so that the estimates of c_r and c_l are obtained from

$$\widehat{c_r} = \frac{\widehat{m_r c_r}}{\widehat{m_r}},\ \widehat{c_l} = \frac{\widehat{m_l c_l}}{\widehat{m_l}},$$

and introducing the hysteresis parameter vectors

$$\theta_h^* = (m_t, c_t, m_b, c_b, m_r, m_r c_r, m_l, m_l c_l)^T \tag{3.69}$$

$$\theta_h = (\widehat{m_t}, \widehat{c_t}, \widehat{m_b}, \widehat{c_b}, \widehat{m_r}, \widehat{m_r c_r}, \widehat{m_l}, \widehat{m_l c_l})^T \tag{3.70}$$

and the corresponding regressor vector

$$\begin{aligned}\omega_h(t) = (&-(\widehat{\chi_t}(t) + \widehat{\chi_w}(t))v(t), -\widehat{\chi_t}(t), -(\widehat{\chi_b}(t) + \widehat{\chi_e}(t))v(t),\\ &-\widehat{\chi_b}(t), -\widehat{\chi_r}(t)v(t), \widehat{\chi_r}(t), -\widehat{\chi_l}(t)v(t), \widehat{\chi_l}(t))^T,\end{aligned} \tag{3.71}$$

we obtain a compact expression for the control error

$$u(t) - u_d(t) = (\theta_h - \theta_h^*)^T \omega_h(t) + d_h(t), \tag{3.72}$$

where

$$d_h(t) = d_1(t) + \widehat{\chi_w}(t)(c_d - \widehat{c_w}) + \widehat{\chi_e}(t)(c_u - \widehat{c_e}). \tag{3.73}$$

We have expressed the control error $u(t) - u_d(t)$ as a sum of a parametrizable part and an unparametrizable part $d_h(t)$ which acts as an unknown disturbance. This disturbance has the following properties.

Proposition 3.3 *The unparametrizable part $d_h(t)$ of the control error $u(t) - u_d(t)$ is bounded for any $t \geq 0$ if the parameter estimates $\widehat{c_t}$, $\widehat{c_b}$, $\widehat{c_l}$, and $\widehat{c_r}$ are bounded, and it reduces to zero for $t \geq t_0$ when the hysteresis parameter error $\theta_h - \theta_h^*$ is zero and the hysteresis inverse is correctly initialized at t_0:*

$$H(HI(u_d(t_0))) = u_d(t_0).$$

In the adaptive inverse case, the estimates $\widehat{c_t}$, $\widehat{c_b}$, $\widehat{c_l}$, and $\widehat{c_r}$ are obtained from an adaptive update law with parameter projection which ensures the boundedness of $\widehat{c_t}$, $\widehat{c_b}$, $\widehat{c_l}$, and $\widehat{c_r}$. Thus, the boundness of the "disturbance" $d_h(t)$ can be guaranteed.

The signals $\chi_t(t)$, $\chi_b(t)$, $\chi_r(t)$, $\chi_l(t)$, $\chi_d(t)$, and $\chi_u(t)$, which describe the motion of the hysteresis output $u(t) = H(v(t))$ in (3.60), are not available for measurement. We have expressed the parametrizable part of the control error $u(t) - u_d(t)$ given by (3.72) in terms of the measured signals $\widehat{\chi_t}(t)$, $\widehat{\chi_b}(t)$, $\widehat{\chi_r}(t)$, $\widehat{\chi_l}(t)$, $\widehat{\chi_w}(t)$, and $\widehat{\chi_e}(t)$ which appear in $\omega_h(t)$.

To derive an expression of the hysteresis inverse estimate, as similar to (3.37) for the backlash inverse estimate, using (3.58), we write

$$\begin{aligned} u_d(t) = {} & \widehat{\chi_t}(t)u_d(t) + \widehat{\chi_b}(t)u_d(t) + \widehat{\chi_r}(t)u_d(t) + \widehat{\chi_l}(t)u_d(t) \\ & + \widehat{\chi_w}(t)u_d(t) + \widehat{\chi_e}(t)u_d(t). \end{aligned} \tag{3.74}$$

Substituting (3.63) - (3.68) in (3.74) and using (3.70) and (3.71), we obtain

$$u_d(t) = -\theta_h^T \omega_h(t) + \widehat{\chi_w}(t)\widehat{c_w}(t) + \widehat{\chi_e}(t)\widehat{c_e}(t), \tag{3.75}$$

which describes the motion of the hysteresis inverse output $v(t) = \widehat{HI}(u_d(t))$ and will be useful for our adaptive controller design. We note that the term $\widehat{\chi_w}(t)\widehat{c_w}(t) + \widehat{\chi_e}(t)\widehat{c_e}(t)$, which does not appear in the similar dead-zone or backlash inverse expression (3.7) or (3.37), is due to the signal motion inside the hysteresis inverse loop.

3.4 Unified Inverse Expression

To develop a control design approach unifying the dead-zone, backlash, and hysteresis inverses $\widehat{DI}(\cdot)$, $\widehat{BI}(\cdot)$, $\widehat{HI}(\cdot)$, we introduce

$$\{\theta_N, \theta_N^*, \omega_N, d_N\} = \begin{cases} \{\theta_d, \theta_d^*, \omega_d, d_d\} & \text{for } \widehat{DI}(\cdot) \\ \{\theta_b, \theta_b^*, \omega_b, d_b\} & \text{for } \widehat{BI}(\cdot) \\ \{\theta_h, \theta_h^*, \omega_h, d_h\} & \text{for } \widehat{HI}(\cdot), \end{cases}$$

where θ_d^*, θ_d, $\omega_d(t)$, $d_d(t)$ (θ_b^*, θ_b, $\omega_b(t)$, $d_b(t)$, or θ_h^*, θ_h, $\omega_h(t)$, $d_h(t)$) are as in (3.4) - (3.6), (3.9) ((3.34) - (3.36), (3.33), or (3.69) - (3.71), (3.73)). Then the control error expressions (3.8), (3.38), and (3.72) are unified as

$$u(t) - u_d(t) = (\theta_N - \theta_N^*)^T \omega_N(t) + d_N(t), \tag{3.76}$$

where $d_N(t)$ is bounded for any $t \geq 0$ and $d_N(t) = 0$ when $\theta_N(t) = \theta_N^*$, $t \geq t_0 \geq 0$, under the initialization conditions:

$$u_d(\tau) = N(NI(u_d(\tau)) \begin{cases} \text{not needed} & \text{for } \widehat{DI}(\cdot) \\ \tau = t_0 & \text{for } \widehat{BI}(\cdot) \\ \tau = t_0 & \text{for } \widehat{HI}(\cdot). \end{cases}$$

In view of (3.7), (3.37), and (3.75), the input nonlinearity inverse $v(t) = \widehat{NI}(u_d(t))$ can be expressed as

$$u_d(t) = -\theta_N^T(t)\omega_N(t) + a_N(t), \tag{3.77}$$

where

$$a_N(t) = \begin{cases} 0 & \text{for } \widehat{DI}(\cdot) \\ 0 & \text{for } \widehat{BI}(\cdot) \\ a_h(t) = \widehat{\chi_w}(t)\widehat{c_w}(t) + \widehat{\chi_e}(t)\widehat{c_e}(t) & \text{for } \widehat{HI}(\cdot). \end{cases}$$

While they have the unified expressions (3.76) and (3.77), the dead-zone inverse characteristic (3.2) is static whereas the backlash inverse characteristic (3.14) or (3.22) and the hysteresis inverse characteristic (3.44) or (3.45) are dynamic. Furthermore, to implement the continuous-time backlash inverse (3.14) or the continuous-time hysteresis inverse (3.44), the knowledge of the sign of $\dot{u}_d(t)$ is needed; however, implementation of their discrete-time counterparts (3.22) or (3.45) does not need such knowledge. Next in Chapters 5 - 7 we will employ these inverses fixed or adaptive to design control schemes for plants with input (actuator) nonlinearities.

Chapter 4

Fixed Inverse Compensation

Since dead-zone, backlash, and hysteresis characteristics are usually poorly known and vary with time, our main goal is to develop adaptive inverse compensation schemes. However, in some situations when these nonlinearities are approximately known, compensation schemes based on fixed approximate inverses may lead to significant performance improvements. Our task in this chapter is to design and analyze compensation schemes with fixed inverses for plants with nonlinearities such as dead-zone, backlash, or hysteresis at the input of a linear part—that is, to address fixed inverse compensation for the input nonlinearity. We will consider four different cases:

(i) Both the nonlinear part and the linear part are known.

(ii) The nonlinear part is known, the linear part is unknown.

(iii) The nonlinear part is unknown, the linear part is known.

(iv) Both the nonlinear part and the linear part are unknown.

For case (i) and case (ii), our compensation scheme employs an exact inverse to cancel the nonlinearity and a fixed or adaptive linear controller structure to achieve the output tracking to a given reference signal. For case (iii) and case (iv), our compensation scheme uses a detuned inverse, whose parameters are fixed inaccurate estimates of the unknown exact inverse parameters, and a fixed or adaptive linear controller structure. Such a compensation scheme ensures closed-loop signal boundedness if the inverse slope parameter errors are small. Although a detuned inverse cannot completely cancel the nonlinearity effect, simulation results show that an approximate inverse may improve system tracking performance considerably.

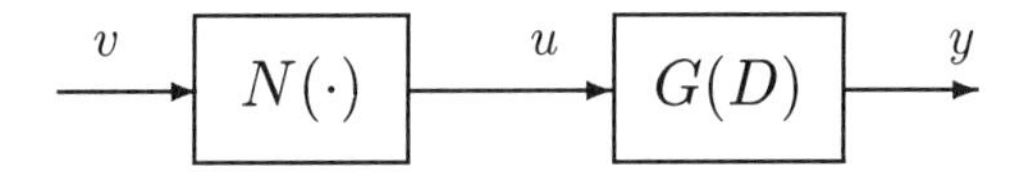

Figure 4.1: Plant with an input nonlinearity.

4.1 Control Objective

We consider the plant with a nonlinear part $N(\cdot)$ at the input of a linear time-invariant part $G(D)$, which is shown in Figure 4.1 and described by

$$y(t) = G(D)[u](t), \; u(t) = N(v(t)), \tag{4.1}$$

where $G(D) = k_p \frac{Z(D)}{P(D)}$, k_p is a constant gain, and $Z(D)$ and $P(D)$ are monic polynomials of degrees n and m, respectively. The symbol D denotes, as the case may be, (1) the Laplace transform variable s or the time differentiation operator $D[x](t) = \dot{x}(t)$ in continuous time, and (2) the z-transform variable z or the time advance operator $D[x](t) = x(t+1)$ in discrete time. With this notation, we use a unified framework to present both the continuous-time and the discrete-time adaptive controller designs.

The design objective is output tracking with stability in the bounded-input and bounded-output sense. The tracking performance will be measured by the error between the plant output $y(t)$ and the output $y_m(t)$ of a given reference model:

$$y_m(t) = W_m(D)[r](t), \tag{4.2}$$

where $r(t)$ is a bounded reference input and $W_m(D)$ is a stable rational transfer function.

Our approach is to use the model reference control strategy (see Appendix A) to design a linear controller structure combined with a fixed inverse $\widehat{NI}(\cdot)$ to compensate for the nonlinearity $N(\cdot)$. We make the following assumptions for the plant and reference model transfer functions $G(D)$ and $W_m(D)$:

(**A1**) $Z(D)$ is a stable polynomial.

(**A2**) The degree n of $P(D)$ is known.

(**A3**) The relative degree $n^* = n - m$ of $G(D)$ is known.

(**A4**) The sign of the gain k_p is known.

(**A5**) $W_m(D) = P_m^{-1}(D)$, for a stable polynomial $P_m(D)$ of degree n^* (for a discrete-time design, $P_m(D) = D^{n^*}$).

We emphasize that the signal $u(t)$ is unavailable for measurement and inaccessible for control.

4.2 Exact Inverse Control

Before we address the control problems when the nonlinear part $N(\cdot)$ is unknown, it is useful to illustrate our inverse control approach for the case when the nonlinear part $N(\cdot)$ is known. As shown in Chapter 3, for dead-zone, backlash, and hysteresis, exact inverses $NI(\cdot)$ exist such that

$$u(t) = N(NI(u_d(t)) = u_d(t),\ t \geq t_0 \tag{4.3}$$

provided that the backlash and hysteresis inverses are correctly initialized:

$$u(t_0) = B(BI(u_d(t_0)),\ \text{or}\ u(t_0) = H(HI(u_d(t_0)). \tag{4.4}$$

This means that the inverse $NI(\cdot)$ cancels the effect of $N(\cdot)$ so that $u_d(t)$ can be designed from any linear controller for $G(D)$ which ensures desired system performance. We will present this exact inverse control design for two cases: one for $G(D)$ known and the other for $G(D)$ unknown.

4.2.1 Designs for $G(D)$ Known

In the exact inverse controller structure in Figure 4.2, the signal $u_d(t)$ is linearly parametrized as

$$u_d(t) = \theta_1^{*T}\omega_1(t) + \theta_2^{*T}\omega_2(t) + \theta_{20}^* y(t) + \theta_3^* r(t), \tag{4.5}$$

where $\omega_1(t)$, $\omega_2(t)$ are the signals from two identical filters:

$$\omega_1(t) = \frac{a(D)}{\Lambda(D)}[u_d](t),\ \omega_2(t) = \frac{a(D)}{\Lambda(D)}[y](t) \tag{4.6}$$

and θ_1^*, θ_2^*, θ_{20}^*, θ_3^* are constant parameters. The signal $u_d(t)$ is then used as the input of $NI(\cdot)$ to generate the control $v(t)$:

$$v(t) = NI(u_d(t)),\ NI(\cdot) \in \{DI(\cdot), BI(\cdot), HI(\cdot)\}, \tag{4.7}$$

where $DI(\cdot)$ (or $BI(\cdot)$, $HI(\cdot)$) denotes the exact dead-zone inverse (or backlash inverse, hysteresis inverse) constructed in Chapter 3.

Dead-zone inverse control. Since a dead-zone inverse $DI(\cdot)$ only needs the knowledge of $u_d(t)$ and not of its derivatives, the filter numerator in (4.6) is an (n − 1)-dimensional vector:

$$a(D) = (1, D, \ldots, D^{n-2})^T \tag{4.8}$$

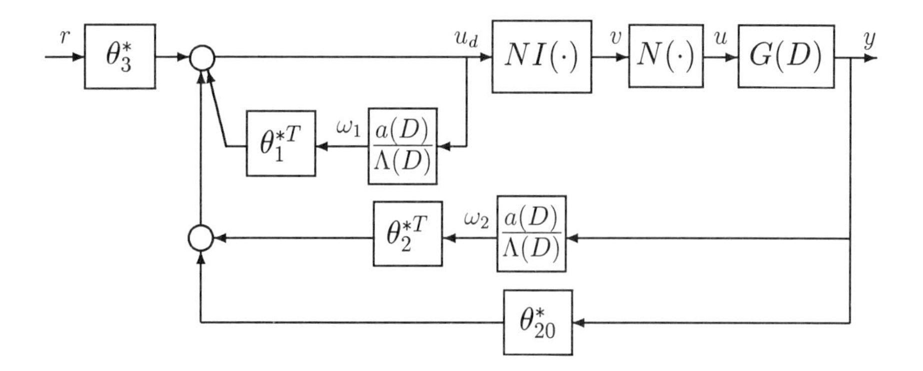

Figure 4.2: Exact inverse controller for $G(D)$ known.

and $\Lambda(D)$ is a stable polynomial of degree $n-1$. The controller parameters

$$\theta_1^* \in R^{n-1},\ \theta_2^* \in R^{n-1},\ \theta_{20}^* \in R,\ \theta_3^* \in R \tag{4.9}$$

are chosen to satisfy the matching equation

$$\begin{aligned}&\theta_1^{*T}a(D)P(D) + (\theta_2^{*T}a(D) + \theta_{20}^*\Lambda(D))k_pZ(D)\\ &= \Lambda(D)(P(D) - \theta_3^*k_pZ(D)P_m(D)).\end{aligned} \tag{4.10}$$

This condition implies that $\theta_3^* = k_p^{-1}$.

A discrete-time dead-zone inverse control scheme has the same controller structure, controller filters and parameters as above but with $\Lambda(z) = z^{n-1}$. An alternative is given below for the backlash case.

Backlash inverse control. A continuous-time backlash inverse $BI(\cdot)$ needs not only the knowledge of $u_d(t)$ but also that of the sign of $\dot{u}_d(t)$. When such knowledge is not available, we can employ a simple approximation of $\dot{u}_d(t)$, such as

$$\dot{u}_d(t) \approx \frac{s}{\tau s + 1}[u_d](t),$$

where $\tau > 0$ is a small constant. If both $\dot{y}(t)$ and $\dot{r}(t)$ are available and $\dot{r}(t)$ is bounded, then $\dot{u}_d(t)$ can be calculated by differentiating (4.5):

$$\dot{u}_d(t) = \theta_1^* \frac{sa(s)}{\Lambda(s)}[u_d](t) + \theta_2^{*T}\frac{sa(s)}{\Lambda(s)}[y](t) + \theta_{20}^*\dot{y}(t) + \theta_3^*\dot{r}(t).$$

If $\dot{y}(t)$ is not available but $\dot{r}(t)$ is measured and bounded, then an alternative to (4.8) - (4.10) is

$$a(D) = (1, D, \ldots, D^{n-1})^T \tag{4.11}$$

with $\Lambda(D)$ a stable polynomial of degree n and with the controller parameters

$$\theta_1^* \in R^n,\ \theta_2^* \in R^n,\ \theta_{20}^* = 0,\ \theta_3^* = k_p^{-1} \tag{4.12}$$

where θ_1^* and θ_2^* satisfy the matching equation

$$\begin{aligned} &\theta_1^{*T} a(D)P(D) + \theta_2^{*T} a(D)k_p Z(D) \\ &\quad = \Lambda(D)(P(D) - Z(D)P_m(D)). \end{aligned} \tag{4.13}$$

Hence, we have a different controller structure

$$u_d(t) = \theta_1^{*T}\omega_1(t) + \theta_2^{*T}\omega_2(t) + \theta_3^* r(t). \tag{4.14}$$

For this choice it follows that

$$\dot{u}_d(t) = \theta_1^* \frac{sa(s)}{\Lambda(s)}[u_d](t) + \theta_2^{*T}\frac{sa(s)}{\Lambda(s)}[y](t) + \theta_3^* \dot{r}(t),$$

which does not depend on $\dot{y}(t)$. The controller (4.11) - (4.13) has $2n+1$ parameters and its order is $2n$, while (4.8) - (4.10) has $2n-1$ parameters and its order is $2n-2$.

We can also use

$$\dot{y}(t) \approx \frac{s}{\tau s + 1}[y](t),\ \dot{r}(t) \approx \frac{s}{\tau s + 1}[r](t)$$

in (4.5) or (4.14) to approximate $\dot{u}_d(t)$ for adaptive backlash and hysteresis inverse controllers when $\dot{y}(t)$ and $\dot{r}(t)$ are not available.

Since a discrete-time backlash inverse $BI(\cdot)$ does not need $\dot{u}_d(t)$ for its implementation, a backlash inverse controller can be designed with either (4.8) - (4.10) with $\Lambda(z) = z^{n-1}$ or (4.11) - (4.13) with $\Lambda(z) = z^n$.

Hysteresis inverse control. For a continuous-time hysteresis inverse $HI(\cdot)$, which also needs the sign of $\dot{u}_d(t)$, the controller structures for the continuous-time backlash inverse design can be used.

As in the design of a discrete-time backlash inverse controller, a discrete-time hysteresis inverse controller can be designed with either (4.8) - (4.10) with $\Lambda(z) = z^{n-1}$, or (4.11) - (4.13) with $\Lambda(z) = z^n$.

Hybrid inverse control. In addition to the designs which provide the knowledge of the sign of $\dot{u}_d(t)$ exactly or approximately, we can also use a hybrid scheme consisting of a continuous-time linear controller and a discrete-time inverse, as shown in Figure 4.3.

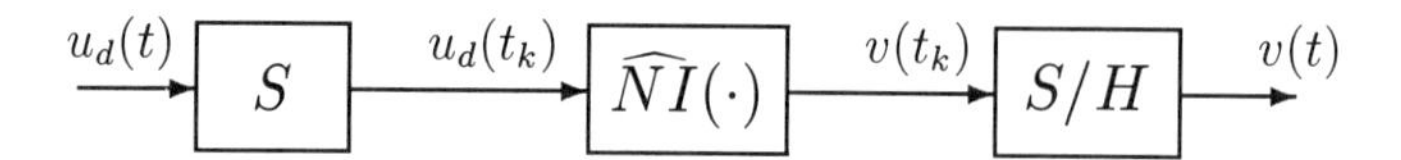

Figure 4.3: A hybrid inverse scheme.

In this scheme, $u_d(t)$ is a continuous-time signal; the block S is a sampler which generates the discrete-time version $u_d(t_k)$ of $u_d(t)$ as the input to $\widehat{NI}(\cdot)$, the backlash or hysteresis inverse designed in discrete time. The discrete-time output $v(t_k)$ of $\widehat{NI}(\cdot)$ is sampled and held by the sample-and-hold block S/H to generate the continuous-time signal $v(t)$ to be applied to the plant.

For hybrid inverse control, the backlash inverse is

$$v(t_k) = \widehat{BI}(u_d(t_k)) = \begin{cases} v(t_{k-1}) & \text{if } u_d(t_k) = u_d(t_{k-1}) \\ \frac{u_d(t_k)}{\widehat{m}} + \widehat{c}_l & \text{if } u_d(t_k) < u_d(t_{k-1}) \\ \frac{u_d(t_k)}{\widehat{m}} + \widehat{c}_r & \text{if } u_d(t_k) > u_d(t_{k-1}) \end{cases}$$

where t_k is the sampling instant and $\widehat{c}_r$, $\widehat{m}$, and $\widehat{c}_l$ are the estimates of the backlash parameters c_r, m, and c_l. The hysteresis inverse is

$$\begin{aligned} v(t_k) &= \widehat{HI}(u_d(t_k)) \\ &= \begin{cases} v(t_{k-1}) & \text{if } u_d(t_k) = u_d(t_{k-1}) \\ \frac{1}{\widehat{m}_t}(u_d(t_k) - \widehat{c}_t) & \text{if } u_d(t_k) \geq u_3 \\ \frac{1}{\widehat{m}_b}(u_d(t_k) - \widehat{c}_b) & \text{if } u_d(t_k) \leq u_4 \\ \frac{1}{\widehat{m}_r}u_d(t_k) + \widehat{c}_r & \text{if } u_3 \geq u_d(t_k) \geq u_e \\ \frac{1}{\widehat{m}_l}u_d(t_k) + \widehat{c}_l & \text{if } u_4 \leq u_d(t_k) \leq u_w \\ \frac{1}{\widehat{m}_t}(u_d(t_k) - \widehat{c}_w) & \text{if } u_w < u_d(t_k) < u_d(t_{k-1}) \\ \frac{1}{\widehat{m}_b}(u_d(t_k) - \widehat{c}_e) & \text{if } u_d(t_{k-1}) < u_d(t_k) < u_e, \end{cases} \end{aligned}$$

where $\widehat{c}_r$, $\widehat{m}_r$, $\widehat{c}_l$, $\widehat{m}_l$, $\widehat{c}_t$, $\widehat{m}_t$, $\widehat{c}_b$, and $\widehat{m}_b$ are the estimates of the hysteresis parameters c_r, m_r, c_l, m_l, c_t, m_t, c_b, and m_b, and other parameters have the same meanings as their counterparts defined in Section 3.3 for the discrete-time hysteresis inverse (3.45).

This inverse scheme can be implemented without the knowledge of the sign of $\dot{u}_d(t)$, because it approximates $\dot{u}_d(t)$ intervally. When the sampling interval $t_k - t_{k-1}$ is small, the approximation error is expected to be small.

The following result for the proposed exact inverse controller for $G(D)$

known is given in a unified form for both the continuous-time and discrete-time cases.

Theorem 4.1 *When applied to the plant (4.1) with an input nonlinearity $N(\cdot)$, the fixed linear controller (4.5) and the exact inverse (4.7) initialized by (4.4) ensure that all closed-loop signals are bounded and the tracking error $y(t) - y_m(t)$ converges to zero exponentially.*

Proof: The existence of θ_1^*, θ_2^*, θ_{20}^*, and θ_3^* which satisfy (4.10) or (4.13) is ensured by the Bezout identity, as shown in [80].

To prove the closed-loop stability and plant-model matching, using (4.3), we rewrite the plant (4.1) with the exact inverse $NI(\cdot)$, initialized by (4.4) so that $u(t) = u_d(t)$, as

$$P(D)[y](t) = k_p Z(D)[u_d](t). \tag{4.15}$$

Operating both sides of (4.10) on $y(t)$ and using (4.15), we obtain

$$\begin{aligned}\theta_1^{*T} a(D) k_p Z(D)[u_d](t) + (\theta_2^{*T} a(D) + \theta_{20}^* \Lambda(D)) k_p Z(D)[y](t) \\ = \Lambda(D) k_p Z(D)[u_d](t) - \Lambda(D) Z(D) P_m(D)[y](t).\end{aligned} \tag{4.16}$$

Because $\Lambda(D)$ and $Z(D)$ are stable polynomials, (4.16) can be expressed as

$$u_d(t) = \theta_1^{*T}\omega_1(t) + \theta_2^{*T}\omega_2(t) + \theta_{20}^* y(t) + \theta_3^* P_m(D)[y](t) + \eta_1(t) \tag{4.17}$$

for some exponentially decaying $\eta_1(t)$. Substituting (4.17) in (4.5) and using (4.2), we obtain

$$\theta_3^* P_m(D)[y - y_m](t) + \eta_1(t) = 0. \tag{4.18}$$

Using (4.15) and (4.18), we have

$$\begin{aligned}P_m(D) Z(D)[u_d](t) &= k_p^{-1} P(D) P_m(D)[y](t) \\ &= P(D)(\theta_3^* r(t) - \eta_1(t)).\end{aligned} \tag{4.19}$$

Since $P_m(D)$ is a stable polynomial and $y_m(t)$ is bounded, (4.18) implies that $y(t)$ is bounded and $y(t) - y_m(t) = \eta_2(t)$ is exponentially decaying. Since $r(t)$ is bounded and $P_m(D)Z(D)$ is a stable polynomial of degree n equal to the degree of $P(D)$, (4.19) implies that $u_d(t)$ is bounded and so are the signals $v(t)$ and $u(t)$ because of the characteristics of the nonlinearity $N(\cdot)$ and its inverse $NI(\cdot)$. Hence all closed-loop signals are bounded. ∇

Since the inverse control scheme (4.11) - (4.13) has the same structure as (4.8) - (4.10) with $\theta_{20}^* = 0$, except for the difference in the orders of $\Lambda(D)$ and $a(D)$, we conclude that Theorem 4.1 holds for both schemes, and so do our results in the next subsection for $G(D)$ unknown.

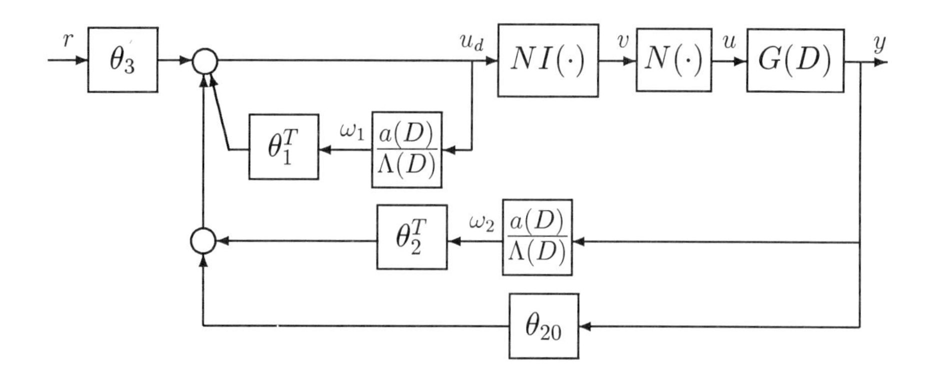

Figure 4.4: Exact inverse controller for $G(D)$ unknown.

4.2.2 Designs for $G(D)$ Unknown

The parameters θ_1^*, θ_2^*, θ_{20}^*, θ_3^* of the linear controller (4.5) depend on the parameters of $G(D)$; see (4.10) or (4.13). When $G(D)$ is unknown, we use $\theta_1(t)$, $\theta_2(t)$, $\theta_{20}(t)$, and $\theta_3(t)$ as the adaptive estimates of θ_1^*, θ_2^*, θ_{20}^*, and θ_3^*. Then the adaptive version of (4.5) is

$$u_d(t) = \theta_1^T(t)\omega_1(t) + \theta_2^T(t)\omega_2(t) + \theta_{20}(t)y(t) + \theta_3(t)r(t), \tag{4.20}$$

where $\omega_1(t)$ and $\omega_2(t)$ are the same as in (4.6):

$$\omega_1(t) = \frac{a(D)}{\Lambda(D)}[u_d](t), \ \omega_2(t) = \frac{a(D)}{\Lambda(D)}[y](t).$$

As shown in Figure 4.4, the signal $u_d(t)$ is applied to $NI(\cdot)$ to generate the control $v(t)$:

$$v(t) = NI(u_d(t)), \ NI(\cdot) \in \{DI(\cdot), BI(\cdot), HI(\cdot)\}. \tag{4.21}$$

For a compact notation, we introduce the parameter vectors θ^* and $\theta(t)$ and the regressor vector $\omega(t)$:

$$\theta^* = (\theta_1^{*T}, \theta_2^{*T}, \theta_{20}^*, \theta_3^*)^T \tag{4.22}$$

$$\theta(t) = (\theta_1^T(t), \theta_2^T(t), \theta_{20}(t), \theta_3(t))^T \tag{4.23}$$

$$\omega(t) = (\omega_1^T(t), \omega_2^T(t), y(t), r(t))^T. \tag{4.24}$$

Next we develop an error model relating the tracking error $e(t)$ and the parameter error $\tilde{\theta}(t)$:

$$e(t) = y(t) - y_m(t),\ \tilde{\theta}(t) = \theta(t) - \theta^*. \tag{4.25}$$

Substituting (4.17) in (4.20) and using (4.22) - (4.25), we obtain

$$\theta_3^* P_m(D)[e](t) + \eta_1(t) = \tilde{\theta}^T(t)\omega(t). \tag{4.26}$$

Ignoring the effect of the exponentially decaying term $\eta_1(t)$ and using $\theta_3^* = k_p^{-1}$, from (4.26), we arrive at the error model

$$e(t) = \frac{k_p}{P_m(D)}[\tilde{\theta}^T\omega](t), \tag{4.27}$$

where $P_m(D)$ is a stable polynomial of degree n^*. In discrete time $P_m(D) = D^{n^*}$. The error model (4.27), familiar from the adaptive control literature [2], [3], [30], [39], [80], [96], will be the basis for a choice of an adaptive update law for $\theta(t)$.

Continuous-Time Design

To design a continuous-time adaptive update law for $\theta(t)$, we introduce the auxiliary signals

$$\xi(t) = \theta^T(t)\zeta(t) - \frac{1}{P_m(s)}[\theta^T\omega](t) \tag{4.28}$$

$$\zeta(t) = \frac{1}{P_m(s)}[\omega](t) \tag{4.29}$$

and define the estimation error as

$$\epsilon(t) = e(t) + \rho(t)\xi(t), \tag{4.30}$$

where $\rho(t)$ is the estimate of $\rho^* = k_p$. With these signals we implement the gradient-type adaptive update law for $\theta(t)$ and $\rho(t)$:

$$\dot{\theta}(t) = -\frac{sign(k_p)\Gamma_\theta\epsilon(t)\zeta(t)}{m^2(t)} \tag{4.31}$$

$$\dot{\rho}(t) = -\frac{\gamma_\rho\epsilon(t)\xi(t)}{m^2(t)}, \tag{4.32}$$

where the normalizing signal $m(t)$ is

$$m(t) = \sqrt{1 + \zeta^T(t)\zeta(t) + \xi^2(t)}$$

and the adaptation gains $\Gamma_\theta \in R^{n_\theta \times n_\theta}$ for $\theta^* \in R^{n_\theta}$, and $\gamma_\rho \in R$ are constant and satisfy

$$\Gamma_\theta = \Gamma_\theta^T > 0, \ \gamma_\rho > 0.$$

With this update law the resulting adaptive system has the following properties.[1]

Lemma 4.1 *For the error model (4.27), the adaptive law (4.31) - (4.32) guarantees that* $\theta(t), \rho(t) \in L^\infty$, *and* $\frac{\epsilon(t)}{m(t)}, \dot{\theta}(t), \dot{\rho}(t) \in L^2 \cap L^\infty$.

Proof: Substituting (4.27) - (4.29) in (4.30) gives

$$\epsilon(t) = k_p \tilde{\theta}^T(t)\zeta(t) + \tilde{\rho}(t)\xi(t), \tag{4.33}$$

where $\tilde{\rho}(t) = \rho(t) - \rho^*$. Using (4.33), we compute the time derivative of the positive definite function

$$V(\tilde{\theta}, \tilde{\rho}) = |k_p|\tilde{\theta}^T \Gamma_\theta^{-1} \tilde{\theta} + \gamma_\rho^{-1} \tilde{\rho}^2$$

along the trajectories of (4.31) - (4.32). The result

$$\dot{V} = \frac{-2\epsilon^2(t)}{m^2(t)} \leq 0 \tag{4.34}$$

proves that $V(\tilde{\theta}(t), \tilde{\rho}(t))$ is a nonincreasing function of t, which implies that $V(\tilde{\theta}(t), \tilde{\rho}(t))$ is bounded so that $\theta(t), \rho(t) \in L^\infty$. From (4.34), it follows that

$$\int_0^\infty \frac{\epsilon^2(t)}{m^2(t)} dt = \frac{1}{2}(V(\tilde{\theta}(0), \tilde{\rho}(0)) - V(\tilde{\theta}(\infty), \tilde{\rho}(\infty))), \tag{4.35}$$

which means $\frac{\epsilon(t)}{m(t)} \in L^2$. By definition, $\frac{\zeta(t)}{m(t)}, \frac{\xi(t)}{m(t)} \in L^\infty$, hence, $\frac{\epsilon(t)}{m(t)} \in L^\infty$ because of (4.33), and $\dot{\theta}(t), \dot{\rho}(t) \in L^2 \cap L^\infty$ because of (4.31), (4.32), and (4.35). ∇

The tracking performance is described by the following theorem.

Theorem 4.2 *When applied to the plant (4.1) with an input nonlinearity* $N(\cdot)$, *the adaptive linear controller (4.20) updated by (4.31) - (4.32) and the exact inverse (4.21) initialized by (4.4) ensure that all closed-loop signals are bounded and the tracking error* $y(t) - y_m(t)$ *converges to zero asymptotically.*

---

[1] A scalar signal $x(t) \in L^\infty$ if $x(t)$ is bounded: $|x(t)| \leq \bar{x}$ for all $t \geq 0$ and some constant $\bar{x} > 0$; a scalar signal $x(t) \in L^2$ if $x(t)$ is square-integrable: $\int_0^\infty x^2(t)dt < \bar{x}$ for some constant $\bar{x} > 0$.

A vector signal $x(t) \in L^\infty$ if each component of $x(t)$, $x_i(t) \in L^\infty$; a vector signal $x(t) \in L^2$ if each component of $x(t)$, $x_i(t) \in L^2$.

Except for the use of the exact inverse $NI(\cdot)$ which cancels the effect of the nonlinearity $N(\cdot)$ so that $u(t) = u_d(t)$, the proof of Theorem 4.2 is standard in adaptive control theory; see, for example, [71], [80], [96]. For completeness, it is given in Appendix B.

Discrete-Time Design

As in the continuous-time case, introducing the auxiliary signals

$$\xi(t) = \theta^T(t)\omega(t-n^*) - \theta^T(t-n^*)\omega(t-n^*) \tag{4.36}$$

we define the estimation error

$$\epsilon(t) = e(t) + \rho(t)\xi(t), \tag{4.37}$$

where $\rho(t)$ is the estimate of $\rho^* = k_p$. Then we choose the gradient-type adaptive update law for $\theta(t)$ and $\rho(t)$:

$$\theta(t+1) = \theta(t) - \frac{sign(k_p)\Gamma_\theta \epsilon(t)\omega(t-n^*)}{m^2(t)} \tag{4.38}$$

$$\rho(t+1) = \rho(t) - \frac{\gamma_\rho \epsilon(t)\xi(t)}{m^2(t)}, \tag{4.39}$$

where the normalizing signal $m(t)$ is

$$m(t) = \sqrt{1 + \omega^T(t-n^*)\omega(t-n^*) + \xi^2(t)}$$

and the adaptation gains $\Gamma_\theta \in R^{n_\theta x n_\theta}$ for $\theta^* \in R^{n_\theta}$, and $\gamma_\rho \in R$ are constant and satisfy

$$0 < \Gamma_\theta = \Gamma_\theta^T < \frac{\gamma_\theta}{k_p^0} I, \; 0 < \gamma_\theta < 2, \; k_p^0 \geq |k_p|, \; 0 < \gamma_\rho < 2.$$

In this way, the knowledge of an upper bound k_p^0 on $|k_p|$ is needed for the discrete-time adaptive law which has the following properties.[2]

Lemma 4.2 *For the error model (4.27), the adaptive law (4.38) - (4.39) guarantees that* $\theta(t), \rho(t) \in l^\infty$, $\frac{\epsilon(t)}{m(t)} \in l^2 \cap l^\infty$, *and* $\theta(t+1)-\theta(t)$, $\rho(t+1)-\rho(t) \in l^2$.

[2] A scalar signal $x(t) \in l^\infty$ if $x(t)$ is bounded: $|x(t)| \leq \bar{x}$ for all $t \geq 0$ and some constant $\bar{x} > 0$; a scalar signal $x(t) \in l^2$ if $x(t)$ is square-summable: $\sum_{t=0}^{\infty} x^2(t) < \bar{x}$ for some constant $\bar{x} > 0$.

A vector signal $x(t) \in l^\infty$ if each component of $x(t)$, $x_i(t) \in l^\infty$; a vector signal $x(t) \in l^2$ if each component of $x(t)$, $x_i(t) \in l^2$.

Proof: Substituting (4.36) and the discrete-time version of (4.27),

$$e(t) = k_p \tilde{\theta}^T(t-n^*)\omega(t-n^*), \tag{4.40}$$

in (4.37), we obtain the error equation

$$\epsilon(t) = k_p \tilde{\theta}^T(t)\omega(t-n^*) + \tilde{\rho}(t)\xi(t),\ \tilde{\rho}(t) = \rho(t) - \rho^*. \tag{4.41}$$

Letting $\gamma_m = \max\{\gamma_\theta, \gamma_\rho\} < 2$ and using (4.41), we express the time increment of the positive definite function

$$V(\tilde{\theta}, \tilde{\rho}) = |k_p|\tilde{\theta}^T \Gamma_\theta^{-1}\tilde{\theta} + \gamma_\rho^{-1}\tilde{\rho}^2$$

along the trajectories of (4.38) - (4.39) as

$$\begin{aligned} & V(\tilde{\theta}(t+1), \tilde{\rho}(t+1)) - V(\tilde{\theta}(t), \tilde{\rho}(t)) \\ & \quad = -(2 - \frac{|k_p|\omega^T(t-n^*)\Gamma_\theta \omega(t-n^*) + \gamma_\rho \xi^2(t)}{m^2(t)})\frac{\epsilon^2(t)}{m^2(t)} \\ & \quad \leq -(2-\gamma_m)\frac{\epsilon^2(t)}{m^2(t)} \leq 0; \end{aligned} \tag{4.42}$$

that is, $V(\tilde{\theta}(t), \tilde{\rho}(t)) \leq V(\tilde{\theta}(0), \tilde{\rho}(0))$ for all $t > 0$ so that $\theta(t),\ \rho(t) \in l^\infty$. From (4.42), it follows that

$$\sum_{t=0}^{\infty} \frac{\epsilon^2(t)}{m^2(t)} \leq \frac{1}{2-\gamma_m}(V(\tilde{\theta}(0), \tilde{\rho}(0)) - V(\tilde{\theta}(\infty), \tilde{\rho}(\infty))), \tag{4.43}$$

which means that $\frac{\epsilon(t)}{m(t)} \in l^2$. By definition, $\frac{\zeta(t)}{m(t)}, \frac{xi(t)}{m(t)} \in L^\infty$; hence $\frac{\epsilon(t)}{m(t)} \in l^\infty$ because of (4.40), and $\theta(t+1) - \theta(t),\ \rho(t+1) - \rho(t) \in l^2$ because of (4.38), (4.39), and (4.43). ∇

The tracking performance of the discrete-time adaptive system is described in the following theorem.

Theorem 4.3 *When applied to the plant (4.1) with an input nonlinearity* $N(\cdot)$, *the adaptive linear controller (4.20) updated by (4.38) - (4.39) and the exact inverse (4.21) initialized by (4.4) ensure that all closed-loop signals are bounded and the tracking error* $y(t) - y_m(t)$ *goes to zero asymptotically.*

The proof of Theorem 4.3 is given in Appendix C.

The adaptive linear controller (4.20) (or its fixed version (4.5)) is able to achieve the desired tracking performance because an exact inverse $NI(\cdot)$ is implemented to cancel the effect of $N(\cdot)$.

4.3 Schemes with Detuned Inverses

Our main task in this book is to develop controller schemes for the plant (4.1) with an unknown $N(\cdot)$. Such schemes will employ the controller structure (4.5) or (4.20), but with some estimates $\widehat{NI}(\cdot)$ rather than with the exact inverses $NI(\cdot)$. An inverse estimate $\widehat{NI}(\cdot)$ is implemented with an estimate θ_N of its true parameter value θ_N^*. Before we proceed to adaptive schemes, we first explore the possibility of using a fixed "detuned" estimate θ_N.

When a dead-zone inverse $\widehat{DI}(\cdot)$, a backlash inverse $\widehat{BI}(\cdot)$, or a hysteresis inverse $\widehat{HI}(\cdot)$ is implemented with a fixed "detuned" parameter estimate $\theta_N \neq \theta_N^*$, it is of interest to examine to what extent the resulting detuned inverse will cancel the effect of $N(\cdot)$.

As was shown in Chapter 3, the control error $u(t) - u_d(t)$ due to an inverse estimate $v(t) = \widehat{NI}(u_d(t))$ can be expressed as

$$u(t) - u_d(t) = (\theta_N(t) - \theta_N^*)^T \omega_N(t) + d_N(t), \tag{4.44}$$

where

$$\{\theta_N, \theta_N^*, \omega_N, d_N\} = \begin{cases} \{\theta_d, \theta_d^*, \omega_d, d_d\} & \text{for } \widehat{DI}(\cdot) \\ \{\theta_b, \theta_b^*, \omega_b, d_b\} & \text{for } \widehat{BI}(\cdot) \\ \{\theta_h, \theta_h^*, \omega_h, d_h\} & \text{for } \widehat{HI}(\cdot) \end{cases}$$

as defined in (3.4) - (3.6) and (3.9), (3.34) - (3.36) and (3.33), or (3.69) - (3.71) and (3.73), respectively.

With a fixed estimate θ_N, the control error $u(t) - u_d(t)$ is due to two factors. One is the parameter error $\theta_N - \theta_N^*$ and the other is the unparametrizable error $d_N(t)$ which has an important property: $d_N(t)$ is bounded for any $t \geq 0$. We will present two control schemes with a detuned inverse $\widehat{NI}(\cdot)$: one for $G(D)$ known and the other for $G(D)$ unknown.

4.3.1 Control Error

We first derive control error expressions for the dead-zone, backlash, and hysteresis inverses, which are easier to use than (4.44).

Dead-zone inverse. Using (3.3) and the fact that

$$\widehat{\chi_i}(t)\widehat{\chi_i}(t) = \widehat{\chi_i}(t),\ i \in \{r, l\},\ \widehat{\chi_r}(t)\widehat{\chi_l}(t) = 0,$$

we have

$$\widehat{\chi_r}(t)u_d(t) = \widehat{\chi_r}(t)\widehat{m_r}v(t) - \widehat{\chi_r}(t)\widehat{m_r b_r} \tag{4.45}$$

$$\widehat{\chi_l}(t)u_d(t) = \widehat{\chi_l}(t)\widehat{m_l}v(t) - \widehat{\chi_l}(t)\widehat{m_l b_l}. \tag{4.46}$$

From (4.45) and (4.46), we have

$$(\widehat{m_r} - m_r)\widehat{\chi_r}(t)v(t) = \frac{\widehat{m_r} - m_r}{\widehat{m_r}}\widehat{\chi_r}(t)(u_d(t) + \widehat{m_r b_r}) \tag{4.47}$$

$$(\widehat{m_l} - m_l)\widehat{\chi_l}(t)v(t) = \frac{\widehat{m_l} - m_l}{\widehat{m_l}}\widehat{\chi_l}(t)(u_d(t) + \widehat{m_l b_l}). \tag{4.48}$$

Introducing

$$\tilde{k}_d(t) = -\left(\frac{\widehat{m_r} - m_r}{\widehat{m_r}}\widehat{\chi_r}(t) + \frac{\widehat{m_l} - m_l}{\widehat{m_l}}\widehat{\chi_l}(t)\right)$$

$$\begin{aligned}\bar{d}_d(t) = & -\frac{\widehat{m_r} - m_r}{\widehat{m_r}}\widehat{\chi_r}(t)\widehat{m_r b_r} + (\widehat{m_r b_r} - m_r b_r)\widehat{\chi_r}(t) \\ & -\frac{\widehat{m_l} - m_l}{\widehat{m_l}}\widehat{\chi_l}(t)\widehat{m_l b_l} + (\widehat{m_l b_l} - m_l b_l)\widehat{\chi_l}(t) + d_d(t)\end{aligned}$$

and using (4.47) and (4.48), we express (3.8) as

$$u(t) - u_d(t) = \tilde{k}_d(t)u_d(t) + \bar{d}_d. \tag{4.49}$$

In this control error expression, $\bar{d}_d(t)$ is bounded for all $t \geq 0$, and $|\tilde{k}_d(t)|$ is small if the slope errors $|\widehat{m_r} - m_r|$, $|\widehat{m_l} - m_l|$ are small.

Backlash inverse. Using (3.25) and (3.31), we obtain

$$u_d(t) = \widehat{m}v(t) - \widehat{\chi_r}(t)\widehat{mc_r} - \widehat{\chi_l}(t)\widehat{mc_l}.$$

From this identity, we have

$$(\widehat{m} - m)v(t) = \frac{\widehat{m} - m}{\widehat{m}}(u_d(t) + \widehat{\chi_r}(t)\widehat{mc_r} + \widehat{\chi_l}(t)\widehat{mc_l}). \tag{4.50}$$

Introducing

$$\tilde{k}_b(t) = -\frac{\widehat{m} - m}{\widehat{m}}$$

$$\begin{aligned}\bar{d}_b(t) = & -\frac{\widehat{m} - m}{\widehat{m}}\widehat{\chi_r}(t)\widehat{mc_r} + (\widehat{mc_r} - mc_r)\widehat{\chi_r}(t) \\ & -\frac{\widehat{m} - m}{\widehat{m}}\widehat{\chi_l}(t)\widehat{mc_l} + (\widehat{mc_l} - mc_l)\widehat{\chi_l}(t) + d_b(t)\end{aligned}$$

and using (4.50), we express (3.38) as

$$u(t) - u_b(t) = \tilde{k}_b(t)u_b(t) + \bar{d}_b. \tag{4.51}$$

Again, $\bar{d}_b(t)$ is bounded for all $t \geq 0$, and $|\tilde{k}_b|$ is small if the slope error $|\widehat{m} - m|$ is small.

Hysteresis inverse. Using (3.63) and (3.64), we obtain

$$(\widehat{m_t} - m_t)\widehat{\chi_t}(t)v(t) = \frac{\widehat{m_t} - m_t}{\widehat{m_t}}\widehat{\chi_t}(t)(u_d(t) - \widehat{c_t}) \tag{4.52}$$

$$(\widehat{m_b} - m_b)\widehat{\chi_b}(t)v(t) = \frac{\widehat{m_b} - m_b}{\widehat{m_b}}\widehat{\chi_b}(t)(u_d(t) - \widehat{c_b}). \tag{4.53}$$

Introducing

$$\tilde{k}_h(t) = -(\frac{\widehat{m_t} - m_t}{\widehat{m_t}}\widehat{\chi_t}(t) + \frac{\widehat{m_b} - m_b}{\widehat{m_b}}\widehat{\chi_b}(t))$$

$$\begin{aligned}\bar{d}_h(t) = & \frac{\widehat{m_t} - m_t}{\widehat{m_t}}\widehat{\chi_t}(t)\widehat{c_t} - (\widehat{m_t} - m_t)\widehat{\chi_w}(t)v(t) - (\widehat{c_t} - c_t)\widehat{\chi_t}(t) \\ & + \frac{\widehat{m_b} - m_b}{\widehat{m_b}}\widehat{\chi_b}(t)\widehat{c_b} - (\widehat{m_b} - m_b)\widehat{\chi_e}(t)v(t) - (\widehat{c_b} - c_b)\widehat{\chi_b}(t) \\ & -(\widehat{m_r} - m_r)\widehat{\chi_r}(t)v(t) + (\widehat{m_r c_r} - m_r c_r)\widehat{\chi_r}(t) \\ & -(\widehat{m_l} - m_l)\widehat{\chi_l}(t)v(t) + (\widehat{m_l c_l} - m_l c_l)\widehat{\chi_l}(t) + d_h(t)\end{aligned}$$

and using (4.52) and (4.53), we express (3.72) as

$$u(t) - u_d(t) = \tilde{k}_h(t)u_h(t) + \bar{d}_h. \tag{4.54}$$

Now $\bar{d}_h(t)$ is bounded for all $t \geq 0$ because

$$\widehat{\chi_w}(t) = \widehat{\chi_e}(t) = \widehat{\chi_r}(t) = \widehat{\chi_l}(t) = 0, \text{ for } v(t) \text{ large.}$$

Furthermore, the gain variation $|\tilde{k}_h(t)|$ is small if the slope errors $|\widehat{m_t} - m_t|$, $|\widehat{m_b} - m_b|$ are small.

To arrive at a unified control error expression for the dead-zone, backlash, and hysteresis inverses, we introduce

$$\tilde{k}_N(t) = \begin{cases} \tilde{k}_d(t) & \text{for } \widehat{DI}(\cdot) \\ \tilde{k}_b(t) & \text{for } \widehat{BI}(\cdot) \\ \tilde{k}_h(t) & \text{for } \widehat{HI}(\cdot) \end{cases} \tag{4.55}$$

$$\bar{d}_N(t) = \begin{cases} \bar{d}_d(t) & \text{for } \widehat{DI}(\cdot) \\ \bar{d}_b(t) & \text{for } \widehat{BI}(\cdot) \\ \bar{d}_h(t) & \text{for } \widehat{HI}(\cdot) \end{cases} \tag{4.56}$$

and express (4.49), (4.51), and (4.54) as

$$u(t) - u_d(t) = \tilde{k}_N(t)u_d(t) + \bar{d}_N(t), \tag{4.57}$$

where $\bar{d}_N(t)$ defined in (4.56) is bounded for all $t \geq 0$, and $\tilde{k}_N(t)$ depends on the dead-zone slope errors $\widehat{m_r} - m_r$, $\widehat{m_l} - m_l$, or the backlash slope error $\widehat{m} - m$, or the hysteresis slope errors $\widehat{m_t} - m_t$, $\widehat{m_b} - m_b$. The size of $\tilde{k}_N(t)$ is proportional to the quantity

$$\mu_N = \begin{cases} \max\{|\widehat{m_r} - m_r|, |\widehat{m_l} - m_l|\} & \text{for } \widehat{DI}(\cdot) \\ |\widehat{m} - m| & \text{for } \widehat{BI}(\cdot) \\ \max\{|\widehat{m_t} - m_t|, |\widehat{m_b} - m_b|\} & \text{for } \widehat{HI}(\cdot), \end{cases} \tag{4.58}$$

which will be used in analyzing the control schemes with a fixed detuned inverse estimate $\widehat{NI}(\cdot)$.

4.3.2 Design for $G(D)$ Known

For $G(D)$ known, the detuned inverse controller structure is shown in Figure 4.5. The signal $u_d(t)$ is generated from the same linear structure as (4.5):

$$u_d(t) = \theta_1^{*T}\omega_1(t) + \theta_2^{*T}\omega_2(t) + \theta_{20}^* y(t) + \theta_3^* r(t), \tag{4.59}$$

where

$$\omega_1(t) = \frac{a(D)}{\Lambda(D)}[u_d](t),\ \omega_2(t) = \frac{a(D)}{\Lambda(D)}[y](t).$$

The linear controller parameters θ_1^*, θ_2^*, θ_{20}^*, θ_3^*, and the controller filter $\frac{a(D)}{\Lambda(D)}$ are designed in the same ways as in Section 4.2. The inverse for the nonlinearity $N(\cdot)$ now is

$$v(t) = \widehat{NI}(u_d(t)),\ NI(\cdot) \in \{\widehat{DI}(\cdot), \widehat{BI}(\cdot), \widehat{HI}(\cdot)\}. \tag{4.60}$$

Since the inverse estimate $\widehat{NI}(\cdot)$ may not completely cancel the effect of the nonlinearity $N(\cdot)$, there is a control error (4.57), characterized by the "uncertainty level" μ_N defined in (4.58). Does the control scheme in Figure 4.5 ensure the closed-loop signal boundedness in the presence of such an error? The following theorem provides an answer.

Theorem 4.4 *There exists a constant $\mu^* > 0$ such that for any nonlinearity $N(\cdot)$ and its detuned inverse $\widehat{NI}(\cdot)$ satisfying $\mu_N \in [0, \mu^*]$ the fixed linear controller (4.59) and the detuned inverse (4.60), applied to the plant (4.1), ensure that all closed-loop signals are bounded.*

Proof: Using the control error expression (4.57), we rewrite the plant (4.1) with the detuned inverse $\widehat{NI}(\cdot)$ as

$$P(D)[y](t) = k_p Z(D)[u_d + \tilde{k}_N u_d + \bar{d}_N](t). \tag{4.61}$$

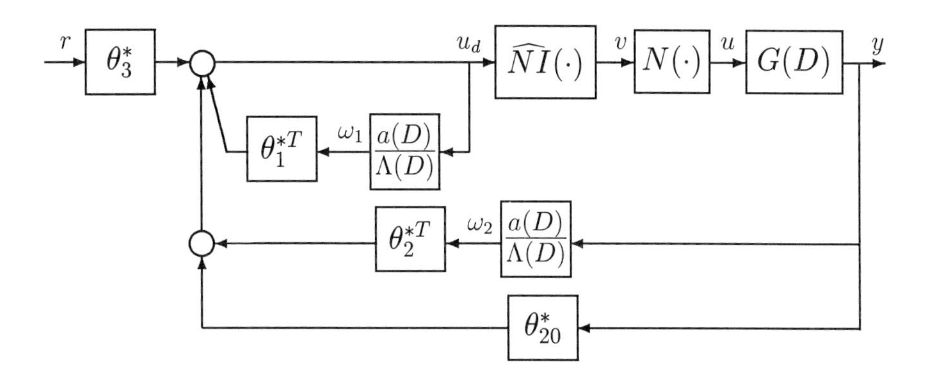

Figure 4.5: Detuned inverse controller for $G(D)$ known.

Operating both sides of (4.10) on $y(t)$ and using (4.61), we obtain

$$\begin{aligned}&\theta_1^{*T}a(D)k_pZ(D)[u_d+\tilde{k}_Nu_d+\bar{d}_N](t)\\&+(\theta_2^{*T}a(D)+\theta_{20}^*\Lambda(D))k_pZ(D)[y](t)\\=\ &\Lambda(D)k_pZ(D)[u_d+\tilde{k}_Nu_d+\bar{d}_N](t)-\Lambda(D)Z(D)P_m(D)[y](t).\end{aligned}\tag{4.62}$$

Because $\Lambda(D)$ and $Z(D)$ are stable, we rewrite (4.62) as

$$\begin{aligned}&u_d(t)+(1-\theta_1^{*T}\frac{a(D)}{\Lambda(D)})[\tilde{k}_Nu_d+\bar{d}_N](t)\\=\ &\theta_1^{*T}\omega_1(t)+\theta_2^{*T}\omega_2(t)+\theta_{20}^*y(t)+\theta_3^*P_m(D)[y](t)+\eta_1(t)\end{aligned}\tag{4.63}$$

for some exponentially decaying $\eta_1(t)$. Substituting (4.63) in (4.59) and using (4.2), we obtain

$$\theta_3^*P_m(D)[y-y_m](t)+\eta_1(t)=(1-\theta_1^{*T}\frac{a(D)}{\Lambda(D)})[\tilde{k}_Nu_d+\bar{d}_N](t).\tag{4.64}$$

Operating both sides of (4.64) by $P(D)$ and using (4.61), (4.2) and the definition $\theta_3^*=k_p^{-1}$, we have

$$\begin{aligned}&P_m(D)Z(D)[u_d+\tilde{k}_Nu_d+\bar{d}_N](t)-\theta_3^*P(D)[r](t)+P(D)[\eta_1](t)\\&=P(D)(1-\theta_1^{*T}\frac{a(D)}{\Lambda(D)})[\tilde{k}_Nu_d+\bar{d}_N](t).\end{aligned}\tag{4.65}$$

Since $P_m(D)$ and $Z(D)$ are stable, we rewrite (4.65) as

$$u_d=(\frac{P(D)}{P_m(D)Z(D)}(1-\theta_1^{*T}\frac{a(D)}{\Lambda(D)})-1)[\tilde{k}_Nu_d](t)+\frac{\theta_3^*P(D)}{P_m(D)Z(D)}[r](t)$$

$$+ (\frac{P(D)}{P_m(D)Z(D)}(1 - \theta_1^{*T} \frac{a(D)}{\Lambda(D)}) - 1)[\bar{d}_N](t) + \eta_2(t) \tag{4.66}$$

for some exponentially decaying $\eta_2(t)$.

Since $r(t)$ and $\bar{d}_N(t)$ are bounded, and $\frac{P(D)}{P_m(D)Z(D)}(1 - \theta_1^{*T} \frac{a(D)}{\Lambda(D)}) - 1$ and $\frac{\theta_3^* P(D)}{P_m(D)Z(D)}$ are stable and proper transfer functions, it follows from (4.66) that there is a constant $k_0 > 0$ such that for $|\tilde{k}_N(t)| \leq k_0$, the signal $u_d(t)$ is bounded. The existence of such a k_0 is equivalent to the existence of the constant $\mu^* > 0$ in Theorem 4.4. The boundedness of $u_d(t)$ and (4.64) imply that $y(t)$ is bounded so all closed-loop signals are bounded. ∇

The tracking performance of the detuned inverse controller is described by the following corollary.

Corollary 4.1 *Under the conditions of Theorem 4.4, the tracking error $y(t) - y_m(t)$ converges exponentially to the the set $\{e : |e| \leq k_1\mu_N + k_2 d_0\}$ for some constants $k_1 > 0$, $k_2 > 0$, where d_0 is the upper bound for $|\bar{d}_N(t)|$.*

This corollary follows from the closed-loop signal boundedness and the tracking error expression (4.64). We note that Theorem 4.4 and Corollary 4.1 hold for both the continuous-time case and the discrete-time case.

4.3.3 Designs for $G(D)$ Unknown

For $G(D)$ unknown, as in Section 4.2.2, the detuned inverse controller also uses the adaptive linear controller structure

$$u_d(t) = \theta_1^T(t)\omega_1(t) + \theta_2^T(t)\omega_2(t) + \theta_{20}(t)y(t) + \theta_3(t)r(t), \tag{4.67}$$

where

$$\omega_1(t) = \frac{a(D)}{\Lambda(D)}[u_d](t),\ \omega_2(t) = \frac{a(D)}{\Lambda(D)}[y](t)$$

and $\theta_1(t)$, $\theta_2(t)$, $\theta_{20}(t)$, and $\theta_3(t)$ are the estimates of θ_1^*, θ_2^*, θ_{20}^* and θ_3^* defined by (4.8) - (4.10) or (4.11) - (4.13) for different inverses; in particular, $\theta_{20} = 0$ if $\theta_{20}^* = 0$ is chosen as in (4.12).

As shown in Figure 4.6, the signal $u_d(t)$ is applied to $\widehat{NI}(\cdot)$ to generate the control $v(t)$:

$$v(t) = \widehat{NI}(u_d(t)),\ \widehat{NI}(\cdot) \in \{\widehat{DI}(\cdot), \widehat{BI}(\cdot), \widehat{HI}(\cdot)\}. \tag{4.68}$$

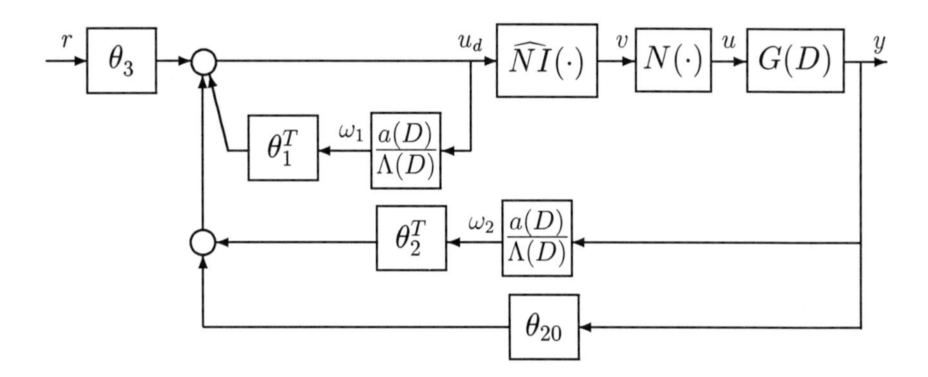

Figure 4.6: Detuned inverse controller for $G(D)$ unknown.

Using the familiar notation

$$\theta^* = (\theta_1^{*T}, \theta_2^{*T}, \theta_{20}^*, \theta_3^*)^T$$
$$\theta(t) = (\theta_1^T(t), \theta_2^T(t), \theta_{20}(t), \theta_3(t))^T$$
$$\omega(t) = (\omega_1^T(t), \omega_2^T(t), y(t), r(t))^T$$
$$e(t) = y(t) - y_m(t),\ \tilde{\theta}(t) = \theta(t) - \theta^*$$

and substituting (4.63) in (4.67), we obtain

$$\theta_3^* P_m(D)[e](t) + \eta_1(t)$$
$$= \tilde{\theta}^T(t)\omega(t) + (1 - \theta_1^{*T}\frac{a(D)}{\Lambda(D)})[\tilde{k}_N u_d + \bar{d}_N](t). \tag{4.69}$$

Ignoring the effect of $\eta_1(t)$ and using $\theta_3^* = k_p^{-1}$, from (4.69), we obtain the tracking error equation

$$e(t) = \frac{k_p}{P_m(D)}[\tilde{\theta}^T\omega](t) + \eta(t) + d(t), \tag{4.70}$$

where

$$\eta(t) = \Delta(D)[\tilde{k}_N u_d](t) \tag{4.71}$$
$$\Delta(D) = \frac{k_p}{P_m(D)}(1 - \theta_1^{*T}\frac{a(D)}{\Lambda(D)})$$
$$d(t) = \frac{k_p}{P_m(D)}(1 - \theta_1^{*T}\frac{a(D)}{\Lambda(D)})[\bar{d}_N](t). \tag{4.72}$$

In this error model the terms $\eta(t)$ and $d(t)$ are due to the inaccuracy of the detuned inverse $\widehat{NI}(\cdot)$. It is important to note that $\eta(t) = \Delta(D)[\tilde{k}_N u_d](t)$

represents an "unmodeled dynamics" whose size of $\tilde{k}_N$ is proportional to μ_N as defined in (4.58), while $d(t)$ is bounded and its size is proportional to d_0, the upper bound of $|\bar{d}_N(t)|$: $|\bar{d}_N(t)| \leq d_0$, for all $t \geq 0$. The error model (4.70) is also familiar from the adaptive control literature [3], [30], [39], [80], [96], based on which a robust adaptive law can be developed to update $\theta(t)$.

Continuous-Time Adaptive Design

We introduce the familiar auxiliary signals

$$\xi(t) = \theta^T(t)\zeta(t) - \frac{1}{P_m(s)}[\theta^T\omega](t) \tag{4.73}$$

$$\zeta(t) = \frac{1}{P_m(s)}[\omega](t) \tag{4.74}$$

and define the estimation error

$$\epsilon(t) = e(t) + \rho(t)\xi(t), \tag{4.75}$$

where $\rho(t)$ is the estimate of $\rho^* = k_p$. We choose the modified gradient-type adaptive update law for $\theta(t) \in R^{n_\theta}$ and $\rho(t)$:

$$\dot{\theta}(t) = -\frac{sign(k_p)\Gamma_\theta\epsilon(t)\zeta(t)}{m^2(t)} + f_\theta(t) \tag{4.76}$$

$$\dot{\rho}(t) = -\frac{\gamma_\rho\epsilon(t)\xi(t)}{m^2(t)} + f_\rho(t), \tag{4.77}$$

where Γ_θ is a constant gain matrix, and γ_ρ is a constant gain:

$$\Gamma_\theta = diag\{\gamma_1, \ldots, \gamma_{n_\theta}\},\ \gamma_j > 0,\ 1 \leq j \leq n_\theta,\ \gamma_\rho > 0.$$

The the normalizing signal $m(t)$ is

$$m(t) = \sqrt{1 + \zeta^T(t)\zeta(t) + \xi^2(t) + m_a^2(t)}, \tag{4.78}$$

where $m_a(t)$ satisfies the differential equation

$$\dot{m}_a(t) = -\delta_m m_a(t) + |u_d(t)|,\ m_a(0) = 0 \tag{4.79}$$

with $\delta_m > 0$ being a constant such that both $\Lambda(s - \delta_m)$, $P_m(s - \delta_m)$ are stable polynomials.

With the modifying terms $f_\theta(t)$, $f_\rho(t)$ we will achieve robustness of the adaptive law with respect to the unmodeled dynamics $\eta(t)$ in (4.71) and the bounded disturbance $d(t)$ in (4.72). Such modifications are available in the robust adaptive control literature—for example, the dead-zone modifications [54], [68], [86], σ-modification [35], [37], switching σ-modification [33], [42], [43], parameter projection modifications [53], [55], [77], ϵ_1-modification [79], and normalization techniques [42], [43], [54], [68], [90]. The switching σ-modification [42] and its discrete-time version [43] have the advantage that the asymptotic tracking is achieved if the terms $\eta(t)$ and $d(t)$ disappear.

The parameter projection algorithms which we will present next not only have the same advantages for output tracking but also ensure that the parameter estimates stay in a prescribed region. For parameter projection, we need to know a convex region in the parameter space, which contains the true parameter θ^*. To simplify notation, we denote θ_j^* as the jth component of θ^* and denote $\theta_j(t)$ as the jth component of $\theta(t)$. Then, to specify the convex region, we assume:

> (**A6**) The jth component of $\theta^* \in R^{n_\theta}$ belongs to a known interval: $\theta_j^* \in [\theta_j^a, \theta_j^b]$, $j = 1, \ldots, n_\theta$, and so does ρ^*: $\rho^* \in [\rho^a, \rho^b]$.

This assumption is used to modify the estimates $\theta(t)$ and $\rho(t)$ according to the following projection algorithm.

Parameter projection. We denote $f_{\theta j}(t)$ as the jth component of $f_\theta(t)$, $g_{\theta j}(t)$ as the jth component of $g_\theta(t)$,

$$g_\theta(t) = -\frac{sign(k_p)\Gamma_\theta \zeta(t)\epsilon(t)}{m^2(t)}, \tag{4.80}$$

choose $\theta_j(0) \in [\theta_j^a, \theta_j^b]$, $\rho(0) \in [\rho^a, \rho^b]$, and let

$$g_\rho(t) = -\frac{\gamma_\rho \xi(t)\epsilon(t)}{m^2(t)}. \tag{4.81}$$

Then, we set

$$f_{\theta j}(t) = \begin{cases} 0 & \text{if } \theta_j(t) \in (\theta_j^a, \theta_j^b), \text{ or} \\ & \text{if } \theta_j(t) = \theta_j^a,\ g_{\theta j}(t) \geq 0, \text{ or} \\ & \text{if } \theta_j(t) = \theta_j^b,\ g_{\theta j}(t) \leq 0 \\ -g_{\theta j}(t) & \text{otherwise} \end{cases} \tag{4.82}$$

$$f_\rho(t) = \begin{cases} 0 & \text{if } \rho(t) \in (\rho^a, \rho^b), \text{ or} \\ & \text{if } \rho(t) = \rho^a,\ g_\rho(t) \geq 0, \text{ or} \\ & \text{if } \rho(t) = \rho^b,\ g_\rho(t) \leq 0 \\ -g_\rho(t) & \text{otherwise.} \end{cases} \tag{4.83}$$

The modified adaptive law (4.76) with (4.82), (4.77), and (4.83) has the following properties.

Lemma 4.3 *For the error model (4.70), the adaptive law (4.76) - (4.77) with the parameter projection (4.80) - (4.83) guarantees that* $\theta(t)$, $\rho(t)$, $\frac{\epsilon(t)}{m(t)}$, $\dot{\theta}(t)$, $\dot{\rho}(t) \in L^\infty$, *and* $\frac{\epsilon(t)}{m(t)}$, $\dot{\theta}(t)$, $\dot{\rho}(t)$ *satisfy*

$$\int_{t_1}^{t_2} \frac{\epsilon^2(t)}{m^2(t)} dt \leq a_1 + b_1 \mu_N^2 (t_2 - t_1) + c_1 \int_{t_1}^{t_2} \frac{d^2(t)}{m^2(t)} dt \tag{4.84}$$

$$\int_{t_1}^{t_2} \|\dot{\theta}(t)\|_2^2 dt \leq a_2 + b_2 \mu_N^2 (t_2 - t_1) + c_2 \int_{t_1}^{t_2} \frac{d^2(t)}{m^2(t)} dt \tag{4.85}$$

$$\int_{t_1}^{t_2} \dot{\rho}^2(t) dt \leq a_3 + b_3 \mu_N^2 (t_2 - t_1) + c_3 \int_{t_1}^{t_2} \frac{d^2(t)}{m^2(t)} dt \tag{4.86}$$

for some constants $a_i > 0$, $b_i > 0$, $c_i > 0$, $i = 1, 2, 3$, *and* $\forall t_2 > t_1 \geq 0$.

Proof: We consider $\eta(t) = \Delta(s)[\tilde{k}_N u_d](t)$ in (4.71), where

$$\Delta(s) = \frac{k_p}{P_m(s)} (1 - \theta_1^{*T} \frac{a(s)}{\Lambda(s)}), \tag{4.87}$$

and introduce the signal $x(t) = e^{\delta_m t} \tilde{k}_N(t) u_d(t)$, with $\delta_m > 0$ as in (4.79), and $k_m > 0$ such that $|\tilde{k}_N(t)| \leq k_m \mu_N$ with $\tilde{k}_N$, μ_N as in (4.55), (4.58). Then, denoting the impulse response of $s\Delta(s - \delta_m)$ by $h_\delta(t, \delta_m)$, and its L^1 signal norm by $\|h_\delta(\cdot, \delta_m)\|_1 \triangleq \int_0^\infty |h_\delta(t, \delta_m)| dt$, we have

$$\begin{aligned}
|\eta(t)| &= |\frac{1}{s + \delta_m} [(s + \delta_m) \Delta(s) [\tilde{k}_N u_d]](t)| \\
&\leq \int_0^t e^{-\delta_m (t - \tau)} |(s + \delta_m) \Delta(s) [\tilde{k}_N u_d](\tau)| d\tau \\
&= e^{-\delta_m t} \int_0^t e^{\delta_m \tau} |(s + \delta_m) \Delta(s) [\tilde{k}_N u_d](\tau)| d\tau \\
&= e^{-\delta_m t} \int_0^t |(s \Delta(s - \delta_m) [x(\tau)]| d\tau \\
&\leq e^{-\delta_m t} \|h_\delta(\cdot, \delta_m)\|_1 \int_0^t |x(\tau)| d\tau \\
&= \|h_\delta(\cdot, \delta_m)\|_1 \int_0^t e^{-\delta_m (t - \tau)} |\tilde{k}_N u_d(\tau)| d\tau \\
&\leq \|h_\delta(\cdot, \delta_m)\|_1 k_m \mu_N \frac{1}{s + \delta_m} [|u_d|](t) \\
&= \|h_\delta(\cdot, \delta_m)\|_1 k_m \mu_N m_a(t),
\end{aligned} \tag{4.88}$$

where the second inequality follows from the definition of $\|h_\delta(\cdot,\delta_m)\|_1$.

From the definition (4.87) of $\Delta(s)$, we see that $\|h_\delta(\cdot,\delta_m)\|_1$ is finite. Hence, (4.78) and (4.88) imply that

$$|\eta(t)| \leq k_\delta \mu_N m(t) \tag{4.89}$$

for some constant $k_\delta > 0$ and $m(t)$ in (4.78).

Substituting (4.70), (4.73), and (4.74) into (4.75), we obtain the estimation error $\epsilon(t)$ in terms of the parameter errors $\tilde{\theta}(t)$, $\tilde{\rho}(t)$:

$$\epsilon(t) = k_p \tilde{\theta}^T(t)\zeta(t) + \tilde{\rho}(t)\xi(t) + \eta(t) + d(t). \tag{4.90}$$

Introducing

$$p_\theta(t) = 2|k_p|\tilde{\theta}^T(t)\Gamma_\theta^{-1} f_\theta(t)$$
$$p_\rho(t) = 2\tilde{\rho}(t)\gamma_\rho^{-1} f_\rho(t)$$

we express the time derivative of

$$V(\tilde{\theta},\tilde{\rho}) = |k_p|\tilde{\theta}^T\Gamma_\theta^{-1}\tilde{\theta} + \gamma_\rho^{-1}\tilde{\rho}^2$$

along the trajectories of (4.76) - (4.77) as

$$\begin{aligned}\dot{V} &= -\frac{2\epsilon^2(t)}{m^2(t)} + \frac{2\epsilon(t)(\eta(t)+d(t))}{m^2(t)} + p_\theta(t) + p_\rho(t)\\ &\leq -\frac{\epsilon^2(t)}{m^2(t)} + \frac{2\eta^2(t)}{m^2(t)} + \frac{2d^2(t)}{m^2(t)} + p_\theta(t) + p_\rho(t).\end{aligned} \tag{4.91}$$

From the projection algorithm (4.80) - (4.83), we see

$$p_\theta(t) = 2|k_p|\sum_{j=1}^{n_\theta}\tilde{\theta}_j(t)\gamma_j^{-1} f_{\theta j}(t),\ \tilde{\theta}_j(t) = \theta_j(t) - \theta_j^*$$
$$\theta_j(t) \in [\theta_j^a, \theta_j^b],\ (\theta_j(t) - \theta_j^*) f_{\theta j}(t) \leq 0,\ 1 \leq j \leq n_\theta$$
$$\rho(t) \in [\rho^a, \rho^b],\ (\rho(t) - \rho^*) f_\rho(t) \leq 0.$$

Thus, $\theta(t)$ and $\rho(t)$ are bounded and

$$p_\theta(t) \leq 0,\ p_\rho(t) \leq 0. \tag{4.92}$$

Using (4.90), we conclude that $\frac{\epsilon(t)}{m(t)}$, $\dot{\theta}(t)$, and $\dot{\rho}(t)$ are bounded. In view of (4.89), (4.91), (4.92), and the boundedness of $d(t)$, we have (4.84). From (4.80) - (4.83), we obtain

$$\dot{\theta}_j^2(t) \leq g_{\theta j}^2(t) \leq \frac{k_\theta \epsilon^2(t)}{m^2(t)},\ 1 \leq j \leq n_\theta \tag{4.93}$$

$$\dot{\rho}^2(t) \leq g_\rho^2(t) \leq \frac{k_\rho \epsilon^2(t)}{m^2(t)} \tag{4.94}$$

for some constant $k_\theta > 0$, $k_\rho > 0$. From (4.84), (4.93), and (4.94), we have (4.85) and (4.86). ∇

Now we present the main property of the continuous-time detuned inverse control scheme (4.67) - (4.68) with the adaptive update law (4.76) - (4.77).

Theorem 4.5 *There exists a constant $\mu^* > 0$ such that for any nonlinearity $N(\cdot)$ and its detuned inverse $\widehat{NI}(\cdot)$ satisfying $\mu_N \in [0, \mu^*]$ the adaptive linear controller (4.67), updated by the adaptive law (4.76) - (4.77), and the detuned inverse (4.68), applied to the plant (4.1), ensure that all closed-loop signals are bounded.*

To prove Theorem 4.5, we note that the estimation error expression (4.90) is standard in the robust adaptive control literature; see, for example, [39]. The error in cancelling the nonlinearity $N(\cdot)$ by a detuned inverse $\widehat{NI}(\cdot)$ consists of the "unmodeled dynamics" $\eta(t) = \Delta(s)[\tilde{k}_N u_d](t)$ and the bounded "disturbance" $d(t)$. Lemma 4.3 shows that, in addition to the boundedness of the parameter estimates $\theta(t)$, $\rho(t)$, the proposed adaptive scheme ensures the smallness of $\frac{\epsilon(t)}{m(t)}$ and $\dot{\theta}(t)$ in the mean sense (4.84), (4.85). These properties are sufficient for the closed-loop signal boundedness if the size μ_N of $\tilde{k}_N$ in the "unmodeled dynamics" $\eta(t)$ is small. The proof of Theorem 4.5 is given in Appendix B.

The tracking performance of the proposed control scheme is described by the following corollary.

Corollary 4.2 *Under the conditions of Theorem 4.5, the tracking error $e(t) = y(t) - y_m(t)$ satisfies*

$$\int_{t_1}^{t_2} e^2(t)dt \leq a_0 + b_0\mu_N^2(t_2 - t_1) + c_0 d_0^2(t_2 - t_1)$$

for an upper bound d_0 on $|\bar{d}_N(t)|$ and some constants a_0, b_0, $c_0 > 0$.

This corollary follows from the closed-loop signal boundedness, (4.75), (4.84), (4.85).

As an alternative, we can also choose to use the switching-σ design [42] for the modification terms $f_\theta(t)$ and $f_\rho(t)$.

Switching σ-modification. Introduce the switching functions

$$\sigma_\theta(t) = \begin{cases} 0 & \text{if } \|\theta(t)\|_2 < M_\theta \\ \sigma_0(\frac{\|\theta(t)\|_2}{M_\theta} - 1) & \text{if } M_\theta \leq \|\theta(t)\|_2 < 2M_\theta \\ \sigma_0 & \text{if } \|\theta(t)\|_2 \geq 2M_\theta \end{cases}$$

$$\sigma_\rho(t) = \begin{cases} 0 & \text{if } |\rho(t)| < M_\rho \\ \sigma_0(\frac{|\rho(t)|}{M_\rho} - 1) & \text{if } M_\rho \leq |\rho(t)| < 2M_\rho \\ \sigma_0 & \text{if } |\rho(t)| \geq 2M_\rho, \end{cases}$$

where $\sigma_0 > 0$ is a design parameter, and M_θ and M_ρ are determined using *a priori* knowledge of upper bounds of $\|\theta^*\|_2$ and $|\rho^*|$:

$$M_\theta > \|\theta^*\|_2, \; M_\rho > |\rho^*|$$

with $\|\cdot\|_2$ being the Euclidean (l^2) vector norm.

Then the switching σ-modifications are

$$f_\theta(t) = -\Gamma_\theta \sigma_\theta(t)\theta(t) \tag{4.95}$$

$$f_\rho(t) = -\gamma_\rho \sigma_\rho(t)\rho(t). \tag{4.96}$$

The adaptive law (4.76) - (4.77) with the modification (4.95) - (4.96) also has the desired properties stated in Lemma 4.3, Theorem 4.5, and Corollary 4.2. To see this fact, we introduce

$$p_\theta(t) = -2|k_p|\sigma_\theta(t)\tilde{\theta}^T(t)\theta(t)$$

$$p_\rho(t) = -2\sigma_\rho(t)\tilde{\rho}(t)\rho(t)$$

and again obtain the expression (4.91):

$$\begin{aligned} \dot{V} &= -\frac{2\epsilon^2(t)}{m^2(t)} + \frac{2\epsilon(t)(\eta(t) + d(t))}{m^2(t)} + p_\theta(t) + p_\rho(t) \\ &\leq -\frac{\epsilon^2(t)}{m^2(t)} + \frac{2\eta^2(t)}{m^2(t)} + \frac{2d^2(t)}{m^2(t)} + p_\theta(t) + p_\rho(t). \end{aligned} \tag{4.97}$$

From the definitions of $\sigma_\theta(t)$, $\sigma_\rho(t)$, it follows that

$$p_\theta(t) \leq 0, \; p_\rho(t) \leq 0$$

and that the term $|p_\theta(t)|$ grows unbounded if $\theta(t)$ grows unbounded. Similarly, the term $|p_\rho(t)|$ grows unbounded if $\rho(t)$ grows unbounded. Therefore, there exist constants $\theta_0 > 0$, $\rho_0 > 0$ such that $\dot{V} < 0$ whenever $\|\theta(t)\|_2 \geq \theta_0$ or/and $|\rho(t)| \geq \rho_0$; that is, $V(\tilde{\theta}(t), \tilde{\rho}(t))$ is bounded, and so are $\theta(t)$ and $\rho(t)$. Then we have the boundedness of $\frac{\epsilon(t)}{m(t)}$, $\dot{\theta}(t)$, $\dot{\rho}(t)$ as well as (4.84). To obtain (4.85) and (4.86), we use (4.98) - (4.99), and, introducing constants k_θ, k_ρ such that

$$\frac{\sigma_\theta(t)\|\Gamma_\theta\|_2^2\|\theta(t)\|_2}{M_\theta - \|\theta^*\|_2} \leq k_\theta \tag{4.98}$$

$$\frac{\sigma_\rho(t)\gamma_\rho^2|\rho(t)|}{M_\rho - |\rho^*|} \leq k_\rho \tag{4.99}$$

we derive the inequalities

$$\begin{aligned}\sigma_\theta^2(t)\theta^T(t)\Gamma_\theta^2\theta(t) &\leq \sigma_\theta^2(t)\|\Gamma_\theta\|_2^2\|\theta(t)\|_2^2 \\ &\leq k_\theta\sigma_\theta(t)\|\theta(t)\|_2(M_\theta - \|\theta^*\|_2) \\ &\leq k_\theta\sigma_\theta(t)\|\theta(t)\|_2(\|\theta(t)\|_2 - \|\theta^*\|_2) \\ &\leq k_\theta\sigma_\theta(t)\tilde{\theta}^T(t)\theta(t)\end{aligned} \tag{4.100}$$

$$\begin{aligned}\sigma_\rho^2(t)\rho(t)\gamma_\rho^2\rho(t) &\leq \sigma_\rho^2(t)\Gamma_\rho^2|\rho(t)|^2 \\ &\leq k_\rho\sigma_\rho(t)|\rho(t)|(M_\rho - |\rho^*|) \\ &\leq k_\rho\sigma_\rho(t)|\rho(t)|(|\rho(t)| - |\rho^*|) \\ &\leq k_\rho\sigma_\rho(t)\tilde{\rho}(t)\rho(t).\end{aligned} \tag{4.101}$$

We thus obtain

$$\|\dot{\theta}(t)\|_2^2 \leq a_4\frac{\epsilon^2(t)}{m^2(t)} + a_5|p_\theta(t)| \tag{4.102}$$

$$|\dot{\rho}(t)|^2 \leq a_6\frac{\epsilon^2(t)}{m^2(t)} + a_7|p_\rho(t)| \tag{4.103}$$

for some constants $a_i > 0$, $i = 4, 5, 6, 7$. Then, (4.85) and (4.86) follow from (4.97), (4.102), and (4.103).

Discrete-Time Adaptive Design

As in the continuous-time case, we introduce the auxiliary signals

$$\xi(t) = \theta^T(t)\omega(t - n^*) - \theta^T(t - n^*)\omega(t - n^*) \tag{4.104}$$

and define the estimation error

$$\epsilon(t) = e(t) + \rho(t)\xi(t), \tag{4.105}$$

where $\rho(t)$ is the estimate of $\rho^* = k_p$. Then we choose the modified gradient-type adaptive update law for $\theta(t) \in R^{n_\theta}$ and $\rho(t)$:

$$\theta(t+1) = \theta(t) - \frac{sign(k_p)\Gamma_\theta\epsilon(t)\omega(t - n^*)}{m^2(t)} + f_\theta(t) \tag{4.106}$$

$$\rho(t+1) = \rho(t) - \frac{\gamma_\rho \epsilon(t)\xi(t)}{m^2(t)} + f_\rho(t), \tag{4.107}$$

where Γ_θ is a diagonal constant gain matrix: $\Gamma_\theta = diag\{\gamma_1, \ldots, \gamma_{n_\theta}\}$ and γ_ρ is a constant gain, which satisfy

$$0 < \gamma_j < \frac{2}{k_p^0},\ 1 \le j \le n_\theta,\ 0 < \gamma_\rho < 2,$$

where k_p^0 is an upper bound on $|k_p|$: $k_p^0 \ge |k_p|$.

The normalizing signal is

$$m(t) = \sqrt{1 + \omega^T(t-n^*)\omega(t-n^*) + \xi^2(t) + m_a^2(t)}, \tag{4.108}$$

where $m_a(t)$ satisfies the difference equation

$$m_a(t+1) = \delta_m m_a(t) + |u_d(t)|,\ m_a(0) = 0 \tag{4.109}$$

with a constant $\delta_m \in (0,1)$ such that both $\Lambda(\delta_m z)$, $P_m(\delta_m z)$ are stable polynomials.

Using the assumption (A6), we update the estimates $\theta(t)$ and $\rho(t)$ using the following projection algorithm.

Parameter projection. Define

$$g_\theta(t) = -\frac{sign(k_p)\Gamma_\theta \omega(t-n^*)\epsilon(t)}{m^2(t)} \tag{4.110}$$

$$g_\rho(t) = -\frac{\gamma_\rho \xi(t)\epsilon(t)}{m^2(t)} \tag{4.111}$$

denote the jth components of $\theta(t)$, $f_\theta(t)$, and $g_\theta(t)$ as $\theta_j(t)$, $f_{\theta j}(t)$, and $g_{\theta j}(t)$, respectively, $j = 1, 2, \ldots, n_\theta$, for $\theta(t)$, $f_\theta(t) \in R^{n_\theta}$, choose

$$\theta_j(0) \in [\theta_j^a, \theta_j^b],\ \rho(0) \in [\rho^a, \rho^b],$$

and let

$$f_{\theta j}(t) = \begin{cases} 0 & \text{if } \theta_j(t) + g_{\theta j}(t) \in [\theta_j^a, \theta_j^b] \\ \theta_j^b - \theta_j(t) - g_{\theta j}(t) & \text{if } \theta_j(t) + g_{\theta j}(t) > \theta_j^b \\ \theta_j^a - \theta_j(t) - g_{\theta j}(t) & \text{if } \theta_j(t) + g_{\theta j}(t) < \theta_j^a \end{cases} \tag{4.112}$$

$$f_\rho(t) = \begin{cases} 0 & \text{if } \rho(t) + g_\rho(t) \in [\rho^a, \rho^b] \\ \rho^b - \rho(t) - g_\rho(t) & \text{if } \rho(t) + g_\rho(t) > \rho^b \\ \rho^a - \rho(t) - g_\rho(t) & \text{if } \rho(t) + g_\rho(t) < \rho^a. \end{cases} \tag{4.113}$$

This modified adaptive law has the following properties.

Lemma 4.4 *For the error model (4.70), the adaptive law (4.106) - (4.107) with the parameter projection (4.110) - (4.113) guarantees that* $\theta(t)$, $\rho(t)$, $\frac{\epsilon(t)}{m(t)} \in l^\infty$, *and* $\frac{\epsilon(t)}{m(t)}$, $\theta(t+1)-\theta(t)$, $\rho(t+1)-\rho(t)$ *satisfy*

$$\sum_{t=t_1}^{t_2} \frac{\epsilon^2(t)}{m^2(t)} \leq a_1 + b_1\mu_N^2(t_2-t_1) + c_1 \sum_{t=t_1}^{t_2} \frac{d^2(t)}{m^2(t)} \tag{4.114}$$

$$\sum_{t=t_1}^{t_2} \|\theta(t+1)-\theta(t)\|_2^2 \leq a_2 + b_2\mu_N^2(t_2-t_1) + c_2 \sum_{t=t_1}^{t_2} \frac{d^2(t)}{m^2(t)} \tag{4.115}$$

$$\sum_{t=t_1}^{t_2} (\rho(t+1)-\rho(t))^2 \leq a_3 + b_3\mu_N^2(t_2-t_1) + c_3 \sum_{t=t_1}^{t_2} \frac{d^2(t)}{m^2(t)} \tag{4.116}$$

for some constants $a_i, b_i, c_i > 0$, $i = 1,2,3$, *and all* $t_2 > t_1 \geq 0$.

Proof: Let us consider $\eta(t) = \Delta(z)[\tilde{k}_N u_d](t)$ in (4.71), where

$$\Delta(z) = \frac{k_p}{P_m(z)}(1 - \theta_1^{*T}\frac{a(z)}{\Lambda(z)}). \tag{4.117}$$

To establish the discrete-time version of (4.89),

$$|\eta(t)| \leq k_\delta \mu_N m(t), \tag{4.118}$$

for some constant $k_\delta > 0$ and $m(t)$ in (4.108), we denote the impulse response of $(z-1)\delta_m\Delta(\delta_m z)$ by $h_\delta(t,\delta_m)$ and denote its l^1 signal norm by $\|h_\delta(\cdot,\delta_m)\|_1 \triangleq \sum_{t=0}^\infty |h_\delta(t,\delta_m)|$. With $\delta_m \in (0,1)$ as in (4.109), we see from (4.117) and the choice of δ_m that $\|h_\delta(\cdot,\delta_m)\|_1$ is finite.

We then introduce the signal $x(t) = \delta_m^{-t}\tilde{k}_N(t)u_d(t)$ for the above δ_m and the constant $k_m > 0$ such that $|\tilde{k}_N(t)| \leq k_m\mu_N$ where $\tilde{k}_N$, μ_N are defined in (4.55), (4.58). We denote the unit step function by $u_s(t)$ and obtain

$$\begin{aligned}
|\eta(t)| &= |\frac{1}{z-\delta_m}[(z-\delta_m)\Delta(z)[\tilde{k}_N u_d]](t)| \\
&\leq \sum_{\tau=0}^{t} \delta_m^{t-\tau-1} u_s(t-\tau-1)|(z-\delta_m)\Delta(z)[\tilde{k}_N u_d](\tau)| \\
&= \delta_m^{t-1} \sum_{\tau=0}^{t} \delta_m^{-\tau} u_s(t-\tau-1)|(z-\delta_m)\Delta(z)[\tilde{k}_N u_d](\tau)| \\
&= \delta_m^{t-1} \sum_{\tau=0}^{t} u_s(t-\tau-1)|(\delta_m z-\delta_m)\Delta(\delta_m z)[x](\tau)| \\
&= \delta_m^{t-1} \sum_{\tau=0}^{t-1} |(\delta_m z-\delta_m)\Delta(\delta_m z)[x](\tau)| \leq \delta_m^{t-1}\|h_\delta(\cdot,\delta_m)\|_1 \sum_{\tau=0}^{t-1} |x(\tau)|
\end{aligned}$$

$$
\begin{aligned}
&= \delta_m^{t-1}\|h_\delta(\cdot,\delta_m)\|_1 \sum_{\tau=0}^{t} u_s(t-\tau-1)|x(\tau)| \\
&= \delta_m^{t-1}\|h_\delta(\cdot,\delta_m)\|_1 \sum_{\tau=0}^{t} tu_s(t-\tau-1)\delta_m^{-\tau}|\tilde{k}_N(\tau)||u_d(\tau)| \\
&= \|h_\delta(\cdot,\delta_m)\|_1 \sum_{\tau=0}^{t} u_s(t-\tau-1)\delta_m^{t-\tau-1}|\tilde{k}_N(\tau)||u_d(\tau)| \\
&\leq \|h_\delta(\cdot,\delta_m)\|_1 k_m\mu_N \sum_{\tau=0}^{t} u_s(t-\tau-1)\delta_m^{t-\tau-1}|u_d(\tau)| \\
&= \|h_\delta(\cdot,\delta_m)\|_1 k_m\mu_N m_a(t),
\end{aligned}
\tag{4.119}
$$

where the second inequality follows from the definition of $\|h_\delta(\cdot,\delta_m)\|_1$. In view of (4.119) and (4.108), we have (4.118).

Substituting (4.104) and the discrete-time version of (4.70),

$$
e(t) = k_p\tilde{\theta}^T(t-n^*)\omega(t-n^*) + \eta(t) + d(t),
$$

in (4.105), we obtain the error equation

$$
\epsilon(t) = k_p\tilde{\theta}^T(t)\omega(t-n^*) + \tilde{\rho}(t)\xi(t) + \eta(t) + d(t). \tag{4.120}
$$

We now proceed to calculate the time increment of the positive definite function

$$
V(\tilde{\theta},\tilde{\rho}) = |k_p|\tilde{\theta}^T\Gamma_\theta^{-1}\tilde{\theta} + \gamma_\rho^{-1}\tilde{\rho}^2
$$

along the trajectories of (4.106) - (4.107). Letting

$$
\gamma_m = \max\{\gamma_1 k_p^0, \ldots, \gamma_{n_\theta} k_p^0, \gamma_\rho\} < 2
$$

and using (4.120), we get

$$
\begin{aligned}
&V(\tilde{\theta}(t+1),\tilde{\rho}(t+1)) - V(\tilde{\theta}(t),\tilde{\rho}(t)) \\
&\quad = -(2 - \frac{|k_p|\omega^T(t-n^*)\Gamma_\theta\omega(t-n^*) + \gamma_\rho\xi^2(t)}{m^2(t)})\frac{\epsilon^2(t)}{m^2(t)} \\
&\qquad +2|k_p|f_\theta^T(t)\Gamma_\theta^{-1}(\tilde{\theta}(t) + g_\theta(t) + f_\theta(t)) - |k_p|f_\theta^T(t)\Gamma_\theta^{-1}f_\theta(t) \\
&\qquad +2f_\rho(t)\gamma_\rho^{-1}(\tilde{\rho}(t) + g_\rho(t) + f_\rho(t)) - f_\rho(t)\gamma_\rho^{-1}f_\rho(t) \\
&\qquad +\frac{2\epsilon(t)(\eta(t)+d(t))}{m^2(t)} \\
&\quad \leq -\frac{2-\gamma_m}{2}\frac{\epsilon^2(t)}{m^2(t)} + \frac{4}{2-\gamma_m}\frac{\eta^2(t)+d^2(t)}{m^2(t)} \\
&\qquad +2|k_p|f_\theta^T(t)\Gamma_\theta^{-1}(\tilde{\theta}(t) + g_\theta(t) + f_\theta(t)) - |k_p|f_\theta^T(t)\Gamma_\theta^{-1}f_\theta(t) \\
&\qquad +2f_\rho(t)\gamma_\rho^{-1}(\tilde{\rho}(t) + g_\rho(t) + f_\rho(t)) - f_\rho(t)\gamma_\rho^{-1}f_\rho(t).
\end{aligned}
\tag{4.121}
$$

Since Γ_θ is a diagonal matrix, we rewrite

$$2|k_p|f_\theta^T(t)\Gamma_\theta^{-1}(\tilde{\theta}(t)+g_\theta(t)+f_\theta(t))$$
$$= 2|k_p|\sum_{j=1}^{n_\theta}\gamma_j^{-1}f_{\theta j}(t)(\theta_j(t)-\theta_j^*+g_{\theta j}(t)+f_{\theta j}(t)). \quad (4.122)$$

Using (4.112), we have

$$f_{\theta j}(t)(\theta_j(t)-\theta_j^*+g_{\theta j}(t)+f_{\theta j}(t)) \leq 0 \quad (4.123)$$

for all $j = 1, \ldots, n_\theta$. Similarly, using (4.113), we have

$$f_\rho(t)(\rho(t)-\rho^*+g_\rho(t)+f_\rho(t)) \leq 0. \quad (4.124)$$

Hence, with (4.122) - (4.124), (4.121) becomes

$$V(\tilde{\theta}(t+1),\tilde{\rho}(t+1)) - V(\tilde{\theta}(t),\tilde{\rho}(t))$$
$$\leq -\frac{2-\gamma_m}{2}\frac{\epsilon^2(t)}{m^2(t)} + \frac{4}{2-\gamma_m}\frac{\eta^2(t)+d^2(t)}{m^2(t)}$$
$$-|k_p|f_\theta^T(t)\Gamma_\theta^{-1}f_\theta(t) - f_\rho(t)\gamma_\rho^{-1}f_\rho(t). \quad (4.125)$$

Since the projection algorithm (4.110) - (4.113) ensures that $\theta_j(t) \in [\theta_j^a, \theta_j^b]$, $j = 1, \ldots, n_\theta$, and $\rho \in [\rho^a, \rho^b]$, it follows that $\theta(t)$ and $\rho(t)$ are bounded. Then (4.118) and (4.108) show that $\frac{\epsilon(t)}{m(t)}$ is bounded. Using (4.118), (4.125) and the boundedness of $d(t)$, we get (4.114). To prove (4.115) and (4.116), from (4.106) and (4.107), we see

$$\|\theta(t+1)-\theta(t)\|_2^2 \leq k_1\frac{\epsilon^2(t)}{m^2(t)} + k_2 f_\theta^T(t)f_\theta(t) \quad (4.126)$$

$$|\rho(t+1)-\rho(t)|^2 \leq k_3\frac{\epsilon^2(t)}{m^2(t)} + k_4 f_\rho^2(t) \quad (4.127)$$

for some constants $k_i > 0$, $i = 1, 2, 3, 4$. Using (4.125) - (4.127), we have (4.115) and (4.116). ∇

Using Lemma 4.4, we can establish the following main property of the discrete-time detuned inverse control scheme (4.67) - (4.68) with the adaptive law (4.106) - (4.107).

Theorem 4.6 *There exists a constant $\mu^* > 0$ such that for any nonlinearity $N(\cdot)$ and its detuned inverse $\widehat{NI}(\cdot)$ satisfying $\mu_N \in [0, \mu^*]$ the adaptive linear controller (4.67), updated by the adaptive law (4.106) - (4.107), and the detuned inverse (4.68), applied to the plant (4.1), ensure that all closed-loop signals are bounded.*

The proof of Theorem 4.6 is given in Appendix C.

The tracking performance of the discrete-time control scheme is described by the following corollary.

Corollary 4.3 *Under the conditions of Theorem 4.6, the tracking error* $e(t) = y(t) - y_m(t)$ *satisfies*

$$\sum_{t=t_1}^{t_2} e^2(t)dt \leq a_0 + b_0\mu_N^2(t_2 - t_1) + c_0 d_0^2(t_2 - t_1)$$

for an upper bound d_0 *on* $|\bar{d}_N(t)|$ *and some constants* $a_0 > 0$, $b_0 > 0$, $c_0 > 0$.

This corollary follows from (4.105), (4.114), (4.115), and the closed-loop signal boundedness properties.

The above results are obtained with the normalizing signal $m_a(t)$ in (4.109). The same result (4.119) leading to Theorem 4.6 can be obtained with

$$m_a(t) = |u_d(t - n^*)|.$$

Similar to the continuous-time case, for the modification terms $f_\theta(t)$ and $f_\rho(t)$, we can also use the switching-σ design [43].

Switching σ-modification. As in (4.115) - (4.116), the discrete-time adaptive law consists of two equations:

$$\theta(t+1) = \theta(t) - \frac{sign(k_p)\Gamma_\theta\epsilon(t)\omega(t-n^*)}{m^2(t)} + f_\theta(t) \tag{4.128}$$

$$\rho(t+1) = \rho(t) - \frac{\gamma_\rho\epsilon(t)\xi(t)}{m^2(t)} + f_\rho(t), \tag{4.129}$$

where $\rho(t)$ is the estimate of $\rho^* = k_p$ for constructing the estimation error $\epsilon(t)$ (see (4.104) and (4.105)). For a switching σ-modification design, we choose

$$\Gamma_\theta = \gamma_\theta I_{n_N},\ 0 < \gamma_\theta < \frac{1}{k_p^0},\ 0 < \gamma_\rho < 1,$$

where k_p^0 is an upper bound of $|k_p|$: $k_p^0 \geq |k_p|$, and introduce the switching functions

$$\sigma_\theta(t) = \begin{cases} \sigma_0 & \text{if } \|\theta(t)\|_2 > 2M_\theta \\ 0 & \text{otherwise} \end{cases}$$

$$\sigma_\rho(t) = \begin{cases} \sigma_0 & \text{if } |\rho(t)| > 2M_\rho \\ 0 & \text{otherwise,} \end{cases}$$

where

$$0 < \sigma_0 < \frac{1}{2}(1-\gamma_m),\ \gamma_m = \max\{\gamma_\theta k_p^0, \gamma_\rho\} < 1$$

and M_θ, M_ρ are the upper bounds of $\|\theta^*\|_2$, $|\rho^*|$:

$$\|\theta^*\|_2 < M_\theta,\ |k_p| < M_\rho = k_p^0.$$

Then the switching σ-modification design is

$$f_\theta(t) = -\sigma_\theta(t)\theta(t) \tag{4.130}$$

$$f_\rho(t) = -\sigma_\rho(t)\rho(t). \tag{4.131}$$

The adaptive law (4.128) - (4.131) has the desired properties stated in Lemma 4.4, Theorem 4.6, and Corollary 4.3. To show this, we use (4.130) and (4.131) to rewrite (4.121) as

$$\begin{aligned}
&V(\tilde\theta(t+1),\tilde\rho(t+1)) - V(\tilde\theta(t),\tilde\rho(t)) \\
&\quad\le -\frac{\gamma_m}{2}\frac{\epsilon^2(t)}{m^2(t)} + \frac{4}{2-\gamma_m}\frac{\eta^2(t)+d^2(t)}{m^2(t)} - (1-\gamma_m)\frac{\epsilon^2(t)}{m^2(t)} \\
&\quad -2|k_p|\sigma_\theta(t)\theta^T(t)\gamma_\theta^{-1}\tilde\theta(t) + 2k_p\sigma_\theta\theta^T(t)\frac{\epsilon(t)\omega(t-n^*)}{m^2(t)} \\
&\quad +|k_p|\sigma_\theta^2(t)\theta^T(t)\gamma_\theta^{-1}\theta(t) - 2\sigma_\rho(t)\rho(t)\gamma_\rho^{-1}\tilde\rho(t) \\
&\quad +2\sigma_\rho\rho(t)\frac{\epsilon(t)\xi(t)}{m^2(t)} + \sigma_\rho^2(t)\rho(t)\gamma_\rho^{-1}\rho(t).
\end{aligned} \tag{4.132}$$

Using the properties

$$\|(k_p\sigma_\theta(t)\theta^T(t), \sigma_\rho(t)\rho(t))^T\|_2^2 = k_p^2\sigma_\theta^2(t)\theta^T(t)\theta(t) + \sigma_\rho^2(t)\rho^2(t)$$

$$\frac{\|(\omega^T(t-n^*), \xi(t))^T\|_2}{m(t)} < 1$$

we obtain

$$\begin{aligned}
&-(1-\gamma_m)\frac{\epsilon^2(t)}{m^2(t)} + 2k_p\sigma_\theta(t)\theta^T(t)\frac{\epsilon(t)\omega(t-n^*)}{m^2(t)} + 2\sigma_\rho(t)\rho(t)\frac{\epsilon(t)\xi(t)}{m^2(t)} \\
&\quad = -(1-\gamma_m)\frac{\epsilon^2(t)}{m^2(t)} + 2(k_p\sigma_\theta(t)\theta^T(t), \sigma_\rho(t)\rho(t))\frac{(\omega^T(t-n^*), \xi(t))^T\epsilon(t)}{m^2(t)} \\
&\quad \le -(1-\gamma_m)\frac{\epsilon^2(t)}{m^2(t)} + 2\|(k_p\sigma_\theta(t)\theta^T(t), \sigma_\rho(t)\rho(t))^T\|_2\frac{|\epsilon(t)|}{m(t)} \\
&\quad \le -(1-\gamma_m)\left(\frac{|\epsilon(t)|}{m(t)} - \frac{\|(k_p\sigma_\theta(t)\theta^T(t), \sigma_\rho(t)\rho(t))^T\|_2}{1-\gamma_m}\right)^2 \\
&\quad + \frac{\|(k_p\sigma_\theta(t)\theta^T(t), \sigma_\rho(t)\rho(t))^T\|_2^2}{1-\gamma_m} \\
&\quad \le \frac{k_p^2\sigma_\theta^2(t)\theta^T(t)\theta(t)}{1-\gamma_m} + \frac{\sigma_\rho^2(t)\rho^2(t)}{1-\gamma_m}.
\end{aligned}$$

With this inequality, (4.132) becomes

$$\begin{aligned}
&V(\tilde{\theta}(t+1), \tilde{\rho}(t+1)) - V(\tilde{\theta}(t), \tilde{\rho}(t)) \\
&\quad \leq -\frac{\gamma_m}{2}\frac{\epsilon^2(t)}{m^2(t)} + \frac{4}{2-\gamma_m}\frac{\eta^2(t)+d^2(t)}{m^2(t)} \\
&\quad\quad -2|k_p|\sigma_\theta(t)\theta^T(t)\gamma_\theta^{-1}\tilde{\theta}(t) + \frac{k_p^2\sigma_\theta^2(t)\theta^T(t)\theta(t)}{1-\gamma_m} + |k_p|\sigma_\theta^2(t)\theta^T(t)\gamma_\theta^{-1}\theta(t) \\
&\quad\quad -2\sigma_\rho(t)\rho(t)\gamma_\rho^{-1}\tilde{\rho}(t) + \frac{\sigma_\rho^2(t)\rho^2(t)}{1-\gamma_m} + \sigma_\rho^2(t)\rho(t)\gamma_\rho^{-1}\rho(t). \qquad (4.133)
\end{aligned}$$

When $\|\theta(t)\|_2 > M_\theta \geq 2\|\theta^*\|_2$, we have that $\sigma_\theta(t) = \sigma_0$, and

$$\begin{aligned}
&-2|k_p|\sigma_\theta(t)\theta^T(t)\gamma_\theta^{-1}\tilde{\theta}(t) + \frac{k_p^2\sigma_\theta^2(t)\theta^T(t)\theta(t)}{1-\gamma_m} + |k_p|\sigma_\theta^2(t)\theta^T(t)\gamma_\theta^{-1}\theta(t) \\
&\quad = -|k_p|\sigma_\theta(t)\gamma_\theta^{-1}(2\theta^T(t)\tilde{\theta}(t) - \frac{\gamma_\theta|k_p|\sigma_\theta(t)\theta^T(t)\theta(t)}{1-\gamma_m} - \sigma_\theta(t)\theta^T(t)\theta(t)) \\
&\quad \leq -\frac{|k_p|k_p^0\sigma_\theta(t)}{2}\theta^T(t)\theta(t). \qquad (4.134)
\end{aligned}$$

The last inequality is due to the facts that $|k_p|\gamma_\theta \leq k_p^0\gamma_\theta \leq \gamma_m < 1$ and that $\sigma_\theta(t) = \sigma_0 \leq \frac{1-\gamma_m}{2}$ and $\theta^T(t)\tilde{\theta}(t) \geq \frac{1}{2}\theta^T(t)\theta(t)$ for $\|\theta(t)\|_2 > 2\|\theta^*\|_2$. When $\|\theta(t)\|_2 \leq M_\theta$, then $\sigma_\theta(t) = 0$ and (4.134) still holds.

Similarly, observing that $\gamma_\rho \leq \gamma_m < 1$ and that $\sigma_\rho(t) = \sigma_0 \leq \frac{1-\gamma_m}{2}$ or $\sigma_\rho(t) = 0$, we have

$$\begin{aligned}
&-2\sigma_\rho(t)\rho(t)\gamma_\rho^{-1}\tilde{\rho}(t) + \frac{\sigma_\rho^2(t)\rho^2(t)}{1-\gamma_m} + \sigma_\rho^2(t)\rho(t)\gamma_\rho^{-1}\rho(t) \\
&\quad = -\sigma_\rho(t)\gamma_\rho^{-1}(2\rho(t)\tilde{\rho}(t) - \frac{\gamma_\rho\sigma_\rho(t)\rho^2(t)}{1-\gamma_m} - \sigma_\rho(t)\rho^2(t)) \\
&\quad \leq -\frac{\sigma_\rho(t)}{2}\rho^2(t). \qquad (4.135)
\end{aligned}$$

Finally, using (4.133) - (4.135), we obtain

$$\begin{aligned}
&V(\tilde{\theta}(t+1), \tilde{\rho}(t+1)) - V(\tilde{\theta}(t), \tilde{\rho}(t)) \\
&\quad \leq -\frac{\gamma_m}{2}\frac{\epsilon^2(t)}{m^2(t)} + \frac{4}{2-\gamma_m}\frac{\eta^2(t)+d^2(t)}{m^2(t)} \\
&\quad\quad -\frac{|k_p|k_p^0\sigma_\theta(t)}{2}\theta^T(t)\theta(t) - \frac{\sigma_\rho(t)}{2}\rho^2(t). \qquad (4.136)
\end{aligned}$$

From the definitions of $\sigma_\theta(t)$, $\sigma_\rho(t)$, we see that the term $\sigma_\theta(t)\theta^T(t)\theta(t)$ grows unbounded if $\theta(t)$ grows unbounded, and similarly the term $\sigma_\rho(t)\rho^2(t)$ grows

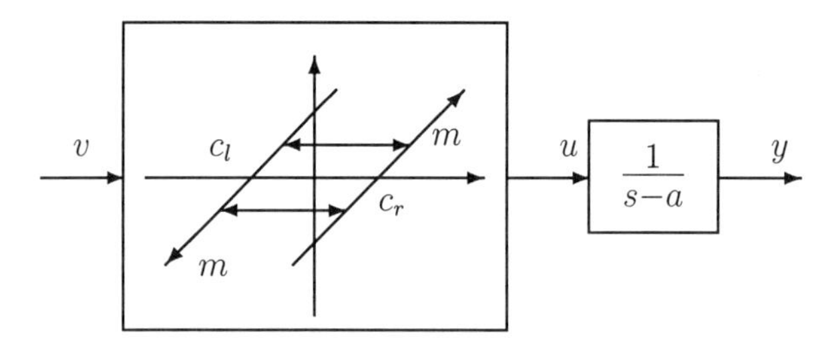

Figure 4.7: Plant with input backlash.

unbounded if $\rho(t)$ grows unbounded. Since $\frac{\eta^2(t)+d^2(t)}{m^2(t)}$ is bounded, there exist constants $\theta_0 > 0$, $\rho_0 > 0$ such that $V(\tilde{\theta}(t+1), \tilde{\rho}(t+1)) - V(\tilde{\theta}(t), \tilde{\rho}(t)) < 0$ whenever $\|\theta(t)\|_2 \geq \theta_0$ or/and $|\rho(t)| \geq \rho_0$. Hence, $V(\tilde{\theta}(t), \tilde{\rho}(t))$ is bounded, and so are $\theta(t)$, $\rho(t)$ and $\frac{\epsilon(t)}{m(t)}$. From (4.136) we have (4.114). To prove (4.115) and (4.116), we see from (4.123), (4.124), (4.130), and (4.131) that

$$\|\theta(t+1) - \theta(t)\|_2^2 \leq k_1 \frac{\epsilon^2(t)}{m^2(t)} + k_2 \sigma_\theta(t) \theta^T(t) \theta(t)$$

$$|\rho(t+1) - \rho(t)|^2 \leq k_3 \frac{\epsilon^2(t)}{m^2(t)} + k_4 \sigma_\rho(t) \rho^2(t)$$

for some constants $k_i > 0$, $i = 1, 2, 3, 4$, which together with (4.136) lead to the inequalities (4.115) and (4.116).

Thus far, we have analyzed the stability of the closed-loop control system. The performance of such a control system, not yet analytically characterized, will be illustrated by simulation results presented in the next section as well as in the subsequent chapters.

4.4 Example: Fixed Backlash Inverse

Consider the first-order linear plant $G(s) = \frac{1}{s-a}$ with input backlash $B(\cdot)$ as in Figure 4.7, where $a = 2$ and $m = 1$ are known, and $c_r = -c_l = 1.3$ is unknown. In this case, the backlash characteristic is defined by the upward line $u(t) = v(t) - 1.3$ and the downward line $u(t) = v(t) + 1.3$.

To specify the system tracking performance, we define a reference signal $y_m(t)$ as the output of the reference model

$$y_m(t) = \frac{1}{s + a_m}[r](t),$$

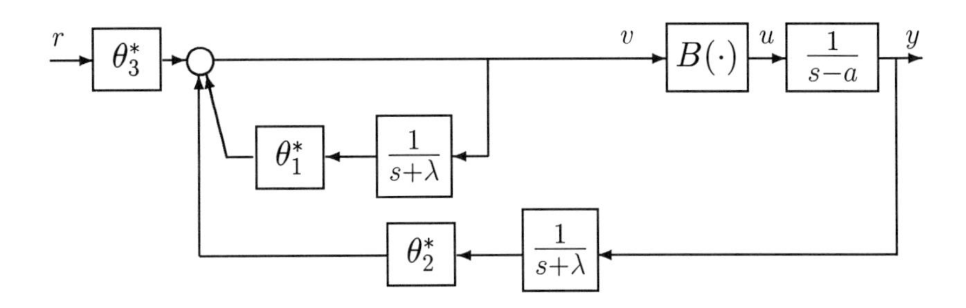

Figure 4.8: Linear controller structure.

where $a_m > 0$ and $r(t)$ is a bounded external input. For simulations we will use $a_m = 1$ and $r(t) = 10\sin 1.3t$.

It can be readily verified that, in the absence of backlash, the linear controller shown in Figure 4.8, namely,

$$v(t) = \frac{\theta_1^*}{s+\lambda}[v](t) + \frac{\theta_2^*}{s+\lambda}[y](t) + \theta_3^* r(t),\ \lambda = 3 \tag{4.137}$$

with

$$\theta_1^* = -(a + a_m),\ \theta_2^* = -(a+\lambda)(a+a_m),\ \theta_3^* = 1, \tag{4.138}$$

ensures stability and output tracking: $\lim_{t\to\infty}(y(t) - y_m(t)) = 0$. However, in the presence of the backlash $B(\cdot)$, this controller cannot achieve output tracking, as shown in Figure 4.9 (a) where the tracking error $e(t) = y(t) - y_m(t)$ is large.

Now we use the fixed compensation scheme shown in Figure 4.10 in which the output $u_d(t)$ of the linear controller is fed to the backlash inverse $\widehat{BI}(\cdot)$ which generates the control $v(t) = \widehat{BI}(u_d(t))$. The linear controller generating $u_d(t)$ has the structure

$$u_d(t) = \frac{\theta_1^*}{s+\lambda}[u_d](t) + \frac{\theta_2^*}{s+\lambda}[y](t) + \theta_3^* r(t).$$

This linear controller is identical to (4.137) and its parameters are calculated from (4.138).

Recall from Section 3.2.1 that the knowledge of $u_d(t)$ and the sign of $\dot{u}_d(t)$ is needed to implement a continuous-time backlash inverse $\widehat{BI}(\cdot)$. In this case, $\dot{u}_d(t)$ is available from the expression

$$\dot{u}_d(t) = \theta_1^* u_d(t) - \frac{\theta_1^*\lambda}{s+\lambda}[u_d](t) + \theta_2^* y(t) - \frac{\theta_2^*\lambda}{s+\lambda}[y](t) + \theta_3^* \dot{r}(t). \tag{4.139}$$

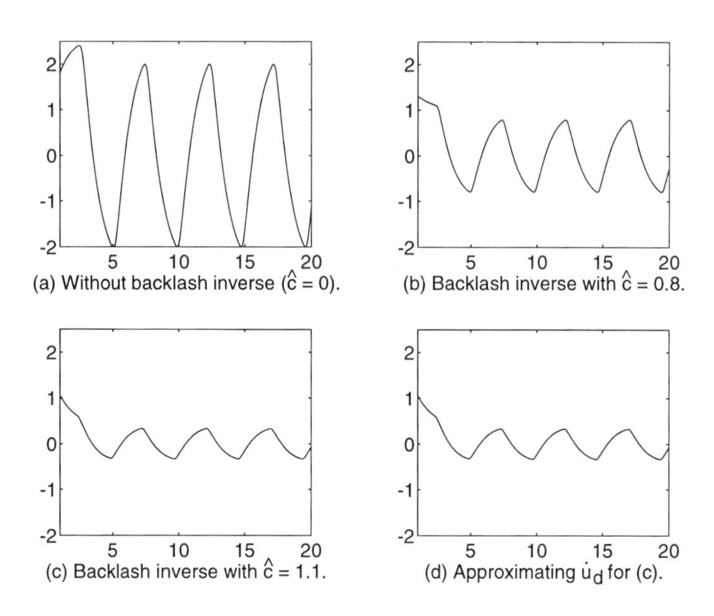

Figure 4.9: Tracking error reduction with inaccurate backlash inverses.

Except for $\dot{r}(t)$, all other signals appearing in (4.139) are available from measurements. We can calculate $\dot{u}_d(t)$ from (4.139) if $\dot{r}(t)$ is also available. When $\dot{r}(t)$ is not available, an approximation is to replace $\dot{r}(t)$ with $\frac{s}{\tau s+1}[r](t)$, for some small $\tau > 0$.

With $u_d(t)$ and $sign(\dot{u}_d(t))$, the backlash inverse (3.15) - (3.16) is implementable. Since $m = 1$ is known and $c_r = -c_l = c$ is unknown, the backlash inverse has only one parameter estimate $\widehat{c} = \widehat{c}_r = -\widehat{c}_l$. Simulation results in-

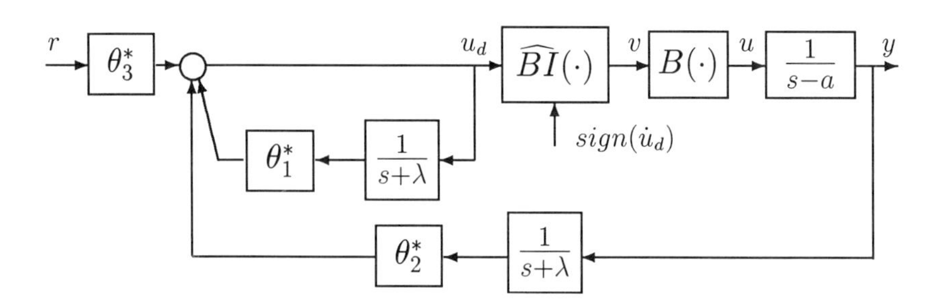

Figure 4.10: Linear controller with backlash compensation.

dicate that using a fixed backlash inverse with parameter estimate $\widehat{c}$ far from its true value $c = 1.3$, the tracking error $e(t)$ is still reduced. The reduction of the tracking error is about the same with $\widehat{c} = 0.8$ (see Figure 4.9 (b)) and with $\widehat{c} = 1.8$ (not shown)—that is, for the parameter error $|\widehat{c} - c| = 0.5$. Further reduction is achieved using a more accurate backlash inverse with $\widehat{c} = 1.1$ such that $|\widehat{c} - c| = 0.2$. The tracking error is shown in Figure 4.9(c) with exact $\dot{r}(t) = 13\cos 1.3t$ for calculating $\dot{u}_d(t)$ from (4.139), and in Figure 4.9(d) with $\frac{15s}{s+15}[r](t)$ instead of $\dot{r}(t)$.

It is clear that a good way to reduce the tracking error is to improve the accuracy of the parameter estimates. This will be the task of the adaptive inverse schemes designed in Chapters 5 - 7.

4.5 Summary

- A plant with an input nonlinearity is modeled as

$$y(t) = G(D)[u](t), \; u(t) = N(v(t)).$$

- Fixed compensation schemes employ a fixed inverse, exact or detuned, for cancelling the nonlinearity $N(\cdot)$ such as a dead-zone, backlash, or hysteresis parametrized by θ_N^*.

- An exact inverse $NI(\cdot)$ implemented with θ_N^*:

$$v(t) = NI(u_d(t))$$

when properly initialized, is able to completely cancel $N(\cdot)$ to result in $u(t) = u_d(t)$, so that the fixed control scheme for the linear part $G(D)$ known:

$$u_d(t) = \theta_1^{*T}\omega_1(t) + \theta_2^{*T}\omega_2(t) + \theta_{20}^* y(t) + \theta_3^* r(t)$$

$$\omega_1(t) = \frac{a(D)}{\Lambda(D)}[u_d](t), \; \omega_2(t) = \frac{a(D)}{\Lambda(D)}[y](t)$$

and the adaptive scheme for $G(D)$ unknown:

$$u_d(t) = \theta_1^T(t)\omega_1(t) + \theta_2^T(t)\omega_2(t) + \theta_{20}(t)y(t) + \theta_3(t)r(t)$$

achieve the desired output tracking:

$$\lim_{t\to\infty}(y(t) - y_m(t)) = 0$$

for a given reference output $y_m(t)$:

$$y_m(t) = W_m(D)[r](t).$$

- A detuned or inaccurate inverse $\widehat{NI}(\cdot)$ implemented with a fixed estimate θ_N of θ_N^*:

$$v(t) = \widehat{NI}(u_d(t))$$

results in a cancellation error

$$u(t) - u_d(t) = \tilde{k}_N(t)u_d(t) + \bar{d}_N(t)$$

with both $\tilde{k}_N(t)$ and $\bar{d}_N(t)$ bounded, so that either the fixed control scheme for $G(D)$ known:

$$u_d(t) = \theta_1^{*T}\omega_1(t) + \theta_2^{*T}\omega_2(t) + \theta_{20}^*y(t) + \theta_3^*r(t)$$

or the adaptive control scheme for $G(D)$ unknown:

$$u_d(t) = \theta_1^T(t)\omega_1(t) + \theta_2^T(t)\omega_2(t) + \theta_{20}(t)y(t) + \theta_3(t)r(t)$$

ensure closed-loop signal boundedness when $\tilde{k}_N$ is small.

- When the parameter error is small, a detuned inverse controller can improve system tracking performance.

Chapter 5

Adaptive Inverse Examples

The models of dead-zone, backlash, and hysteresis in Chapter 2 are nonsmooth and the inverses of dead-zone and backlash contain discontinuities. Examples in this chapter show that such discontinuities need not preclude creative engineering designs.

Many continuous-time systems with discontinuities can be treated using "sliding mode" methods [24], [132], which are well developed in the theory of variable structure systems (VSS). Although very useful, the VSS theory is not a required background for our methodology. All we need our readers to accept is that this theory can be used to legitimize the implementation of purely discontinuous adaptive inverses. Moreover, a "softer" continuous approximation is often a practical alternative.

The purpose of the first example in this chapter is to introduce the discontinuous adaptive dead-zone inverse and to analyze both regular and "sliding" sections of the trajectories of the designed adaptive inverse system. It will be shown that a "soft" inverse, with which the trajectories are continuous, results in a behavior and performance close to that achievable with the discontinuous inverse. The continuous-time analysis in this example is representative for most of the continuous-time results in the subsequent chapters.

The second example illustrates the design of a continuous-time adaptive backlash inverse for the plant with input backlash discussed in Section 3.4 and shown in Figure 3.7. In this example the adaptive version of the backlash compensation scheme of Section 3.4 reduces the tracking error to very small values.

Modern control systems are most frequently implemented with digital controllers so that a discrete-time treatment is closer to practice. An additional benefit is that some of the continuous-time difficulties, such as the need for a time derivative and a discontinuous inverse, are practically eliminated. The

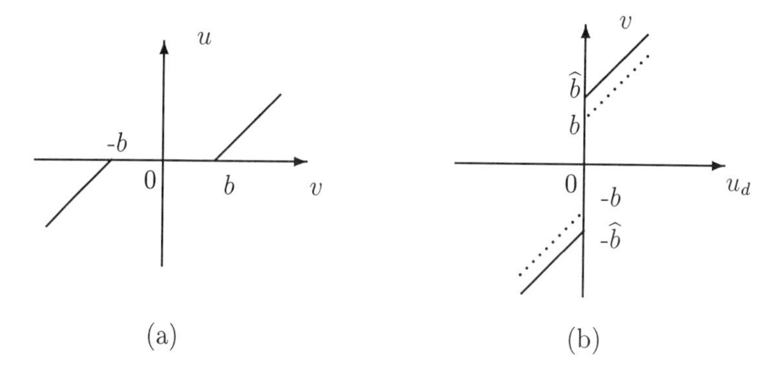

Figure 5.1: (a) Dead-zone; (b) dead-zone inverse.

third example introduces our discrete-time adaptive inverse approach and applies it to the sampled-data model of the plant with input backlash which appeared in Figure 2.20 of Section 2.5.

5.1 Dead-Zone Inverse

We now design an adaptive inverse controller for a plant with an unknown symmetric dead-zone at its input:

$$u(t) = DZ(v(t)) = \begin{cases} v(t) - b & \text{if } v(t) \geq b \\ 0 & \text{if } -b < v(t) < b \\ v(t) + b & \text{if } v(t) \leq -b, \end{cases}$$

where the unknown parameter is the break-point b. This dead-zone and its exact inverse $DI(\cdot)$ (dotted line) and estimated inverse $\widehat{DI}(\cdot)$ (solid line) are shown in Figure 5.1, where $\widehat{b}$ is an estimate of b.

The estimated dead-zone inverse is described by

$$v(t) = \widehat{DI}(\cdot) = u_d(t) + \widehat{b}(t)sgn(u_d(t)). \tag{5.1}$$

It is clear from (5.1) that, when the break-point parameter b is overestimated, $\widehat{b} > b$, as in Figure 5.1(b), the estimated inverse introduces a relay-type discontinuity in $u(t)$. Let us now develop an adaptive inverse controller for the plant with an unstable linear part $G(s) = \frac{1}{s-1}$:

$$\dot{y}(t) = y(t) + u, \; u(t) = DZ(v(t))$$

Suppose that the objective is to place the pole at $s = -1$, and regulate $y(t)$ to a constant set point $r > 0$. Then the desired control $u_d(t)$ is

$$u_d = -2y + r.$$

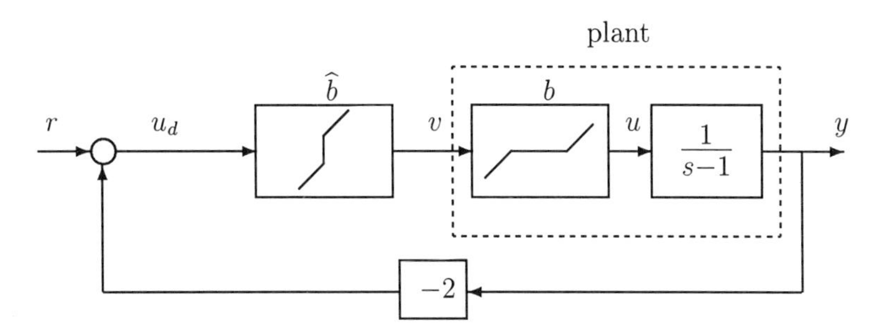

Figure 5.2: Fixed dead-zone compensation scheme.

This $u_d(t)$ is then applied to the dead-zone inverse to generate the plant input $v(t) = \widehat{DI}(u_d(t))$. The nonadaptive version of the dead-zone inverse control system is shown in Figure 5.2.

Our task is to develop an adaptive update law for $\widehat{b}(t)$. A convenient notation is in terms of the tracking error e and the parameter error ϕ:

$$e(t) = y(t) - r, \; \phi(t) = \widehat{b}(t) - b.$$

Since v can be either inside or outside the dead-zone, the equation for the tracking error $e(t)$ has two forms:

$$\dot{e} = -e - u_d, \qquad -b < v(t) < b \tag{5.2}$$

$$\dot{e} = -e + \phi sgn(u_d), \; v(t) \le -b \text{ or } v(t) \ge b. \tag{5.3}$$

The form of (5.3) and $b > 0$ suggest a Lyapunov-type adaptive update law with projection to $\widehat{b}(t) \ge 0$, namely,

$$\dot{\phi} = \dot{\widehat{b}} = \begin{cases} -e \; sgn(u_d) & \text{if } \widehat{b} > 0 \\ 0 & \text{if } \widehat{b} = 0 \text{ and } -e \; sgn(u_d) < 0. \end{cases} \tag{5.4}$$

This update law is based on the Lyapunov function

$$V(e,\phi) = \frac{1}{2}(e^2 + \phi^2) \tag{5.5}$$

with which we can prove that $(e, \phi) = (0, 0)$ is the globally asymptotically stable equilibrium of the adaptive system.

With the adaptive law (5.4), the adaptive dead-zone inverse control system is shown in Figure 5.3.

Let us first ignore the possibility of a sliding mode. Using $V(e, \phi)$ in (5.5) for (5.3), (5.4), we see that $\dot{V} = -e^2$ for all solutions outside the dead-zone.

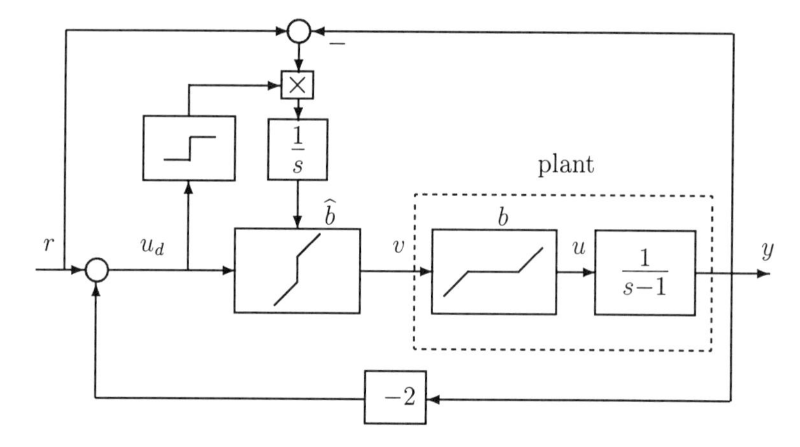

Figure 5.3: Adaptive dead-zone compensation scheme.

This proves that all solutions which start and remain outside the dead-zone converge to the equilibrium $(e, \phi) = (0, 0)$. For the solutions inside the dead-zone we use (5.2) and (5.4) to obtain

$$\dot{V} = -e^2 - e(u_d + \phi sgn(u_d)) = e(e + r - \phi sgn(u_d)).$$

Since $sgn(u_d) = sgn(v)$, we need to examine $\dot{V}$ in the following two cases $u_d > 0$, $0 < v < b$ and $u_d < 0$, $-b < v < 0$. In the first case, $e < -\frac{r}{2} < 0$ and $-e < e + r - \phi$ so that again $\dot{V} < -e^2$. In the second case, $\dot{V} \leq -e^2$ whenever $e \geq 0$, because then $u_d - \phi > 0$. Also, $\dot{V} < 0$ for $-\frac{r}{2} < e < 0$ and $e + r + \phi > 0$. In the remaining region $-\frac{r}{2} < e < 0$ and $e + r + \phi \leq 0$ so that $\dot{V} \geq 0$. We must prove that the state leaves this region in finite time. This region is compact because $-b \leq \phi$, and it suffices to show that $e(t)$ defined by (5.2) becomes positive in finite time. This immediately follows from $\dot{e} = -e - u_d = e + r > \frac{r}{2}$, which completes the proof of the global asymptotic stability of $(e, \phi) = (0, 0)$. Hence both $e(t)$ and $\phi(t)$ converge to zero.

Now we examine the possibility and effects of sliding modes. In the adaptive system represented either by (5.2) and (5.4), or by (5.3) and (5.4), the discontinuity is at $u_d = 0$—that is, at $e = -\frac{r}{2}$. The sliding mode is to occur if the vector field on each side of the line $e = -\frac{r}{2}$ is directed into this line—that is, if $\dot{u}_d u_d < 0$ when $u_d > 0$ as well as when $u_d < 0$. This test applied to (5.2) shows that, as expected, the sliding mode does not occur while $v(t)$ is inside the dead-zone. For the system outside the dead-zone, as described by (5.3) - (5.4), regardless of the sign of $u_d(t)$, we have

$$\dot{u}_d u_d = 2(e\, u_d - \phi\, |u_d|) < 0 \text{ if } \phi > \frac{r}{2}.$$

This proves that the sliding mode occurs only along the half-line

$$e = -\frac{r}{2}, \; \phi > \frac{r}{2}.$$

The sliding mode solution defined in the sense of Filippov [24] is obtained from

$$\dot{e} = 0, \qquad e(t) = -\frac{r}{2} \tag{5.6}$$

$$\dot{\phi} = -\frac{e^2}{\phi} = -\frac{r^2}{4\phi}, \; \phi(t) > \frac{r}{2}. \tag{5.7}$$

Hence, this solution $\phi(t)$ of (5.7) for $\phi(0) > \frac{r}{2}$ reaches the end of the sliding mode $\phi = \frac{r}{2}$ in finite time because of (5.6). Outside the sliding mode the adaptive system has a solution in the usual sense. The solution is continuous, while its derivative experiences a discontinuity at $u_d = 0$. Since the solution can be in the sliding mode only during a finite time interval the above proof of global asymptotic stability of the equilibrium $(e, \phi) = (0, 0)$ remains valid.

The phase portrait of the adaptive system with the adaptive dead-zone inverse (5.1) is shown in Figure 5.4, where $b = 3$ and $r = 2$ so that the discontinuity $u_d = 0$ is at $e = -1$. Note that the sliding ends at $(\phi, e) = (1, -1)$. The phase portrait of the same adaptive system with a "soft" adaptive dead-zone inverse is shown in Figure 5.5. The specific form of the "soft" inverse was the differentiable function

$$v = u_d + \widehat{b}(1 - e^{-10u_d^2})sgn(u_d).$$

The only qualitative difference between the two phase portraits is that in Figure 5.5 the sliding mode is eliminated. Instead, the motion proceeds along an attractive manifold. Asymptotic tracking and parameter convergence are practically the same as with the discontinuous inverse. In general, using a "soft" inverse, sliding modes are avoided with a negligible effect on the tracking performance.

Finally, we note that without a dead-zone inverse the linear controller $v(t) = -2y + r$ results in large tracking errors $y(t) - r$; while the fixed dead-zone compensation scheme with a fixed estimate $\widehat{b}$, as shown in Figure 5.2, can reduce the tracking error but can not make the error zero, as shown by simulations.

5.2 Continuous-Time Backlash Inverse

As we have described in Section 2.2 and Section 3.2, a major difference between the backlash nonlinearity and the dead-zone nonlinearity is that the backlash

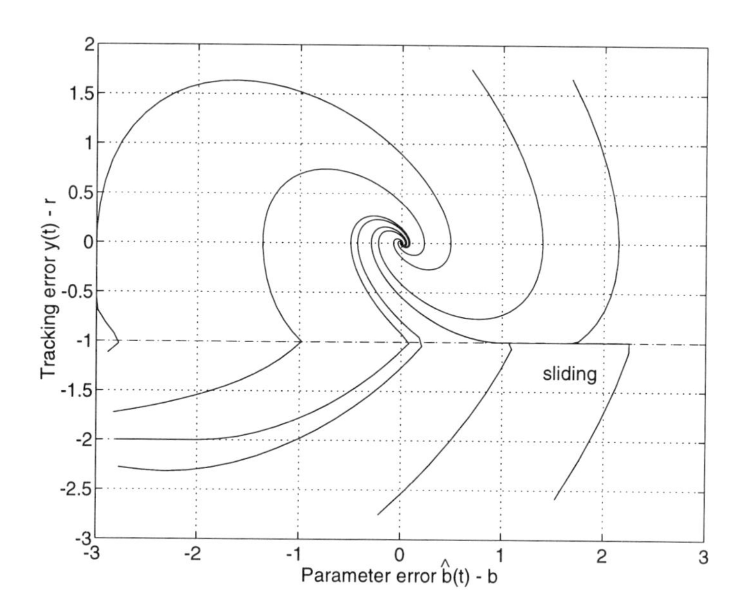

Figure 5.4: Phase portrait of the adaptive dead-zone inverse system.

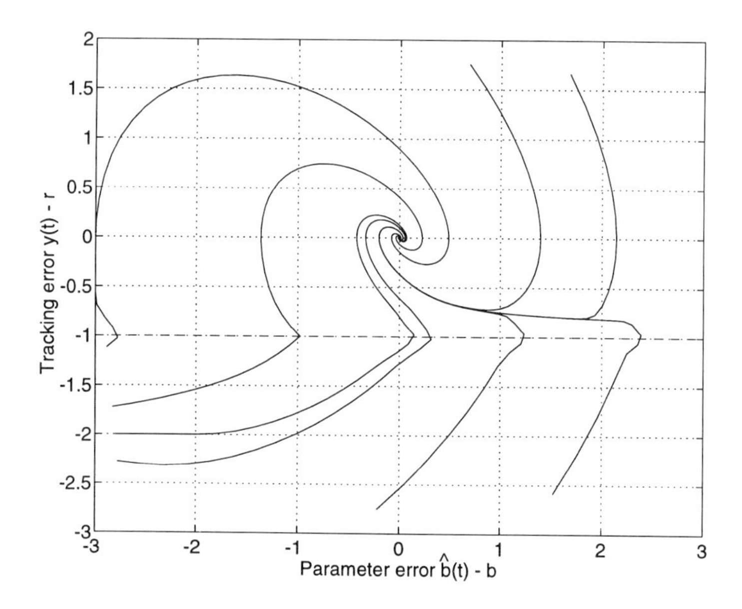

Figure 5.5: Phase portrait with a soft adaptive dead-zone inverse.

characteristic is dynamic and so is its inverse. In particular, to implement the continuous-time backlash inverse characteristic $\widehat{BI}(\cdot)$ as described by (3.15) - (3.16), one needs the knowledge of $sign(\dot{u}_d(t))$ to determine whether the signal motion is on the upward side of $\widehat{BI}(\cdot)$ or on the downward side of $\widehat{BI}(\cdot)$. When the exact knowledge of $sign(\dot{u}_d(t))$ is not available, certain approximation of $\dot{u}_d(t)$ can be used to provide a practical implementation of $\widehat{BI}(\cdot)$. It is the goal of this section to use an example with simulation results to illustrate the design and analysis of a feedback control system with an adaptive backlash inverse implemented in continuous time and to compare the system performance when the backlash inverse is implemented with the exact $sign(\dot{u}_d(t))$ and with an approximate of it. As will be shown by simulation results, both implementation schemes achieve almost the same significant improvement of the system tracking performance.

Let us consider the plant with input backlash discussed in Section 4.4 and shown in Figure 4.7:

$$y(t) = \frac{1}{s-a}[u](t), \ u(t) = B(v(t)), \tag{5.8}$$

where $a = 2$ is known, and the backlash slope $m = 1$ is known, while the crossing parameter $c_r = -c_l = 1.3$ is unknown. We will use the controller structure shown in Figure 4.8 but now with an adaptive backlash inverse $v(t) = \widehat{BI}(u_d(t))$ whose parameter $\widehat{c}(t)$ is updated by an adaptive law. The control objective is to generate the input signal $v(t)$ such that the plant output $y(t)$ tracks the refence signal

$$y_m(t) = \frac{1}{s+a_m}[r](t), \tag{5.9}$$

where $r(t)$ is a bounded signal.

For a linear controller generating the signal $u_d(t)$, we still use (4.137) with $\lambda = 3$:

$$u_d(t) = \frac{\theta_1^*}{s+3}[u_d](t) + \frac{\theta_2^*}{s+3}[y](t) + \theta_3^* r(t) \tag{5.10}$$

whose parameters are calculated from (4.138):

$$\theta_1^* = -(a+a_m), \ \theta_2^* = -(a+3)(a+a_m), \ \theta_3^* = 1.$$

This choice of θ_1^*, θ_2^*, θ_3^* is made based on the matching equation

$$\theta_1^*(s-a) + \theta_2^* = (s+3)((s-a) - \theta_3^*(s+a_m). \tag{5.11}$$

With such a choice of θ_1^*, θ_2^*, θ_3^*, the control law $v(t) = u_d(t)$, in the absence of the backlash $B(\cdot)$: $u(t) = u_d(t)$, would lead to the error equation: $(s+a_m)[y-$

$y_m](t) = 0$, which would imply asymptotic output tracking: $\lim_{t\to\infty}(y(t) - y_m(t)) = 0$.

As shown in Section 4.4, in the presence of backlash $u(t) = B(v(t))$, the linear controller (5.10) alone would result in large tracking error $e(t) = y(t) - y_m(t)$, and a backlash inverse $v(t) = \widehat{BI}(u_d(t))$ whose parameters are fixed can make the tracking error smaller but not convergent to zero. It is the task of an adaptive version of the backlash inverse $v(t) = \widehat{BI}(u_d(t))$ to further reduce the tracking error $e(t)$.

To derive an error model which is crucial for developing an adaptive law to update the backlash inverse $\widehat{BI}(\cdot)$, we divide (5.11) by $s+3$ to obtain

$$\theta_1^* \frac{1}{s+3}(s-a) + \theta_2^* \frac{1}{s+3} = (s-a) - \theta_3^*(s+a_m). \tag{5.12}$$

Operating both sides of (5.12) on $y(t)$ and using (5.8), we have

$$u(t) = \theta_1^* \frac{1}{s+3}[u](t) + \theta_2^* \frac{1}{s+3}[y](t) + \theta_3^*(s+3)[y](t). \tag{5.13}$$

Recall from (3.37) that if a backlash inverse $v(t) = \widehat{BI}(u_d(t))$ is used, then the control error $u(t) - u_d(t)$ can be expressed as

$$u(t) - u_d(t) = (\theta_b(t) - \theta_b^*)^T \omega_b(t) + d_b(t) \tag{5.14}$$

for some bounded error $d_b(t)$. Since $m = 1$ is known and $c_r = -c_l$ in this example, we denote $\theta(t) = \widehat{c}(t)$, $\theta^* = c$, $\phi(t) = \theta(t) - \theta^*$, and $\omega(t) = \widehat{\chi_r}(t) - \widehat{\chi_l}(t)$ and simplify (5.14) as

$$u(t) - u_d(t) = \phi(t)\omega(t) + d_b(t). \tag{5.15}$$

Substituting (5.13) and (5.15) in (5.10) and defining

$$W(s) = \frac{1}{s+a_m}\left(1 - \frac{\theta_1^*}{s+3}\right),\ d(t) = W(s)[d_b](t) \tag{5.16}$$

we have the tracking error equation

$$e(t) = W(s)[\phi\omega](t) + d(t) \tag{5.17}$$

where $d(t)$ is bounded for any $t \geq 0$ because $d_b(t)$ is.

The tracking error equation (5.17), which is familiar from adaptive linear control theory, suggests the adaptive law for updating $\theta(t)$:

$$\dot{\theta}(t) = -\frac{\gamma\zeta(t)\epsilon(t)}{1+\zeta^2(t)} + f(t),\ \gamma > 0, \tag{5.18}$$

where the *estimate error* $\epsilon(t)$ is defined as

$$\epsilon(t) = e(t) + \xi(t) \tag{5.19}$$

with the auxiliary signals

$$\xi(t) = \theta(t)\zeta(t) - W(s)[\theta\omega](t) \tag{5.20}$$

$$\zeta(t) = W(s)[\omega](t). \tag{5.21}$$

The initial parameter $\theta(0)$ is chosen to satisfy $\theta(0) \in [\theta^a, \theta^b]$, where the constants θ^a, θ^b are the lower and upper bounds of the unknown backlash parameter $\theta^* = c$: $\theta^a \leq c \leq \theta^b$, which are determined from *a priori* knowledge of c. A natural constraint is that $\theta^a \geq 0$ since $c \geq 0$. The modifying term $f(t)$ based on parameter projection is designed as

$$f(t) = \begin{cases} 0 & \text{if } \theta(t) \in (\theta^a, \theta^b), \text{ or} \\ & \text{if } \theta(t) = \theta^a,\ g(t) \geq 0, \text{ or} \\ & \text{if } \theta(t) = \theta^b,\ g(t) \leq 0 \\ -g(t) & \text{otherwise,} \end{cases} \tag{5.22}$$

where

$$g(t) = -\frac{\gamma\zeta(t)\epsilon(t)}{1+\zeta^2(t)}. \tag{5.23}$$

This projection of $\theta(t)$ ensures that $\theta(t) = \widehat{c}(t) \geq 0$ so that an adaptive backlash inverse $\widehat{BI}(\cdot)$ is implementable.

Proposition 5.1 *All signals in the closed-loop control system with the adaptive backlash inverse update law (5.18) are bounded and*

$$\int_{t_1}^{t_2} e^2(t)dt \leq c_1 + c_2 \int_{t_1}^{t_2} d_b^2(t)dt \tag{5.24}$$

for some positive constants c_1, c_2, and any $t_2 \geq t_1 \geq 0$.

Proof: Substituting (5.17), (5.20), and (5.21) in (5.19), we obtain

$$\epsilon(t) = \phi(t)\zeta(t) + d(t).$$

Using this equation and (5.18), we express the time derivative of the positive definite function $V(\phi) = \phi^2\gamma^{-1}$ as

$$\begin{aligned} \dot{V} &= \frac{-\zeta(t)\phi(t)\epsilon(t)}{1+\zeta^2(t)} + \frac{f(t)\phi(t)}{\gamma} \\ &= \frac{-\epsilon^2(t)}{2m^2(t)} - \frac{(\epsilon(t)-d(t))^2}{2m^2(t)} + \frac{d^2(t)}{2m^2(t)} + \frac{f(t)\phi(t)}{\gamma}, \end{aligned} \tag{5.25}$$

where $m(t) = \sqrt{1+\zeta^2(t)}$.

From the definition (5.22) of $f(t)$, it follows that

$$\theta(t) \in [\theta^a, \theta^b],\ \phi(t)f(t) \leq 0. \tag{5.26}$$

Since $\omega(t)$ is bounded by definition, from (5.21), $\zeta(t)$ is bounded, and so is $\xi(t)$ in (5.21). Using (5.24) and (5.19), we see that $\epsilon(t)$ and $e(t)$ are both bounded, and so is $y(t)$. Substituting (5.8), (5.9), (5.16), and (5.17) in (5.13), we express $u(t)$ as

$$\begin{aligned} u(t) &= \theta_1^* \frac{s-a}{s+3}[y](t) + \theta_2^* \frac{1}{s+3}[y](t) \\ &\quad + \theta_3^* \frac{s+3}{s+a_m}[(1-\frac{\theta_1^*}{s+3})[\phi\omega + d_b] + r](t), \end{aligned}$$

which means that $u(t)$ is also bounded, and so is $v(t)$ because of the backlash characteristic $u(t) = B(v(t))$.

To prove (5.23), from the boundness of $\zeta(t)$ and $\phi(t)$, using (5.25) and (5.26), we see that

$$\int_{t_1}^{t_2} \epsilon^2(t)dt \leq a_1 + a_2 \int_{t_1}^{t_2} d^2(t)dt \tag{5.27}$$

for some constants $a_1 > 0$, $a_2 > 0$, and any $t_2 \geq t_1 \geq 0$. From (5.18), (5.22), and (5.23), we obtain

$$(\dot{\theta}(t))^2 \leq (\frac{\gamma\zeta(t)\epsilon(t)}{1+\zeta^2(t)})^2 \leq a_3\epsilon^2(t) \tag{5.28}$$

for some constant $a_3 > 0$. From (5.27) and (5.28), we have

$$\int_{t_1}^{t_2} (\dot{\theta}(t))^2 dt \leq a_4 + a_5 \int_{t_1}^{t_2} d^2(t)dt \tag{5.29}$$

for some constants $a_4 > 0$, $a_5 > 0$, and any $t_2 \geq t_1 \geq 0$.

Using the definitions (5.20), (5.16) of $\xi(t)$, $W(s)$, we express

$$\begin{aligned} \xi(t) &= \theta(t)\frac{s+3-\theta_1^*}{(s+a_m)(s+3)}[\omega](t) - \frac{s+3-\theta_1^*}{(s+a_m)(s+3)}[\theta\omega](t) \\ &= \frac{1}{s+a_m}[(s+a_m)[\theta\frac{s+3-\theta_1^*}{(s+a_m)(s+3)}[\omega]] - \frac{s+3-\theta_1^*}{s+3}[\theta\omega]](t) \\ &= \frac{1}{s+a_m}[\dot{\theta}\frac{s+3-\theta_1^*}{(s+a_m)(s+3)}[\omega]](t) \\ &\quad + \frac{1}{s+a_m}[\theta\frac{s+3-\theta_1^*}{s+3}[\omega] - \frac{s+3-\theta_1^*}{s+3}[\theta\omega]](t) \\ &= \frac{1}{s+a_m}[\dot{\theta}\frac{s+3-\theta_1^*}{(s+a_m)(s+3)}[\omega]](t) \\ &\quad + \frac{1}{s+a_m}[\frac{1}{s+3}[\dot{\theta}\frac{s+3-\theta_1^*}{s+3}[\omega] - \dot{\theta}\omega]](t). \end{aligned} \tag{5.30}$$

Here in (5.30) we have used the equality

$$
\begin{aligned}
&\theta\frac{s+3-\theta_1^*}{s+3}[\omega](t) - \frac{s+3-\theta_1^*}{s+3}[\theta\omega](t) \\
&= \frac{1}{s+3}[(s+3)[\theta\frac{s+3-\theta_1^*}{s+3}[\omega]] - (s+3-\theta_1^*)[\theta\omega]](t) \\
&= \frac{1}{s+3}[\dot{\theta}\frac{s+3-\theta_1^*}{s+3}[\omega] + \theta(s+3-\theta_1^*)[\omega] - (s+3-\theta_1^*)[\theta\omega]](t) \\
&= \frac{1}{s+3}[\dot{\theta}\frac{s+3-\theta_1^*}{s+3}[\omega] - \dot{\theta}\omega](t).
\end{aligned}
$$

Since $\omega(t)$ and $\xi(t)$ are bounded and $a_m > 0$, from the dependence of $\xi(t)$ on $\dot{\theta}(t)$ as shown in (5.30) and the property of $\dot{\theta}(t)$ as shown by (5.29), it can be shown that

$$
\int_{t_1}^{t_2} \xi^2(t)dt \leq a_6 + a_7 \int_{t_1}^{t_2} d^2(t)dt \tag{5.31}
$$

for some positive constants a_6, a_7, and any $t_2 \geq t_1 \geq 0$. Then it follows from (5.19), (5.27), and (5.31) that

$$
\int_{t_1}^{t_2} e^2(t)dt \leq a_8 + a_9 \int_{t_1}^{t_2} d^2(t)dt \tag{5.32}
$$

for some positive constants a_8, a_9, and any $t_2 \geq t_1 \geq 0$. Finally, since

$$
d(t) = \frac{1}{s+a_m}(1 - \frac{\theta_1^*}{s+3})[d_b](t) \tag{5.33}
$$

it can be shown from (5.32) and (5.33) that

$$
\int_{t_1}^{t_2} e^2(t)dt \leq c_1 + c_2 \int_{t_1}^{t_2} d_b^2(t)dt
$$

for some positive constants c_1, c_2, and any $t_2 \geq t_1 \geq 0$. ∇

Simulations were performed for the adaptive control system with $c = 1.3$, $r(t) = 10\sin 1.3t$, and $\gamma = 0.5$, $\theta^a = 0.01$, $\theta^b = 3$.

For $\widehat{c}(0) = 0.8$, the simulation results are shown in Figure 4.6(a) - (b) for $\dot{u}_d(t)$ calculated with $\dot{r}(t) = 13\cos 1.3t$ and are shown in Figure 5.6(c) - (d) for $\dot{u}_d(t)$ calculated with $\frac{15s}{s+15}[r](t)$ replacing $\dot{r}(t)$.

For $\widehat{c}(0) = 0$, the simulation results are shown in Figure 5.7(a) - (b) for $\dot{u}_d(t)$ calculated with $\dot{r}(t) = 13\cos 1.3t$ and are shown in Figure 5.7(c) - (d) for $\dot{u}_d(t)$ calculated with $\frac{15s}{s+15}[r](t)$ replacing $\dot{r}(t)$.

These results indicate that with an adaptive backlash inverse the tracking error $e(t) = y(t) - y_m(t)$ converges to very small values, and so does the parameter error $\widehat{c}(t) - c$, and that with $\frac{15s}{s+15}[r](t)$ approximating $\dot{r}(t)$, the performance of the adaptive backlash inverse controller is as good as that with the exact $\dot{r}(t)$.

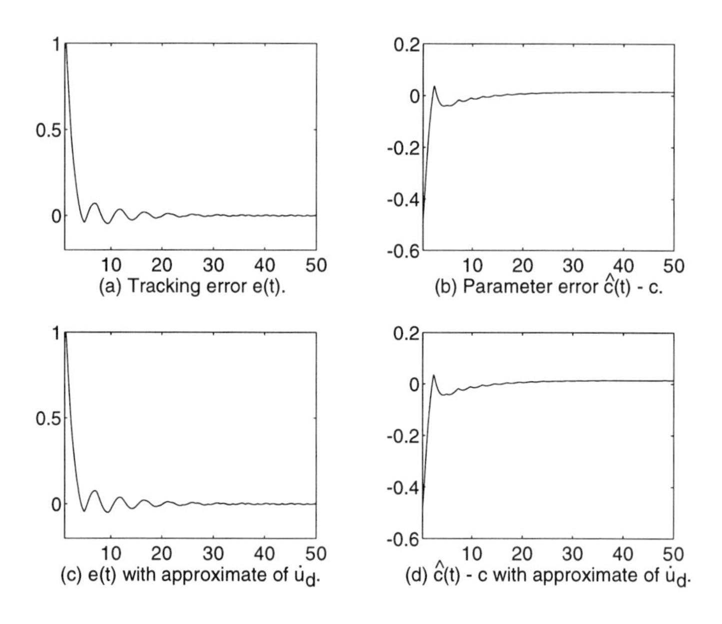

Figure 5.6: Responses of the adaptive $\widehat{BI}(\cdot)$ for $\widehat{c}(0) = 0.8$.

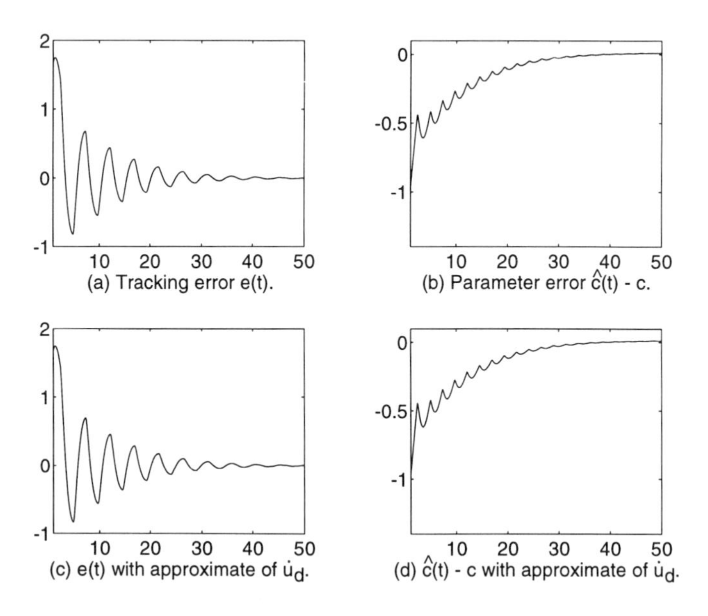

Figure 5.7: Responses of the adaptive $\widehat{BI}(\cdot)$ for $\widehat{c}(0) = 0$.

5.3 Discrete-Time Backlash Inverse

Recall that the example in Section 2.4 exhibited undesirable sustained oscillations and a loss of accuracy for a plant with a backlash at its input, controlled by a PI controller. One of the goals of this section is to design a discrete-time adaptive backlash inverse controller for this plant to achieve asymptotic tracking despite the presence of backlash. Another goal is to show how a discrete-time backlash inverse design can overcome the difficulty of generating $sign(\dot{u}_d(t))$, which has appeared in implementing a continuous-time backlash inverse as shown in the previous section.

Let us consider the same plant whose linear part is $G(s) = \frac{k_p}{s}$, where k_p is a known constant, assuming that the backlash slope $m > 0$ is known, but its width—that is, the parameter $c_r = -c_l = c$—is unknown. For this plant we now design a discrete-time adaptive backlash inverse.

For a discrete-time control design, the linear part is given by

$$y(t_{k+1}) = y(t_k) + k_p \int_{t_k}^{t_{k+1}} u(t)dt. \tag{5.34}$$

With $u(t)$ piecewise constant over $[t_k, t_{k+1}]$, $T \triangleq t_{k+1} - t_k > 0$, scaling $Tk_p = 1$, and changing notation $t_{k+1} = t+1$ and $t_k = t$, the linear part of the plant becomes its discrete-time version:

$$y(t+1) = y(t) + u(t). \tag{5.35}$$

In the absence of backlash our design objective to stabilize the closed-loop system and make the plant output $y(t)$ track a given reference signal $y_m(t)$ which specifies the desired system behavior would be achieved by the controller

$$u_d(t) = -y(t) + y_m(t+1). \tag{5.36}$$

In the presence of backlash we use this controller along with an adaptive scheme designed to update the backlash inverse $v(t) = \widehat{BI}(u_d(t))$ on-line, as shown in Figure 5.8.

Since, by assumption, m is known and $c_r = -c_l = c$, we let $\widehat{m}(t) = m$ and $\widehat{mc_l}(t) = -\widehat{mc_r}(t) = \widehat{mc}(t)$ and introduce

$$\phi(t) = \theta(t) - \theta^*, \ \theta(t) = \widehat{mc}(t), \ \theta^* = mc \tag{5.37}$$

so that the backlash inverse error equation (3.37) becomes

$$u(t) - u_d(t) = \phi(t)\omega(t) + d_b(t), \tag{5.38}$$

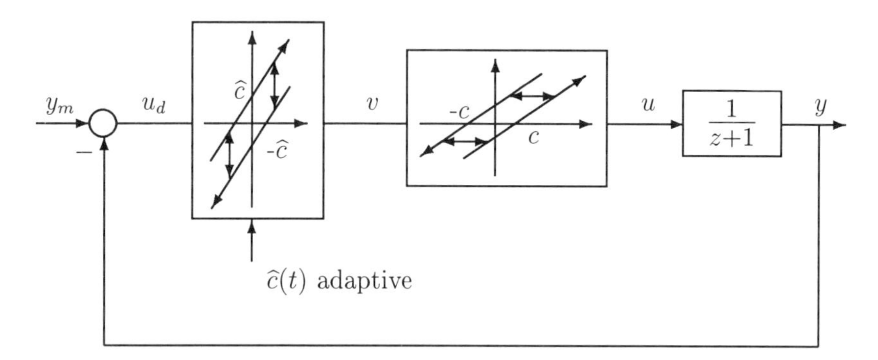

Figure 5.8: Adaptive backlash compensation scheme.

where $\omega(t) = \widehat{\chi_r}(t) - \widehat{\chi_l}(t)$ is the regressor.

For the *tracking error* $e(t) = y(t) - y_m(t)$, from (5.34) - (5.37), we obtain the expression

$$e(t) = \theta(t-1)\omega(t-1) - \theta^*\omega(t-1) + d_b(t-1). \tag{5.39}$$

An important quantity to be used in the discrete-time adaptive update law is the *estimation error* defined as

$$\epsilon(t) = e(t) + \theta(t)\omega(t-1) - \theta(t-1)\omega(t-1). \tag{5.40}$$

This form of $\epsilon(t)$ is implementable, and from (5.38) and (5.39), a simpler form of $\epsilon(t)$ is

$$\epsilon(t) = \phi(t)\omega(t-1) + d_b(t-1). \tag{5.41}$$

This unimplementable form is convenient for analysis.

Using the implementable form of $\epsilon(t)$, our update law for $\theta(t)$ based on a gradient-type algorithm is

$$\theta(t+1) = \theta(t) - \frac{\gamma\omega(t-1)\epsilon(t)}{1+\omega^2(t-1)} + f(t),\ 0 < \gamma < 2, \tag{5.42}$$

where the modifying term $f(t)$ is designed as

$$f(t) = \begin{cases} 0 & \text{if } \theta(t) + g(t) \in [\theta^a, \theta^b] \\ \theta^b - \theta(t) - g(t) & \text{if } \theta(t) + g(t) > \theta^b \\ \theta^a - \theta(t) - g(t) & \text{if } \theta(t) + g(t) < \theta^a \end{cases} \tag{5.43}$$

$$g(t) = -\frac{\gamma\omega(t-1)\epsilon(t)}{1+\omega^2(t-1)} \tag{5.44}$$

with the constants θ^a, θ^b being the lower and upper bounds of the unknown backlash parameter $\theta^* = mc$: $\theta^a \leq mc \leq \theta^b$, which are determined from *a priori* knowledge of mc. A natural constraint is that $\theta^a \geq 0$ since $mc \geq 0$. This projection of $\theta(t)$ ensures that $\widehat{mc}(t) \geq 0$.

The stability and tracking properties of the closed-loop system are summarized as follows.

Proposition 5.2 *All signals in the closed-loop system are bounded and there exist $\alpha_0 > 0$, $\beta_0 > 0$ such that*

$$\sum_{k=k_1}^{k_1+k_2} e^2(t) \leq \alpha_0 \sum_{k=k_1-2}^{k_1+k_2-1} d_b^2(t) + \beta_0 \tag{5.45}$$

for any $k_1 \geq 2$, $k_2 \geq 0$.

Proof: Using (5.41), (5.42) and introducing

$$\bar{\epsilon}(t) = \frac{\epsilon(t)}{m(t)},\ \bar{d}_b(t-1) = \frac{d_b(t-1)}{m(t)},\ m(t) = \sqrt{1+\omega^2(t-1)}$$

we obtain the time increment of $V(\phi) = \phi^2$ as

$$\begin{aligned} & V(\phi(t+1)) - V(\phi(t)) \\ & = -(2 - \frac{\gamma\omega^2(t-1)}{m^2(t)})\frac{\epsilon^2(t)}{m^2(t)} + \frac{2\epsilon(t)d(t)}{m^2(t)} \\ & \quad +2f(t)\gamma^{-1}(\phi(t)+g(t)+f(t)) - f^2(t)\gamma^{-1} \\ & \leq -\frac{2-\gamma}{2}\bar{\epsilon}^2(t) + \frac{2}{2-\gamma}\bar{d}^2(t) \\ & \quad +2f(t)\gamma^{-1}(\phi(t)+g(t)+f(t)) - f^2(t)\gamma^{-1}. \end{aligned} \tag{5.46}$$

From (5.42) - (5.44), we have

$$\theta(t) \in [\theta^a, \theta^b],\ f(\phi(t)+g(t)+f(t)) \leq 0. \tag{5.47}$$

This proves that $\phi(t)$ is bounded. By definition, $\omega(t)$ is bounded. Hence $e(t)$ in (5.39) is bounded, and so is $y(t)$. Finally, $u_d(t)$ in (5.36), $v(t) = \widehat{BI}(u_d(t))$, and $u(t) = B(v(t))$ are bounded, and thus all closed-loop signals are bounded.

Using (5.40), (5.42), we obtain

$$\begin{aligned} e^2(t) & \leq 2\epsilon^2(t) + 2\omega^2(t-1)(\theta(t)-\theta(t-1))^2 \\ & \leq 2\epsilon^2(t) + \frac{4\gamma^2\omega^2(t-1)\omega^2(t-2)}{1+\omega^2(t-2)}\bar{\epsilon}^2(t-1). \end{aligned} \tag{5.48}$$

Here we have used the fact that

$$\theta(t+1) - \theta(t) = \begin{cases} g(t) & \text{if } \theta(t) + g(t) \in [\theta^a, \theta^b] \\ \theta^b - \theta(t) & \text{if } \theta(t) + g(t) > \theta^b \\ \theta^a - \theta(t) & \text{if } \theta(t) + g(t) < \theta^a, \end{cases}$$

which implies that $|\theta(t+1) - \theta(t)| \leq |g(t)|$. Since $\omega(t)$ is bounded, from (5.46) - (5.48), we have (5.45). ∇

It should be pointed out that in this discrete-time design we have not encountered the time derivative and discontinuity difficulties present in the continuous-time designs.

To evaluate the system performance improvement achieved by the proposed adaptive backlash inverse, simulations were performed for the first-order plant (5.34). The backlash parameters were taken as $m = 0.4625$ known and $c_r = -c_l = c = 1.25$ unknown to the adaptive backlash inverse.

Two cases were studied for comparison: (a) Only the controller (5.35) is applied—that is, no backlash inverse is implemented; (b) the controller (5.35) and an adaptive backlash inverse are applied.

The tracking error $e(t) = y(t) - y_m(t)$ for $y_m(t) = 10 \sin 0.126t$ with $\widehat{c}(0) = 1.91$ is shown in Figure 5.9, and the tracking error for $y_m(t) = 10\, sgn(\sin 0.22t)$ (square wave) with $\widehat{c}(0) = 0.59$ is shown in Figure 5.10. The parameter error $\widehat{c}(t) - c$ is shown in Figure 5.11, where $\widehat{c}(t) = \frac{\widehat{mc}(t)}{\widehat{m}(t)}$ and $\widehat{m}(t) \equiv m$. In these simulations, $\gamma = 0.5$, $\theta^a = 0$, $\theta^b = 5$ were used.

The simulation results show that the adaptive backlash inverse (case (b)) leads to major system performance improvements in all the cases of different initial conditions and different reference signals: In addition to the signal boundedness the adaptive scheme achieves convergence to zero of both tracking error and parameter error. Along with the parameter error the control error also converges to zero.

Compared with the continuous-time backlash inverse design of Section 4.2, the discrete-time backlash inverse of this section does not need the knowledge of $sign(\dot{u}_d(t))$ for implementation. This property is the main feature of a discrete-time backlash inverse design, which is of major practical significance because very often such signal derivative knowledge is not available in applications.

The adaptive inverse examples presented in this chapter carry an important message: An adaptive inverse can cancel the effect of the unknown nonlinearity

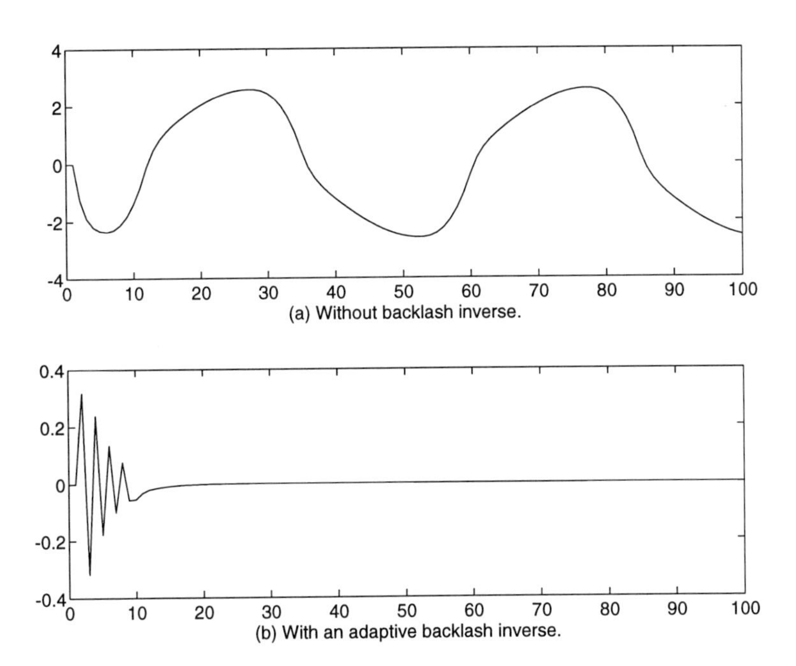

Figure 5.9: Tracking error for $y_m(t) = 10 \sin 0.126t$.

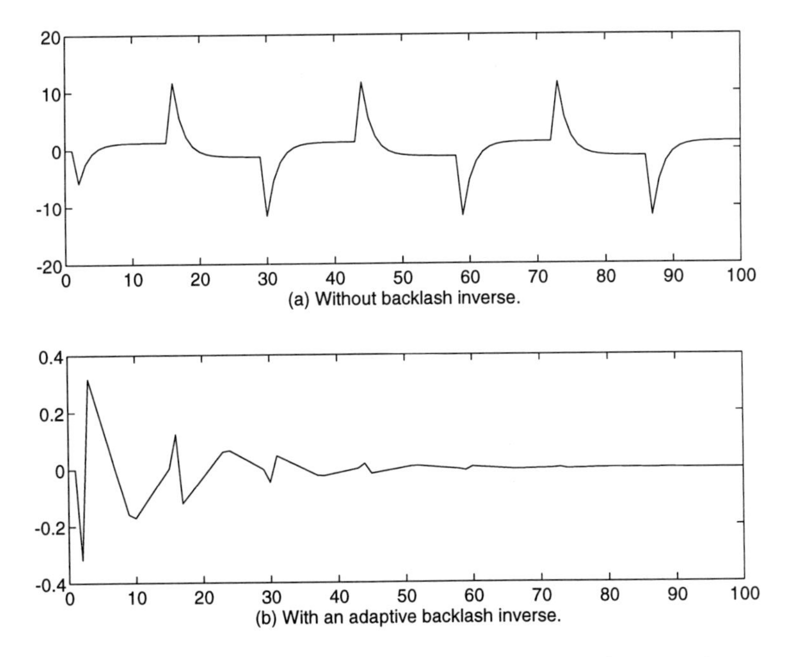

Figure 5.10: Tracking error for $y_m(t) = 10\, sgn(\sin 0.22t)$.

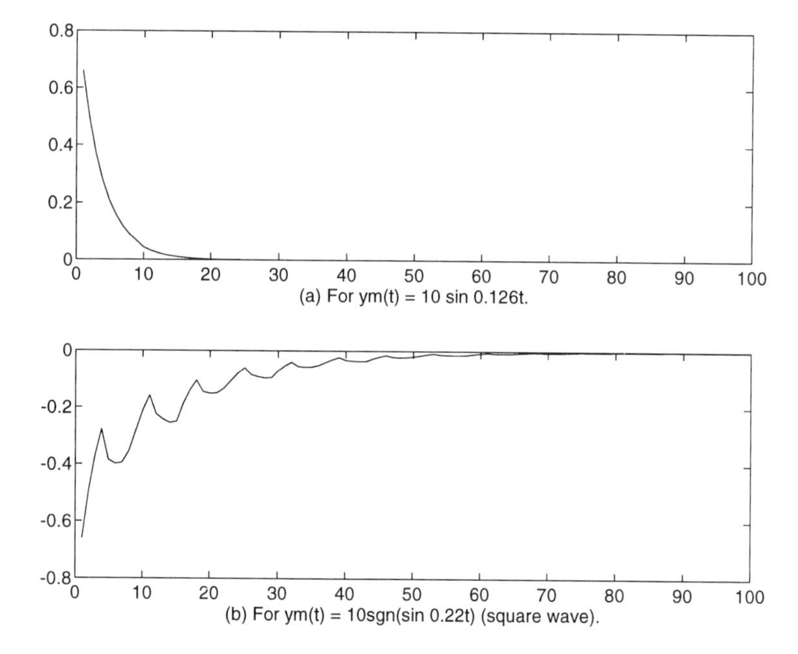

Figure 5.11: Parameter error of the adaptive backlash inverse.

and thus improve system tracking performance. Our main task in the subsequent chapters is to design adaptive controllers with such adaptive inverses for plants with a general linear part and a general nonlinear characteristic.

Chapter 6

Continuous-Time Adaptive Inverse Control

We have seen in Chapter 4 that a fixed inverse compensation can improve the system tracking performance. However, when the fixed inverse is implemented with inaccurate parameter estimates, then the error in cancelling the effect of nonlinearity can be significant. As indicated by the *control error equation* of Chapter 3, the magnitude of such an error is proportional to the magnitude of the parameter error. Therefore, to further reduce the tracking error between the plant output and a given reference signal, we will use adaptive laws to update the parameter estimates of the inverse. Our examples in Chapter 5 demonstrated that an adaptive inverse has the potential to improve the system tracking performance. In this chapter we will develop adaptive inverse schemes for continuous-time plants with an unknown dead-zone, backlash, or hysteresis. The unknown nonlinearity is assumed to be "in the actuator"—that is, at the input of a linear part described by a known or unknown transfer function.

6.1 Control Objective

Our objective is to control a continuous-time plant with a nonlinearity $N(\cdot)$ at the input of a linear time-invariant part $G(s)$, which is shown in Figure 6.1 and described by

$$y(t) = G(s)[u](t), \; u(t) = N(v(t)), \tag{6.1}$$

where $v(t)$ is the control input and $y(t)$ is the measured output. The nonlinear part $N(\cdot)$ represents either a dead-zone, a backlash, or a hysteresis. The linear part is $G(s) = k_p \frac{Z(s)}{P(s)}$, where k_p is a constant gain and $Z(s)$, $P(s)$ are monic polynomials in s. The symbol s denotes, as the case may be, the Laplace transform variable or the time differentiation operator: $s[x](t) = \dot{x}(t)$.

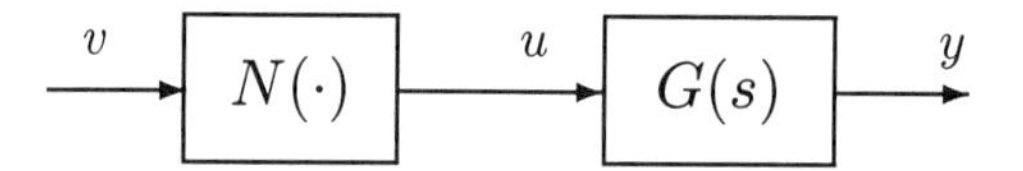

Figure 6.1: Plant with an input nonlinearity.

We will consider the plant (6.1) with two key features:

- The output $u(t)$ of $N(\cdot)$ is not accessible for measurement.
- The parameters of the nonlinear part $N(\cdot)$ are unknown.

We want to generate a feedback control signal $v(t)$ which ensures closed-loop signal boundedness and reduced tracking error between the plant output $y(t)$ and the output $y_m(t)$ of a given reference model:

$$y_m(t) = W_m(s)[r](t), \tag{6.2}$$

where $r(t)$ is bounded and $W_m(s)$ is stable.

We will develop two adaptive inverse controllers: one for plants with a known linear part and an unknown nonlinear part, and the other for both parts unknown. Our approach is to use the model reference control strategy (see Appendix A) to design a linear controller structure combined with an adaptive inverse $\widehat{NI}(\cdot)$ to compensate for the nonlinearity $N(\cdot)$. We assume that the linear part $G(s)$ and the reference model $W_m(s)$ satisfy the assumptions (A1) - (A5) stated in Section 4.1.

Recall that the inverse $\widehat{NI}(\cdot)$ is implemented with parameter estimates, and $u_d(t)$ is a signal from a linear controller structure to be designed:

$$v(t) = \widehat{NI}(u_d(t)),\ NI(\cdot) \in \{\widehat{DI}(\cdot), \widehat{BI}(\cdot), \widehat{HI}(\cdot)\}. \tag{6.3}$$

In Chapter 3, we have developed the dead-zone inverse $\widehat{DI}(\cdot)$, backlash inverse $\widehat{BI}(\cdot)$, and hysteresis inverse $\widehat{HI}(\cdot)$. When implemented with true parameters θ_d^*, θ_b^*, or θ_h^*, these inverses result in the exact cancellation of the input nonlinearity $N(\cdot)$—that is, $u(t) = u_d(t)$. However, when implemented with parameter estimates, they result in a control error of the form (3.76):

$$u(t) - u_d(t) = (\theta_N(t) - \theta_N^*)^T \omega_N(t) + d_N(t) \tag{6.4}$$

for some measured signal $\omega_N(t)$, where

$$\theta_N^* \in \{\theta_d^*, \theta_b^*, \theta_h^*\},\ \theta_N(t) \in \{\theta_b(t), \theta_b(t), \theta_h(t)\}$$

with $\theta_N(t)$ being the estimate of θ_N^*. An important property of the inverses is that the unknown term $d_N(t)$ is bounded.[1] Furthermore, this term has the property that $d_N(t) = 0$ when $\theta_N = \theta_N^*$ and the backlash or hysteresis inverse is properly initialized:

$$u_d(t_0) = B(BI(u_d(t_0))), \quad \text{or } u_d(t_0) = H(HI(u_d(t_0))).$$

We have shown in Chapter 3 (see (3.80)) that the inverse scheme (6.3) can be parametrized as

$$\begin{aligned} u_d(t) &= -\theta_N^T(t)\omega_N(t) + a_h(t) \\ &= \begin{cases} -\theta_d^T \omega_d(t) & \text{for } \widehat{DI}(\cdot) \\ -\theta_b^T \omega_b(t) & \text{for } \widehat{BI}(\cdot) \\ -\theta_h^T \omega_h(t) + \widehat{\chi_w}(t)\widehat{c_w}(t) + \widehat{\chi_e}(t)\widehat{c_e}(t) & \text{for } \widehat{HI}(\cdot), \end{cases} \end{aligned} \tag{6.5}$$

where $a_h(t) = \widehat{\chi_w}(t)\widehat{c_w}(t) + \widehat{\chi_e}(t)\widehat{c_e}(t)$ is a known signal used only for the hysteresis inverse $\widehat{NI}(\cdot) = \widehat{HI}(\cdot)$.

The control error expression (6.4) and the inverse parametrization (6.5) will be used to derive adaptive laws for updating the parameters of the adaptive inverse $\widehat{NI}(\cdot)$.

To obtain suitable parameter estimates for implementing the dead-zone inverse (3.2), backlash inverse (3.15) - (3.16), and hysteresis inverse (3.44), we make the following assumptions for the components of the true parameters θ_d^*, θ_b^* and θ_h^*:

(A7a) For *dead-zone*, $m_{r1} \le m_r \le m_{r2}$, $m_{l1} \le m_l \le m_{l2}$, $0 \le b_r \le b_{r0}$, $-b_{l0} \le b_l \le 0$, for some known positive constants m_{r1}, m_{l1}, m_{r2}, m_{l2}, b_{r0}, b_{l0}.

(A7b) For *backlash*, $m_1 \le m \le m_2$, $0 \le c_r \le c_{r0}$, $-c_{l0} \le c_l \le 0$, for some known positive constants m_1, m_2, c_{r0}, c_{l0}.

(A7c) For *hysteresis*, $m_{t1} \le m_t \le m_{t2}$, $\max\{m_{t2}, m_{b2}\} < m_{r1} \le m_r \le m_{r2}$, $m_{b1} \le m_b \le m_{b2}$, $\max\{m_{t2}, m_{b2}\} < m_{l1} \le m_l \le m_{l2}$, $-c_{b0} \le c_b \le 0 \le c_t \le c_{t0}$, $-c_{l0} \le c_l \le 0 \le c_r \le c_{r0}$, for some known positive constants m_{t1}, m_{t2}, m_{r1}, m_{r2}, m_{b1}, m_{b2}, m_{l1}, m_{l2}, c_{b0}, c_{t0}, c_{l0}, c_{r0}.

Assumptions (A7a) - (A7c) will be used to project the components of the parameters estimates $\theta_d(t)$, $\theta_b(t)$ and $\theta_h(t)$ from adaptive update laws. Thus, the same bounds will hold for parameter estimates.

[1] In the case of adaptive hysteresis inverse, $d_N(t)$ is bounded under the condition that parameter estimates $\widehat{c_t}$, $\widehat{c_b}$, $\widehat{c_l}$ and $\widehat{c_r}$, are bounded (see Proposition 3.3). This can be ensured by using parameter projection.

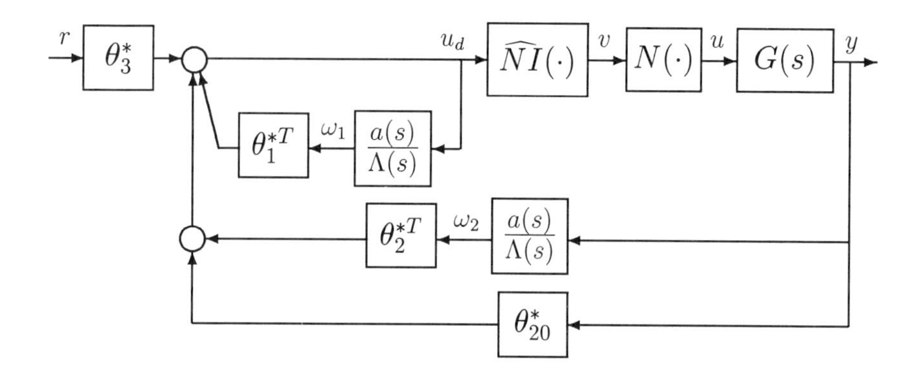

Figure 6.2: Controller structure for $G(s)$ known.

6.2 Designs for $G(s)$ Known

When the linear part $G(s)$ of the plant (6.1) is known, then the controller structure depends on the nature of the adaptive inverse. For the implementation of backlash and hysteresis inverses, which are dynamic, the information about the sign of $\dot{u}_d(t)$ must be available. This is not required for the dead-zone inverse, which is static. The two controller structures derived below differ in the availability of signals which provide the information about $\dot{u}_d(t)$.

With an appropriate interpretation of the individual blocks, the block diagram in Figure 6.2 is representative of both controller structures.

6.2.1 Controller Structure with $\theta_{20}^* \neq 0$

If $\dot{y}(t)$ and $\dot{r}(t)$ are both available and, furthermore, $\dot{r}(t)$ is bounded, then the following controller structure is suitable:

$$u_d(t) = \theta_1^{*T}\omega_1(t) + \theta_2^{*T}\omega_2(t) + \theta_{20}^* y(t) + \theta_3^* r(t), \tag{6.6}$$

where θ_1^*, θ_2^* are vector parameters and θ_{20}^*, θ_3^* are scalar parameters. In the filters

$$\omega_1(t) = \frac{a(s)}{\Lambda(s)}[u_d](t),\ \omega_2(t) = \frac{a(s)}{\Lambda(s)}[y](t) \tag{6.7}$$

the vector $a(s)$ is

$$a(s) = (1, s, \ldots, s^{n-2})^T \tag{6.8}$$

and $\Lambda(s)$ is a stable polynomial of degree $n-1$. As in the fixed inverse compensation schemes of Chapter 4, the parameters

$$\theta_1^* \in R^{n-1},\ \theta_2^* \in R^{n-1},\ \theta_{20}^* \in R,\ \theta_3^* \in R \tag{6.9}$$

are chosen to satisfy the matching equation

$$\begin{aligned}\theta_1^{*T} a(s)P(s) + (\theta_2^{*T} a(s) + \theta_{20}^* \Lambda(s))k_p Z(s) \\ = \Lambda(s)(P(s) - \theta_3^* k_p Z(s) P_m(s)).\end{aligned} \quad (6.10)$$

This condition implies that $\theta_3^* = k_p^{-1}$.

With the controller structure (6.6) a *dead-zone inverse* is directly implementable. The signal $\dot{u}_d(t)$ needed for a *backlash inverse* and a *hysteresis inverse* can be obtained from the implementable expression

$$\dot{u}_d(t) = \theta_1^* \frac{sa(s)}{\Lambda(s)}[u_d](t) + \theta_2^{*T} \frac{sa(s)}{\Lambda(s)}[y](t) + \theta_{20}^* \dot{y}(t) + \theta_3^* \dot{r}(t). \quad (6.11)$$

6.2.2 Controller Structure with $\theta_{20}^* = 0$

An alternative controller structure is now developed for the case when $\dot{y}(t)$ is not available, but $\dot{r}(t)$ is measured and bounded. We increase the order of $\Lambda(s)$ to n, modify $a(s)$ as

$$a(s) = (1, s, \ldots, s^{n-1})^T, \quad (6.12)$$

and choose the parameters

$$\theta_1^* \in R^n, \ \theta_2^* \in R^n, \ \theta_{20}^* = 0, \ \theta_3^* = k_p^{-1} \quad (6.13)$$

so that θ_1^* and θ_2^* satisfy the matching equation

$$\theta_1^{*T} a(s)P(s) + \theta_2^{*T} a(s)k_p Z(s) = \Lambda(s)(P(s) - Z(s)P_m(s)). \quad (6.14)$$

With this choice, the controller structure, alternative to (6.6), is

$$u_d(t) = \theta_1^{*T} \omega_1(t) + \theta_2^{*T} \omega_2(t) + \theta_3^* r(t). \quad (6.15)$$

It is also represented by Figure 6.2, where $\theta_{20}^* = 0$ and $a(s)$, $\Lambda(s)$, $\omega_1(t)$, and $\omega_2(t)$ are accordingly redefined.

Differentiating (6.15) with respect to time we obtain

$$\dot{u}_d(t) = \theta_1^* \frac{sa(s)}{\Lambda(s)}[u_d](t) + \theta_2^{*T} \frac{sa(s)}{\Lambda(s)}[y](t) + \theta_3^* \dot{r}(t), \quad (6.16)$$

which means that the signal $\dot{u}_d(t)$ is available. We note that the controller structure (6.15) can also be used for an adaptive dead-zone inverse controller.

6.2.3 Approximations of $\dot{u}_d(t)$

If $\dot{y}(t)$ or $\dot{r}(t)$ is not available, then we can employ a simple approximation of $\dot{u}_d(t)$, such as

$$\dot{u}_d(t) \approx \frac{s}{\tau s + 1}[u_d](t), \tag{6.17}$$

where $\tau > 0$ is a small constant.

Alternatively, in (6.11) and (6.16), we can use the approximations

$$\dot{y}(t) \approx \frac{s}{\tau s + 1}[y](t)$$

$$\dot{r}(t) \approx \frac{s}{\tau s + 1}[r](t).$$

We can also use the hybrid inverse scheme proposed in Section 4.2, which does not need the knowledge of the sign of $\dot{u}_d(t)$.

6.2.4 Tracking Error Equation

To derive an error equation which will be instrumental for our update law design, we operate both sides of (6.10) on $y(t)$ to obtain

$$\begin{aligned} &\theta_1^{*T} a(s)P(s)[y](t) + (\theta_2^{*T} a(s) + \theta_{20}^* \Lambda(s))k_p Z(s)[y](t) \\ &\quad = \Lambda(s)P(s)[y](t) - \theta_3^* k_p \Lambda(s) Z(s) P_m(s)[y](t). \end{aligned} \tag{6.18}$$

Then, using the plant (6.1):

$$P(s)[y](t) = k_p Z(s)[u](t),$$

recalling that $Z(s)$ and $\Lambda(s)$ are stable polynomials, and ignoring the effect of the initial conditions, from (6.18), we get

$$u(t) = \theta_1^{*T} \frac{a(s)}{\Lambda(s)}[u](t) + \theta_2^{*T} \omega_2(t) + \theta_{20}^* y(t) + \theta_3^* P_m(s)[y](t). \tag{6.19}$$

Substituting the control error (6.4) into (6.19), using the model (6.2), and the controller (6.6), and letting

$$e(t) = y(t) - y_m(t)$$

$$W(s) = \frac{k_p}{P_m(s)}(1 - \theta_1^{*T} \frac{a(s)}{\Lambda(s)}),$$

we obtain the *tracking error equation*

$$e(t) = W(s)[(\theta_N - \theta_N^*)^T \omega_N](t) + d(t), \tag{6.20}$$

where $d(t) = W(s)[d_N](t)$. In an analogous fashion, the same error equation can be obtained for the controller structure (6.15).

It is important to note that in the dead-zone and backlash inverse cases, $d_N(t)$ is bounded. It is also bounded for the hysteresis inverse when its parameter estimates $\widehat{c_t}$, $\widehat{c_b}$, $\widehat{c_l}$, and $\widehat{c_r}$ are bounded, as will be guaranteed by a projection algorithm. It is useful to note that the error model (6.20), familiar from the adaptive control literature [3], [30], [39], [80], [96], is parametrized by the nonlinearity parameter error $\theta_N - \theta_N^*$ only because the controller parameters θ_1^*, θ_2^*, θ_{20}^*, θ_3^* are known when $G(s)$ is known.

6.2.5 Adaptive Update Law

Based on the error equation (6.20), we choose the following update law:

$$\dot{\theta}_N(t) = -\frac{\Gamma_N \zeta_N(t) \epsilon_N(t)}{m_N^2(t)} + f_N(t), \tag{6.21}$$

where $f_N(t)$ is a modification term,

$$\epsilon_N(t) = e(t) + \xi_N(t) \tag{6.22}$$

$$\xi_N(t) = \theta_N^T(t)\zeta_N(t) - W(s)[\theta_N^T \omega_N](t) \tag{6.23}$$

$$\zeta_N(t) = W(s)[\omega_N](t) \tag{6.24}$$

$$m_N(t) = \sqrt{1 + \zeta_N^T(t)\zeta_N(t) + \xi_N^2(t)} \tag{6.25}$$

$$\Gamma_N = diag\{\gamma_1, \ldots, \gamma_{n_N}\},\ \gamma_i > 0,\ i = 1, \ldots, n_N \tag{6.26}$$

and n_N is the dimension of θ_N.

This is a modified gradient algorithm in which $m_N(t)$ is a normalizing signal, $\epsilon_N(t)$ is an estimation error, and $\xi_N(t)$ is an auxiliary signal for a linearly parametrized $\epsilon_N(t)$. The modifying term $f_N(t)$ is introduced to ensure the boundedness of $\theta_N(t)$ and smallness in the mean of $\frac{\epsilon_N(t)}{m_N(t)}$ and $\dot{\theta}_N(t)$ in the presence of the bounded "disturbance" $d(t)$ in (6.20) (see Lemma 6.1 below). Such a task can be fulfilled by a choice of $f_N(t)$ as any one of the modifications proposed in the adaptive control literature, e.g., dead-zone modifications [54], [68], [86], σ-modification [35], [37], switching σ-modification [33], [42], [43], parameter projection modifications [53], [55], [77], ϵ_1-modification [79], or normalization techniques [42], [43], [54], [68], [90]. Among these modifications, the switching σ-modification is particularly desirable because it ensures that the asymptotic tracking is achieved when the "disturbance" $d(t)$ vanishes. Another task of $f_N(t)$ is to ensure that the components of $\theta_N(t)$ stay in a parameter region, required by the adaptive inverse $\widehat{NI}(\cdot)$.

6.2.6 Parameter Projection

A parameter projection technique can be used to ensure the desired properties that the adaptive law (6.21) is robust with respect to the disturbance $d(t)$ and that the components of $\theta_N(t)$ satisfy the assumptions (A7a) - (A7c) in Section 6.1. It also ensures that the asymptotic tracking is achieved when $d(t)$ vanishes.

For parameter projection, we need to know a convex region in the parameter space, which contains the true parameter θ_N^*. Using the assumptions (A7a) - (A7c), such a region is specified as follows.

For the *dead-zone inverse* $\widehat{DI}(\cdot)$ with

$$\begin{aligned}\theta_N^* = \theta_d^* &= (m_r, m_r b_r, m_l, m_l b_l)^T \\ &= (\theta_{N1}^*, \theta_{N2}^*, \theta_{N3}^*, \theta_{N4}^*)^T\end{aligned} \tag{6.27}$$

the assumption (A7a) implies that

$$\theta_{Nj}^* \in [\theta_{Nj}^a, \theta_{Nj}^b],\ j = 1, 2, 3, 4 \tag{6.28}$$

$$\theta_{N1}^a = m_{r1},\ \theta_{N2}^a = 0,\ \theta_{N3}^a = m_{l1},\ \theta_{N4}^a = -m_{l2} b_{l0} \tag{6.29}$$

$$\theta_{N1}^b = m_{r2},\ \theta_{N2}^b = m_{r2} b_{r0},\ \theta_{N3}^b = m_{l2},\ \theta_{N4}^b = 0. \tag{6.30}$$

For the *backlash inverse* $\widehat{BI}(\cdot)$ with

$$\begin{aligned}\theta_N^* = \theta_b^* &= (mc_r, m, mc_l)^T \\ &= (\theta_{N1}^*, \theta_{N2}^*, \theta_{N3}^*)^T\end{aligned} \tag{6.31}$$

the assumption (A7b) implies that

$$\theta_{Nj}^* \in [\theta_{Nj}^a, \theta_{Nj}^b],\ j = 1, 2, 3 \tag{6.32}$$

$$\theta_{N1}^a = 0,\ \theta_{N2}^a = m_1,\ \theta_{N3}^a = -m_2 c_{l0} \tag{6.33}$$

$$\theta_{N1}^b = m_2 c_{r0},\ \theta_{N2}^b = m_2,\ \theta_{N3}^b = 0. \tag{6.34}$$

For the *hysteresis inverse* $\widehat{HI}(\cdot)$ with

$$\begin{aligned}\theta_N^* = \theta_h^* &= (m_t, c_t, m_b, c_b, m_r, m_r c_r, m_l, m_l c_l)^T \\ &= (\theta_{N1}^*, \theta_{N2}^*, \theta_{N3}^*, \theta_{N4}^*, \theta_{N5}^*, \theta_{N6}^*, \theta_{N7}^*, \theta_{N8}^*)^T\end{aligned} \tag{6.35}$$

the assumption (A7c) implies that

$$\theta_{Nj}^* \in [\theta_{Nj}^a, \theta_{Nj}^b],\ j = 1, \ldots, 8 \tag{6.36}$$

$$\theta_{N1}^a = m_{t1},\ \theta_{N2}^a = 0,\ \theta_{N3}^a = m_{b1},\ \theta_{N4}^a = -c_{b0} \tag{6.37}$$

$$\theta_{N5}^a = m_{r1},\ \theta_{N6}^a = 0,\ \theta_{N7}^a = m_{l1},\ \theta_{N8}^a = -c_{l0}m_{l2} \tag{6.38}$$

$$\theta_{N1}^b = m_{t2},\ \theta_{N2}^b = c_{t0},\ \theta_{N3}^b = m_{b2},\ \theta_{N4}^b = 0 \tag{6.39}$$

$$\theta_{N5}^b = m_{r2},\ \theta_{N6}^b = c_{r0}m_{r2},\ \theta_{N7}^b = m_{l2},\ \theta_{N8}^b = 0. \tag{6.40}$$

Using the knowledge of the parameter region $[\theta_{Nj}^a, \theta_{Nj}^b]$, we now let $f_N(t)$ modify the estimates $\theta_N(t)$ in (6.21) according to the following parameter projection algorithm.

Denote $\theta_{Nj}(t)$, $f_{Nj}(t)$, and $g_{Nj}(t)$ as the jth components of $\theta_N(t)$, $f_N(t)$, and

$$g_N(t) = -\frac{\Gamma_N \zeta_N(t) \epsilon_N(t)}{m_N^2(t)}, \tag{6.41}$$

respectively, for $j = 1, 2, \ldots, n_N$. Choose

$$\theta_{Nj}(0) \in [\theta_{Nj}^a, \theta_{Nj}^b]$$

and set

$$f_{Nj}(t) = \begin{cases} 0 & \begin{array}{l}\text{if } \theta_{Nj}(t) \in (\theta_{Nj}^a, \theta_{Nj}^b), \text{ or} \\ \text{if } \theta_{Nj}(t) = \theta_{Nj}^a,\ g_{Nj}(t) \geq 0, \text{ or} \\ \text{if } \theta_{Nj}(t) = \theta_{Nj}^b,\ g_{Nj}(t) \leq 0 \end{array} \\ -g_{Nj}(t) & \text{otherwise.} \end{cases} \tag{6.42}$$

This modified adaptive law has the following properties.

Lemma 6.1 *For the error model (6.20), the adaptive law (6.21) with the parameter projection (6.41) - (6.42) guarantees that $\theta_N(t)$, $\frac{\epsilon_N(t)}{m_N(t)}$, $\dot{\theta}_N(t) \in L^\infty$, and $\frac{\epsilon_N(t)}{m_N(t)}$, $\dot{\theta}_N(t)$ satisfy*

$$\int_{t_1}^{t_2} \frac{\epsilon_N^2(t)}{m_N^2(t)} dt \leq a_1 + b_1 \int_{t_1}^{t_2} \frac{d^2(t)}{m_N^2(t)} dt \tag{6.43}$$

$$\int_{t_1}^{t_2} \|\dot{\theta}_N(t)\|_2^2 dt \leq a_2 + b_2 \int_{t_1}^{t_2} \frac{d^2(t)}{m_N^2(t)} dt \tag{6.44}$$

for some constants $a_i, b_i > 0$, $i = 1, 2$, and all $t_2 > t_1 \geq 0$.

Proof: Substituting (6.20), (6.23), and (6.24) in (6.22) we have

$$\epsilon_N(t) = \tilde{\theta}_N^T(t)\zeta_N(t) + d(t),\ \tilde{\theta}_N(t) = \theta_N(t) - \theta_N^*. \tag{6.45}$$

We then obtain the time derivative of

$$V(\tilde{\theta}_N) = \tilde{\theta}_N^T \Gamma_N^{-1} \tilde{\theta}_N$$

along the trajectories of (6.21), as

$$\begin{aligned}\dot{V} &= -\frac{2\epsilon_N^2(t)}{m_N^2(t)} + \frac{2\epsilon_N(t)d(t)}{m_N^2(t)} + p_N(t) \\ &\leq -\frac{\epsilon_N^2(t)}{m_N^2(t)} + \frac{d^2(t)}{m_N^2(t)} + p_N(t),\end{aligned} \tag{6.46}$$

where

$$p_N(t) = 2\tilde{\theta}_N^T(t)\Gamma_N^{-1} f_N(t).$$

From the projection algorithm (6.41) - (6.42), we see that

$$p_N(t) = 2\sum_{j=1}^{n_N} \tilde{\theta}_{Nj}(t)\gamma_j^{-1} f_{Nj}(t),\ \tilde{\theta}_{Nj}(t) = \theta_{Nj}(t) - \theta_{Nj}^*$$

$$\theta_{Nj}(t) \in [\theta_{Nj}^a, \theta_{Nj}^b]$$

$$(\theta_{Nj}(t) - \theta_{Nj}^*) f_{Nj}(t) \leq 0,\ 1 \leq j \leq n_N.$$

Thus, $\theta_N(t)$ is bounded and

$$p_N(t) \leq 0. \tag{6.47}$$

Using (6.45), we conclude that $\frac{\epsilon(t)}{m_N(t)}$ and $\dot{\theta}(t)$ are bounded. In view of (6.46), (6.47), and the boundedness of $d(t)$, we have (6.43). From (6.41) and (6.42), we obtain

$$\dot{\theta}_{Nj}^2(t) \leq g_{Nj}^2(t) \leq \frac{k_0\epsilon_N^2(t)}{m_N^2(t)} \tag{6.48}$$

for $j = 1, 2, \ldots, n_N$ and some constant $k_0 > 0$. From (6.43) and (6.48), we obtain (6.44). ∇

The property (6.43) (or (6.44)) is the so-called "smallness in the mean" property: If $\frac{1}{t_2-t_1}\int_{t_1}^{t_2} \frac{d^2(t)}{m_N^2(t)} dt$, the mean of $\frac{d^2(t)}{m_N^2(t)}$, is small, then $\frac{1}{t_2-t_1}\int_{t_1}^{t_2} \frac{\epsilon_N^2(t)}{m_N^2(t)} dt$ (or $\frac{1}{t_2-t_1}\int_{t_1}^{t_2} \|\dot{\theta}_N(t)\|_2^2 dt$), the mean of $\frac{\epsilon_N^2(t)}{m_N^2(t)}$ (or $\|\dot{\theta}_N(t)\|_2^2$), is small, because a_i, b_i are constants. These properties, together with the boundedness of $\theta_N(t)$, $\frac{\epsilon_N(t)}{m_N(t)}$, and $\dot{\theta}_N(t)$, are crucial for the closed-loop signal boundedness.

6.2.7 σ-Modification with Parameter Projection

A "softer" tool for achieving boundedness of parameter estimates is the switching σ-modification [42] which can be used even when *a priori* knowledge of the parameter region for θ_N^* is lacking. It is, therefore, practical to combine this tool with parameter projection and select $f_N(t)$ in (6.21) as

$$f_N(t) = -\Gamma_N \sigma_N(t) \theta_N(t) + f_p(t), \tag{6.49}$$

where Γ_N is as in (6.26), and $f_p(t)$ is a projection. The switching-σ term is given by

$$\sigma_N(t) = \begin{cases} 0 & \text{if } \|\theta_N(t)\|_2 < M_N \\ \sigma_0(\frac{\|\theta_N(t)\|_2}{M_N} - 1) & \text{if } M_N \leq \|\theta_N(t)\|_2 < 2M_N \\ \sigma_0 & \text{if } \|\theta_N(t)\|_2 \geq 2M_N \end{cases}$$

with $\sigma_0 > 0$ and $M_N > \|\theta_N^*\|_2$.

Recall from Propositions 3.1 and 3.2 that the error $d_N(t)$ is bounded and so is $d(t) = W(s)[d_N](t)$ in (6.20) or (6.45) for the dead-zone and backlash inverse cases. With $f_p(t) = 0$, the adaptive law (6.21) with $f_N(t) = -\Gamma_N \sigma_N(t) \theta_N(t)$ ensures that $\theta_N(t) \in L^\infty$ in the dead-zone or backlash case. However, the components of $\theta_N(t)$ do not necessarily satisfy both lower and upper bounds in the assumptions (A7a) and (A7b). Thus, a projection $f_p(t)$ is still needed.

For the *dead-zone inverse* $\widehat{DI}(\cdot)$, with

$$\begin{aligned} \theta_N = \theta_d &= (\widehat{m_r}, \widehat{m_r b_r}, \widehat{m_l}, \widehat{m_l b_l})^T \\ &= (\theta_{N1}, \theta_{N2}, \theta_{N3}, \theta_{N4})^T \end{aligned}$$

from the adaptive law with the switching σ, we have

$$\theta_{N1} \leq \bar{\theta}_{N1}^b,\ \theta_{N2} \leq \bar{\theta}_{N2}^b,\ \theta_{N3} \leq \bar{\theta}_{N3}^b,\ \bar{\theta}_{N4}^a \leq \theta_{N4} \tag{6.50}$$

for some constants $\bar{\theta}_{Nj}^b > 0$, $j = 1, 2, 3$, and $\bar{\theta}_{N4}^a < 0$. To obtain parameter estimates suitable for implementing the adaptive dead-zone inverse, we need to use parameter projection to ensure that

$$\bar{\theta}_{N1}^a \leq \theta_{N1},\ 0 \leq \theta_{N2},\ \bar{\theta}_{N3}^a \leq \theta_{N3},\ \theta_{N4} \leq 0$$

for some constants $\bar{\theta}_{Nj}^a > 0$, $j = 1, 3$.

For the *backlash inverse* $\widehat{BI}(\cdot)$, with

$$\begin{aligned} \theta_N = \theta_b &= (\widehat{mc_r}, \widehat{m}, \widehat{mc_l})^T \\ &= (\theta_{N1}, \theta_{N2}, \theta_{N3})^T \end{aligned}$$

we have from the adaptive law with the switching σ-modification that

$$\theta_{N1} \le \bar{\theta}^b_{N1},\ \theta_{N2} \le \bar{\theta}^b_{N2},\ \bar{\theta}^a_{N3} \le \theta_{N3} \tag{6.51}$$

for some constants $\bar{\theta}^b_{Nj} > 0$, $j = 1, 2$, and $\bar{\theta}^a_{N3} < 0$. To obtain parameter estimates suitable for implementing the adaptive backlash inverse, we need to use parameter projection to ensure that

$$0 \le \theta_{N1},\ \bar{\theta}^a_{N2} \le \theta_{N2},\ \theta_{N3} \le 0$$

for some constant $\bar{\theta}^a_{N2} > 0$.

For the *hysteresis inverse* $\widehat{HI}(\cdot)$, the situation is different. Recall from Proposition 3.3 that the error $d_N(t)$ is bounded and so is $d(t) = W(s)[d_N](t)$ if the parameter estimates $\widehat{c}_t$, $\widehat{c}_b$, $\widehat{c}_l$, and $\widehat{c}_r$ are bounded. To generate parameter estimates for the adaptive hysteresis inverse, we need to use parameter projection to ensure that $\widehat{c}_t$, $\widehat{c}_b$, $\widehat{c}_l$, and $\widehat{c}_r$ are bounded so that $d(t)$ is bounded. On the other hand, to implement the adaptive hysteresis inverse, we need to ensure that the hysteresis inverse parameter estimates satisfy the assumption (A7c). This is achieved by a parameter projection algorithm combined with the switching σ-modification.

For the *hysteresis inverse* $\widehat{HI}(\cdot)$, with

$$\begin{aligned}\theta_N = \theta_h &= (\widehat{m_t}, \widehat{c}_t, \widehat{m_b}, \widehat{c}_b, \widehat{m_r}, \widehat{m_r c_r}, \widehat{m_l}, \widehat{m_l c_l})^T \\ &= (\theta_{N1}, \theta_{N2}, \theta_{N3}, \theta_{N4}, \theta_{N5}, \theta_{N6}, \theta_{N7}, \theta_{N8})^T\end{aligned}$$

the estimates of c_r and c_l are obtained from

$$\widehat{c}_r = \frac{\widehat{m_r c_r}}{\widehat{m_r}},\ \widehat{c}_l = \frac{\widehat{m_l c_l}}{\widehat{m_l}}.$$

For the boundedness of $\widehat{c}_t$, $\widehat{c}_b$, $\widehat{c}_l$, and $\widehat{c}_r$, we need to use parameter projection to ensure that

$$0 \le \theta_{N2} \le \bar{\theta}^b_{N2},\ \bar{\theta}^a_{N4} \le \theta_{N4} \le 0,\ \bar{\theta}^a_{N5} \le \theta_{N5} \tag{6.52}$$

$$0 \le \theta_{N6} \le \bar{\theta}^b_{N6},\ \bar{\theta}^a_{N7} \le \theta_{N7},\ \bar{\theta}^a_{N8} \le \theta_{N8} \le 0 \tag{6.53}$$

for some constants $\bar{\theta}^a_{N4} < 0$, $\bar{\theta}^a_{N8} < 0$, $\bar{\theta}^a_{Nj} > 0$, $j = 5, 7$, and $\bar{\theta}^b_{Nk} > 0$, $k = 2, 6$. This also ensures that $\widehat{c}_t \ge 0$, $\widehat{c}_b \le 0$, $\widehat{c}_l \le 0$, and $\widehat{c}_r \ge 0$ for implementing the hysteresis inverse because we have

$$0 \le \theta_{N2},\ \theta_{N4} \le 0,\ 0 \le \theta_{N6},\ \theta_{N8} \le 0.$$

We also need to use parameter projection to meet one of the requirements of the assumption (A7c):

$$\bar{\theta}^a_{N1} \leq \theta_{N1} \leq \bar{\theta}^b_{N1},\ \bar{\theta}^a_{N3} \leq \theta_{N3} \leq \bar{\theta}^b_{N3}$$

for some constants $\bar{\theta}^a_{Nj} > 0,\ \bar{\theta}^b_{Nj} > 0,\ j = 1, 3$ such that

$$\bar{\theta}^b_{N1} < min\{\bar{\theta}^a_{N5}, \bar{\theta}^a_{N7}\},\ \bar{\theta}^b_{N3} < min\{\bar{\theta}^a_{N5}, \bar{\theta}^a_{N7}\}.$$

When (6.52) and (6.53) are met, $d_N(t)$ is bounded so that the switching σ-modification ensures that $\theta_N(t) \in R^\infty$, which implies that

$$\theta_{N5} \leq \bar{\theta}^b_{N5},\ \theta_{N7} \leq \bar{\theta}^b_{N7} \tag{6.54}$$

for some constants $\bar{\theta}^b_{Nj} > 0,\ j = 5, 7$.

This analysis has established the parameter bounds achieved by (6.49) with $f_p(t) = 0$. To ensure that the components of θ_N satisfy the assumptions (A7a) - (A7c), the remaining bounds are to be guaranteed by the projection $f_p(t)$.

Using the assumption (A7a), we define

$$\theta^a_{N1} = m_{r1},\ \theta^a_{N2} = 0,\ \theta^a_{N3} = m_{l1},\ \theta^b_{N4} = 0 \tag{6.55}$$

for the *dead-zone inverse* parameter estimation.

Using the assumption (A7b), we define

$$\theta^a_{N1} = 0,\ \theta^a_{N2} = m_1,\ \theta^b_{N3} = 0 \tag{6.56}$$

for the *backlash inverse* parameter estimation.

Using the assumption (A7c), we define

$$\theta^a_{N1} = m_{t1},\ \theta^a_{N2} = 0,\ \theta^a_{N3} = m_{b1},\ \theta^a_{N4} = -c_{b0} \tag{6.57}$$

$$\theta^a_{N5} = m_{r1},\ \theta^a_{N6} = 0,\ \theta^a_{N7} = m_{l1},\ \theta^a_{N8} = -c_{l0}m_{l2} \tag{6.58}$$

$$\theta^b_{N1} = m_{t2},\ \theta^b_{N2} = c_{t0},\ \theta^b_{N3} = m_{b2} \tag{6.59}$$

$$\theta^b_{N4} = 0,\ \theta^b_{N6} = c_{r0}m_{r2},\ \theta^b_{N8} = 0 \tag{6.60}$$

for the *hysteresis inverse* parameter estimation.

For all other θ^a_{Nj}, θ^b_{Nj} in the *dead-zone*, *backlash*, and *hysteresis* cases, we define

$$\theta^a_{Nj} = -\infty,\ \theta^b_{Nj} = \infty \tag{6.61}$$

as the corresponding bounds $\bar{\theta}^a_{Nj}$, $\bar{\theta}^b_{Nj}$ in (6.50), (6.51), and (6.54) are guaranteed by the switching σ-modification. This also means that only a part of the assumption (A7a) (or (A7b), (A7c)) is needed when $f_N(t)$ in (6.21) is the switching σ-modification combined with parameter projection.

Then, for parameter projection, we denote $\theta_{Nj}(t)$, $f_{pj}(t)$, and $\bar{g}_{Nj}(t)$ as the jth components of $\theta_N(t)$, $f_p(t)$, and $\bar{g}_N(t)$:

$$\bar{g}_N(t) = -\frac{\Gamma_N \zeta_N(t)\epsilon_N(t)}{m_N^2(t)} - \Gamma_N \sigma_N(t)\theta_N(t), \tag{6.62}$$

respectively, choose

$$\theta_{Nj}(0) \in [\theta_{Nj}^a, \theta_{Nj}^b],$$

and set

$$f_{pj}(t) = \begin{cases} 0 & \text{if } \theta_{Nj}(t) \in (\theta_{Nj}^a, \theta_{Nj}^b), \text{ or} \\ & \text{if } \theta_{Nj}(t) = \theta_{Nj}^a,\ \bar{g}_{Nj}(t) \geq 0, \text{ or} \\ & \text{if } \theta_{Nj}(t) = \theta_{Nj}^b,\ \bar{g}_{Nj}(t) \leq 0 \\ -\bar{g}_{Nj}(t) & \text{otherwise} \end{cases} \tag{6.63}$$

for $j = 1, 2, \ldots, n_N$, with n_N being the dimension of θ_N.

The adaptive law (6.21) with the modification $f_N(t)$ defined by (6.49) and (6.63) also has the desired properties stated in Lemma 6.1. To show this result, we note that from the definitions of θ_{Nj}^a and θ_{Nj}^b, it follows that

$$\theta_{Nj}^* \in [\theta_{Nj}^a, \theta_{Nj}^b],\ j = 1, 2, \ldots, n_N. \tag{6.64}$$

With $f_N(t)$ from (6.49), (6.46) becomes

$$\dot{V} \leq -\frac{\epsilon_N^2(t)}{m_N^2(t)} + \frac{d^2(t)}{m_N^2(t)} + p_N(t) \tag{6.65}$$

$$p_N(t) = 2\tilde{\theta}_N^T(t)\Gamma_N^{-1}(-\Gamma_N \sigma_N(t)\theta_N(t) + f_p(t)). \tag{6.66}$$

From (6.64) and the projection algorithm (6.63), we see that

$$2\tilde{\theta}_N^T(t)\Gamma_N^{-1} f_p(t) = 2\sum_{j=1}^{n_N} \tilde{\theta}_{Nj}(t)\gamma_j^{-1} f_{pj}(t) \tag{6.67}$$

$$\tilde{\theta}_{Nj}(t) f_{pj}(t) \leq 0,\ \tilde{\theta}_{Nj}(t) = \theta_{Nj}(t) - \theta_{Nj}^*,\ 1 \leq j \leq n_N. \tag{6.68}$$

On the other hand, from the definition of $\sigma_N(t)$, it follows that

$$2\tilde{\theta}_N^T(t)\Gamma_N^{-1}(-\Gamma_N \sigma_N(t)\theta_N(t)) = -2\sigma_N(t)\tilde{\theta}_N^T(t)\theta_N(t) \leq 0 \tag{6.69}$$

and $2\sigma_N(t)\tilde{\theta}_N^T(t)\theta_N(t)$ grows unbounded if any component of $\theta_N(t)$ grows unbounded. Therefore, $V(\tilde{\theta}_N(t)) = \tilde{\theta}_N^T(t)\Gamma_N^{-1}\tilde{\theta}_N(t)$ is bounded, and so are $\frac{\epsilon_N(t)}{m_N(t)}$ and $\dot{\theta}_N(t)$.

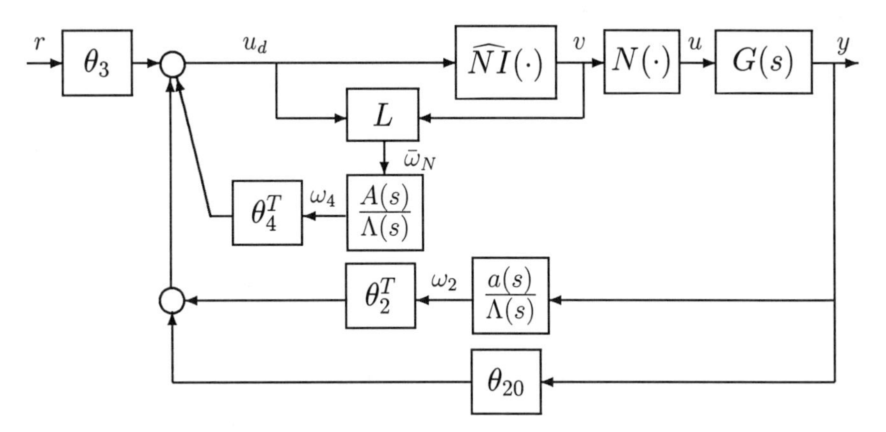

Figure 6.3: Controller structure for $G(s)$ unknown.

Similar to (4.98) - (4.101), it follows that

$$\sigma_N^2(t)\theta_N^T(t)\Gamma_N^2\theta_N(t) \leq k_N\sigma_N(t)\tilde{\theta}_N^T(t)\theta_N(t) \tag{6.70}$$

for some constant $k_N > 0$. Using (6.70), (6.21), (6.49), (6.62) and (6.63), we obtain

$$\|\dot{\theta}_N(t)\|_2^2 \leq \|\bar{g}_N(t)\|_2^2 \leq a_3\frac{\epsilon_N^2(t)}{m_N^2(t)} + a_4\sigma_N(t)\tilde{\theta}_N^T(t)\theta_N(t) \tag{6.71}$$

for some constants $a_i > 0$, $i = 3, 4$. Finally, (6.43) and (6.44) follow from (6.65) - (6.71). Thus, Lemma 6.1 also holds for the switching-σ combined with parameter projection modification.

6.3 Designs for $G(s)$ Unknown

The problem of adaptively controlling the plant (6.1) with both the nonlinear part $N(\cdot)$ and the linear part $G(s)$ unknown is much more complex. It involves estimating two sets of parameters: one from the adaptive inverse $\widehat{NI}(\cdot)$ and the other from a linear controller structure. A new difficulty in this situation is that the parameters of $G(s)$ and $N(\cdot)$ multiply each other and create a bilinear parametrization not suitable for parameter estimation. To achieve a desired linear parametrization, we need new controller structures. Again, we develop two controller structures which are both representable by the block diagram in Figure 6.3. Similar to the case when $G(s)$ is known, these two controller structures differ in the availability of signals which provide the information about $\dot{u}_d(t)$ for implementing an adaptive backlash or hysteresis inverse.

6.3.1 Controller Structure with $\theta_{20} \neq 0$

For a *dead-zone inverse*, using (6.5) with true parameters:

$$u_d(t) = -\theta_d^{*T} \omega_d(t)$$

we reparametrize the term $\theta_1^{*T}\omega_1(t)$ in the controller structure (6.6) as

$$\theta_1^{*T}\omega_1(t) = \theta_1^{*T}\frac{a(s)}{\Lambda(s)}[u_d](t) = \theta_4^{*T}\omega_4(t), \tag{6.72}$$

where $\theta_4^* \in R^{4(n-1)}$ is defined as the Kronecker product of $-\theta_1^*$ and θ_d^*:

$$\theta_4^* = -\theta_1^* \otimes \theta_d^*$$

and the corresponding regressor vector $\omega_4(t) \in R^{4(n-1)}$ is

$$\omega_4(t) = \frac{A(s)}{\Lambda(s)}[\omega_d](t).$$

As before, $\Lambda(s)$ is a stable and monic polynomial of degree $n-1$, while $A(s)$, analogous to the vector

$$a(s) = (1, s, \ldots, s^{n-2})^T,$$

now is a matrix:

$$A(s) = (I_4, sI_4, \ldots, s^{n-2}I_4)^T,$$

where I_4 is the 4×4 identity matrix.

With $\omega_2(t) = \frac{a(s)}{\Lambda(s)}[y](t)$ as defined in (6.7), the reparametrized controller structure is

$$u_d(t) = \theta_2^T\omega_2(t) + \theta_{20}y(t) + \theta_3 r(t) + \theta_4^T\omega_4(t), \tag{6.73}$$

where θ_2, θ_{20}, θ_3, and θ_4 are the estimates of $\theta_2^* \in R^{n-1}$, $\theta_{20}^* \in R$, $\theta_3^* \in R$, and $\theta_4^* \in R^{4(n-1)}$. The price paid to achieve this linear parametrization is the higher dimension of θ_4 and $\omega_4(t)$.

Recall that for the implementation of the backlash and hysteresis inverses, we need the knowledge of the sign of $u_d(t)$. The controller structure (6.73) is suitable if both $\dot{y}(t)$ and $\dot{r}(t)$ are available and $\dot{r}(t)$ is bounded.

For a *backlash inverse*, we obtain (6.72) with

$$\theta_4^* = -\theta_1^* \otimes \theta_b^*$$

$$\omega_4(t) = \frac{A(s)}{\Lambda(s)}[\omega_b](t)$$

with $\Lambda(s)$ as above and

$$A(s) = (I_3, sI_3, \ldots, s^{n-2}I_3)^T.$$

A *hysteresis inverse* with true parameters can be expressed as

$$u_d(t) = -\theta_h^{*T}\omega_h(t) + a_h(t),$$

where

$$a_h(t) = \widehat{\chi_w}(t)\widehat{c_w}(t) + \widehat{\chi_e}(t)\widehat{c_e}(t).$$

To obtain (6.89), we define

$$\theta_4^* = -\theta_1^* \otimes (\theta_h^{*T}, -1)^T$$

$$\omega_4(t) = \frac{A(s)}{\Lambda(s)}[(\omega_h^T, \widehat{\chi_w}\widehat{c_w} + \widehat{\chi_e}\widehat{c_e})^T](t)$$

for $\Lambda(s)$ as above and

$$A(s) = (I_9, sI_9, \ldots, s^{n-2}I_9)^T.$$

In this case, $\dot{u}_d(t)$ is available from the implementable expression

$$\dot{u}_d(t) = \dot{\theta}_2^T(t)\omega_2(t) + \theta_2^T\dot{\omega}_2(t) + \dot{\theta}_{20}(t)y(t) + \theta_{20}(t)\dot{y}(t) + \dot{\theta}_3(t)r(t) \\ +\theta_3(t)\dot{r}(t) + \dot{\theta}_4^T(t)\omega_4(t) + \theta_4^T(t)\dot{\omega}_4(t).$$

Note that $\dot{\theta}_2(t)$, $\dot{\theta}_{20}(t)$, $\dot{\theta}_3(t)$, and $\dot{\theta}_4(t)$ will be available from the update laws which are yet to be designed. Also, $\dot{\omega}_2(t)$ and $\dot{\omega}_4(t)$ are available because both filters $\frac{a(s)}{\Lambda(s)}$ and $\frac{A(s)}{\Lambda(s)}$ are strictly proper.

The controller structure (6.73), applicable to dead-zone, backlash, and hysteresis, is shown in Figure 6.3 where the logic block L specifies the corresponding form of $\bar{\omega}_N(t)$:

$$\bar{\omega}_N(t) = \begin{cases} \omega_d(t) & \text{for } \widehat{DI}(\cdot) \\ \omega_b(t) & \text{for } \widehat{BI}(\cdot) \\ (\omega_h^T(t), \widehat{\chi_w}(t)\widehat{c_w}(t) + \widehat{\chi_e}(t)\widehat{c_e}(t))^T & \text{for } \widehat{HI}(\cdot). \end{cases}$$

6.3.2 Controller Structure with $\theta_{20} = 0$

If $\dot{y}(t)$ is not available but $\dot{r}(t)$ is measured and bounded, then an alternative controller structure is obtained by using (6.12) - (6.15) with $\theta_1^* \in R^n$ and

$$a(s) = (1, s, \ldots, s^{n-1})^T$$

to reparametrize $\theta_1^{*T}\omega(t)$. In this case, the expression (6.89) still holds but now with $\Lambda(s)$ of degree n and

$$A(s) = (I_i, sI_i, \ldots, s^{n-1}I_i)^T,$$

where $i = 4$ for the *dead-zone* case, $i = 3$ for the *backlash* case, and $i = 9$ for the *hysteresis* case. The new controller structure, similar to (6.73), is

$$u_d(t) = \theta_2^T\omega_2(t) + \theta_3 r(t) + \theta_4^T\omega_4(t), \tag{6.74}$$

where $\omega_2(t) = \frac{a(s)}{\Lambda(s)}[y](t)$, and θ_2, θ_3, and θ_4 are the estimates of $\theta_2^* \in R^n$, $\theta_3^* \in R$, and $\theta_4^* \in R^{4n}$ for the *dead-zone* case, $\theta_4^* \in R^{3n}$ for the *backlash* case, and $\theta_4^* \in R^{9n}$ for the *hysteresis* case. With this controller structure, the signal $\dot{u}_d(t)$ is available from the expression

$$\begin{aligned}\dot{u}_d(t) = {} & \dot{\theta}_2^T(t)\omega_2(t) + \theta_2^T(t)\dot{\omega}_2(t) + \dot{\theta}_3(t)r(t) \\ & + \theta_3(t)\dot{r}(t) + \dot{\theta}_4^T(t)\omega_4(t) + \theta_4^T(t)\dot{\omega}_4(t),\end{aligned}$$

which is implementable when $y(t)$, $u_d(t)$, $r(t)$, and $\dot{r}(t)$ are available.

For the case when $\dot{r}(t)$ is not available, we can employ an approximation of $\dot{u}_d(t)$, such as (6.17), for $u_d(t)$ from either (6.73) or (6.74). Approximations of $\dot{y}(t)$ and $\dot{r}(t)$ can also be employed to generate an approximation of $\dot{u}_d(t)$ for $u_d(t)$ from either (6.73) or (6.74). Yet another possibility is to use the hybrid inverse scheme proposed in Section 4.2 to implement discrete-time inverses which do not need the knowledge of the sign of $\dot{u}_d(t)$.

6.3.3 Tracking Error Equation

To obtain a tracking error equation associated with the controller (6.73) or (6.74), we use (6.4) and (6.5) to express

$$u(t) = -\theta_N^{*T}\omega_N(t) + a_h(t) + d_N(t). \tag{6.75}$$

Using (6.4) for $u(t)$ on the left side of (6.19), and (6.75) for $u(t)$ on the right side of (6.19), we obtain

$$\begin{aligned}& u_d(t) + (\theta_N(t) - \theta_N^*)^T\omega_N(t) + d_N(t) \\ & = \theta_2^{*T}\omega_2(t) + \theta_{20}^{*T}y(t) + \theta_3^* r(t) \\ & \quad + \theta_3^* P_m(s)[y - y_m](t) + \theta_4^{*T}\omega_4(t) + \theta_1^{*T}\frac{a(s)}{\Lambda(s)}[d_N](t).\end{aligned} \tag{6.76}$$

Substituting (6.76) in (6.73) or (6.74), with $W_m(s) = P_m^{-1}(s)$, we arrive at the tracking error equation

$$e(t) = k_p W_m(s)[(\theta - \theta^*)^T \omega](t) + d(t), \tag{6.77}$$

where

$$\omega(t) = \begin{cases} (\omega_N^T(t), \omega_2^T(t), y(t), r(t), \omega_4^T(t))^T & \text{for (6.73)} \\ (\omega_N^T(t), \omega_2^T(t), r(t), \omega_4^T(t))^T & \text{for (6.74)} \end{cases}$$

$$\theta(t) = \begin{cases} (\theta_N^T(t), \theta_2^T(t), \theta_{20}(t), \theta_3(t), \theta_4^T(t))^T & \text{for (6.73)} \\ (\theta_N^T(t), \theta_2^T(t), \theta_3(t), \theta_4^T(t))^T & \text{for (6.74)} \end{cases}$$

$$\theta^* = \begin{cases} (\theta_N^{*T}, \theta_2^{*T}, \theta_{20}^*, \theta_3^*, \theta_4^{*T})^T & \text{for (6.73)} \\ (\theta_N^{*T}, \theta_2^{*T}, \theta_3^*, \theta_4^{*T})^T & \text{for (6.74)} \end{cases}$$

$$d(t) = k_p W_m(s)(1 - \theta_1^{*T} \frac{a(s)}{\Lambda(s)})[d_N](t).$$

It is important to note that the error equation (6.77) is linear in the overall system parameter error $\theta - \theta^*$. This form, familiar from the adaptive control literature [3], [30], [39], [80], [96], is ensured by the new controller structures (6.73) and (6.74).

Using the error equation (6.77) we now design two adaptive schemes: one with a gradient-type update law, and the other with a Lyapunov-type update law. Depending on the available signals, each of these schemes can employ either of the two controller structures.

6.3.4 Gradient-Type Adaptive Scheme

For this scheme, we use the following gradient-type update law:

$$\dot{\theta}(t) = -\frac{sign(k_p)\Gamma_\theta \zeta(t)\epsilon(t)}{1 + \zeta^T(t)\zeta(t) + \xi^2(t)} + f_\theta(t) \tag{6.78}$$

$$\dot{\rho}(t) = -\frac{\gamma_\rho \xi(t)\epsilon(t)}{1 + \zeta^T(t)\zeta(t) + \xi^2(t)} + f_\rho(t), \tag{6.79}$$

where $\rho(t)$ is the estimate of $\rho^* = k_p$, and

$$\epsilon(t) = e(t) + \rho(t)\xi(t) \tag{6.80}$$

$$\xi(t) = \theta^T(t)\zeta(t) - W_m(s)[\theta^T \omega](t)$$

$$\zeta(t) = W_m(s)[\omega](t)$$

$$\Gamma_\theta = diag\{\gamma_1, \ldots, \gamma_{n_\theta}\},\ \gamma_j > 0,\ 1 \le j \le n_\theta,\ \gamma_\rho > 0,$$

with n_θ being the dimension of the parameter estimate $\theta(t)$. The terms $f_\theta(t)$ and $f_\rho(t)$ are to be chosen to achieve robustness with respect to $d(t)$ and to implement the adaptive inverse $\widehat{NI}(\cdot)$.

Parameter Projection

For parameter projection, $f_\theta(t)$ and $f_\rho(t)$ are similar to (6.41) - (6.42). To simplify notation, we let θ_j^* be the jth component of θ^* and let $\theta_j(t)$ be the jth component of $\theta(t)$. We assume that a convex region in the parameter space, which contains the true parameter θ^*, is known, as described by the following assumption:

> (**A8**) The jth component of θ^* is $\theta_j^* \in [\theta_j^a, \theta_j^b]$ for some known constants $\theta_j^a \leq \theta_j^b$, for $j = 1, \ldots, n_\theta$; $\rho^* \in [\rho^a, \rho^b]$ for some known constants $\rho^a \leq \rho^b$.

For the parameter θ_N, such an assumption is characterized by (6.27) - (6.40), based on the assumptions (A7a) - (A7c). To design $f_\theta(t)$, $f_\rho(t)$, we let

$$g_\theta(t) = -\frac{sign(k_p)\Gamma_\theta \zeta(t)\epsilon(t)}{1 + \zeta^T(t)\zeta(t) + \xi^2(t)} \tag{6.81}$$

$$g_\rho(t) = -\frac{\gamma_\rho \xi(t)\epsilon(t)}{1 + \zeta^T(t)\zeta(t) + \xi^2(t)} \tag{6.82}$$

and denote the jth component of $g_\theta(t)$, $f_\theta(t)$ as $g_{\theta j}(t)$, $f_{\theta j}(t)$, respectively. Then, we choose

$$\theta_j(0) \in [\theta_j^a, \theta_j^b], \ \rho(0) \in [\rho^a, \rho^b]$$

and set

$$f_{\theta j}(t) = \begin{cases} 0 & \text{if } \theta_j(t) \in (\theta_j^a, \theta_j^b), \text{ or} \\ & \text{if } \theta_j(t) = \theta_j^a, \ g_{\theta j}(t) \geq 0, \text{ or} \\ & \text{if } \theta_j(t) = \theta_j^b, \ g_{\theta j}(t) \leq 0 \\ -g_{\theta j}(t) & \text{otherwise} \end{cases} \tag{6.83}$$

$$f_\rho(t) = \begin{cases} 0 & \text{if } \rho(t) \in (\rho^a, \rho^b), \text{ or} \\ & \text{if } \rho(t) = \rho^a, \ g_\rho(t) \geq 0, \text{ or} \\ & \text{if } \rho(t) = \rho^b, \ g_\rho(t) \leq 0 \\ -g_\rho(t) & \text{otherwise.} \end{cases} \tag{6.84}$$

The resulting gradient adaptive scheme has the following properties.

Lemma 6.2 *The adaptive law (6.78) - (6.79) with the parameter projection (6.81) - (6.84) guarantees that* $\theta(t)$, $\rho(t)$, $\frac{\epsilon(t)}{m(t)}$, $\dot{\theta}(t)$, $\dot{\rho}(t) \in L^\infty$, *and* $\frac{\epsilon(t)}{m(t)}$, $\dot{\theta}(t)$, $\dot{\rho}(t)$ *satisfy*

$$\int_{t_1}^{t_2} \frac{\epsilon^2(t)}{m^2(t)} dt \leq a_1 + b_1 \int_{t_1}^{t_2} \frac{d^2(t)}{m^2(t)} dt \tag{6.85}$$

$$\int_{t_1}^{t_2} \|\dot{\theta}(t)\|_2^2 dt \leq a_2 + b_2 \int_{t_1}^{t_2} \frac{d^2(t)}{m^2(t)} dt \tag{6.86}$$

$$\int_{t_1}^{t_2} \dot{\rho}^2(t) dt \leq a_3 + b_3 \int_{t_1}^{t_2} \frac{d^2(t)}{m^2(t)} dt \tag{6.87}$$

for some constants $a_i > 0,\ b_i > 0,\ i = 1, 2, 3,$ *and all* $t_2 > t_1 \geq 0$, *where*

$$m(t) = \sqrt{1 + \zeta^T(t)\zeta(t) + \xi^2(t)}.$$

Proof: Using (6.77) and (6.80), we obtain

$$\epsilon(t) = k_p \tilde{\theta}^T(t)\zeta(t) + \tilde{\rho}(t)\xi(t) + d(t), \tag{6.88}$$

where

$$\tilde{\theta}(t) = \theta(t) - \theta^*,\ \tilde{\rho}(t) = \rho(t) - \rho^*.$$

Introducing

$$p_\theta(t) = 2|k_p|\tilde{\theta}^T(t)\Gamma_\theta^{-1} f_\theta(t)$$
$$p_\rho(t) = 2\tilde{\rho}(t)\gamma_\rho^{-1} f_\rho(t)$$

we obtain the time derivative of

$$V(\tilde{\theta}, \tilde{\rho}) = |k_p|\tilde{\theta}^T \Gamma_\theta^{-1}\tilde{\theta} + \gamma_\rho^{-1}\tilde{\rho}^2$$

along the trajectories of (6.78) - (6.79) as

$$\begin{aligned} \dot{V} &= -\frac{2\epsilon^2(t)}{m^2(t)} + \frac{2\epsilon(t)d(t)}{m^2(t)} + p_\theta(t) + p_\rho(t) \\ &\leq -\frac{\epsilon^2(t)}{m^2(t)} + \frac{d^2(t)}{m^2(t)} + p_\theta(t) + p_\rho(t). \end{aligned} \tag{6.89}$$

From the projection algorithm (6.81) - (6.84), we see that

$$p_\theta(t) = 2|k_p| \sum_{j=1}^{n_\theta} \tilde{\theta}_j(t)\gamma_j^{-1} f_{\theta j}(t),\ \tilde{\theta}_j(t) = \theta_j(t) - \theta_j^*$$

$$\theta_j(t) \in [\theta_j^a, \theta_j^b],\ (\theta_j(t) - \theta_j^*) f_{\theta j}(t) \leq 0,\ 1 \leq j \leq n_\theta$$

$$\rho(t) \in [\rho^a, \rho^b],\ (\rho(t) - \rho^*) f_\rho(t) \leq 0.$$

Thus, $\theta(t)$ is bounded, and so is $\rho(t)$, and

$$p_\theta(t) \leq 0,\ p_\rho(t) \leq 0. \tag{6.90}$$

Recalling from Propositions 3.1 - 3.3 that the boundedness of $\theta_N(t)$ implies that $d(t) = W(s)[d_N](t)$ is bounded in the hysteresis case (in the dead-zone or backlash case $d(t)$ is bounded regardless of the boundedness of $\theta_N(t)$), and using (6.88), (6.78), and (6.79), we conclude that $\frac{\epsilon(t)}{m(t)}$, $\dot{\theta}(t)$, and $\dot{\rho}(t)$ are bounded. In view of (6.89), (6.90), and the boundedness of $d(t)$, we have (6.85). From (6.81) - (6.84), we obtain

$$\dot{\theta}_j^2(t) \leq g_{\theta j}^2(t) \leq \frac{k_\theta \epsilon^2(t)}{m^2(t)},\ 1 \leq j \leq n_\theta \tag{6.91}$$

$$\dot{\rho}^2(t) \leq g_\rho^2(t) \leq \frac{k_\rho \epsilon^2(t)}{m^2(t)} \tag{6.92}$$

for some constant $k_\theta > 0$, $k_\rho > 0$. From (6.85), (6.91), and (6.92), we have (6.86) and (6.87). ∇

Switching σ-Modification with Parameter Projection

In this case, $f_\theta(t)$ and $f_\rho(t)$ are

$$f_\theta(t) = -\Gamma_\theta \sigma_\theta(t)\theta(t) + f_p(t) \tag{6.93}$$

$$f_\rho(t) = -\gamma_\rho \sigma_\rho(t)\rho(t), \tag{6.94}$$

where

$$\sigma_\theta(t) = \begin{cases} 0 & \text{if } \|\theta(t)\|_2 < M_\theta \\ \sigma_0(\frac{\|\theta(t)\|_2}{M_\theta} - 1) & \text{if } M_\theta \leq \|\theta(t)\|_2 < 2M_\theta \\ \sigma_0 & \text{if } \|\theta(t)\|_2 \geq 2M_\theta \end{cases} \tag{6.95}$$

$$\sigma_\rho(t) = \begin{cases} 0 & \text{if } |\rho(t)| < M_\rho \\ \sigma_0(\frac{|\rho(t)|}{M_\rho} - 1) & \text{if } M_\rho \leq |\rho(t)| < 2M_\rho \\ \sigma_0 & \text{if } |\rho(t)| \geq 2M_\rho \end{cases} \tag{6.96}$$

$\sigma_0 > 0$, and M_θ, M_ρ are the upper bounds of $\|\theta^*\|_2$, $|\rho^*|$:

$$M_\theta > \|\theta^*\|_2,\ M_\rho > |\rho^*|.$$

The parameter projection term $f_p(t)$ is designed to modify only the nonlinearity parameter estimate $\theta_N(t)$ which is a part of the overall system parameter vector $\theta(t)$. With the use of the switching-σ term $-\Gamma_\theta \sigma_\theta(t)\theta(t)$, there is no need to project other parts of $\theta(t)$. The parameter boundaries θ_{Nj}^a, θ_{Nj}^b for θ_{Nj}^* are given in (6.55) - (6.61), which are needed for projecting θ_{Nj}. Denoting the jth component of $\theta(t)$ as $\theta_j(t)$ and that of $f_p(t)$ as $f_{pj}(t)$ (that is, $\theta_j(t) = \theta_{Nj}(t)$, $j = 1, 2, \ldots, n_N$ for $\theta_N(t) \in R^{n_N}$), the parameter projection is

$$f_{pj}(t) = \begin{cases} f_{Nj}(t) & \text{for } j = 1, 2, \ldots, n_N \\ 0 & \text{otherwise,} \end{cases}$$

where $f_{Nj}(t)$ is the projection for $\theta_j(t) = \theta_{Nj}(t)$:

$$f_{Nj}(t) = \begin{cases} 0 & \begin{array}{l} \text{if } \theta_{Nj}(t) \in (\theta_{Nj}^a, \theta_{Nj}^b), \text{ or} \\ \text{if } \theta_{Nj}(t) = \theta_{Nj}^a,\ g_j(t) \geq 0, \text{ or} \\ \text{if } \theta_{Nj}(t) = \theta_{Nj}^b,\ g_j(t) \leq 0 \end{array} \\ -g_j(t) & \text{otherwise} \end{cases}$$

and $g_j(t)$ is the jth component of

$$g(t) = -\frac{\Gamma_\theta \zeta(t)\epsilon(t)}{m^2(t)} - \Gamma_\theta \sigma_\theta(t)\theta(t).$$

The initial parameter estimates are

$$\theta_j(0) \in \begin{cases} [\theta_j^a, \theta_j^b] = [\theta_{Nj}^a, \theta_{Nj}^b] & \text{for } j = 1, 2, \ldots, n_N \\ \text{no constraint} & \text{otherwise.} \end{cases}$$

Similar to the switching σ-modification with parameter projection in Section 6.2, we can show that the desired properties stated in Lemma 6.2 hold for the adaptive law with this combination of the switching-σ and parameter projection modifications.

6.3.5 Lyapunov-Type Adaptive Scheme

In the design of a Lyapunov-type scheme we use a different estimation error based on new auxiliary signals $\zeta(t)$ and $\xi(t)$. This leads to an adaptive system with a dynamic order lower than that of (6.78) - (6.79).

Letting $L(s)$ be a Hurwitz polynomial of degree n^*-1 such that $W_m(s)L(s)$ is strictly positive real [40] and introducing

$$\zeta(t) = L^{-1}(s)[\omega](t)$$

$$\xi(t) = \theta^T(t)\zeta(t) - L^{-1}(s)[\theta^T\omega](t)$$

we define the estimation error $\epsilon(t)$ from

$$\epsilon(t) = e(t) + W_m(s)L(s)[\rho\xi - \alpha\epsilon(\zeta^T\zeta + \xi^2)](t), \tag{6.97}$$

where $\rho(t)$ is the estimate of $\rho^* = k_p$, and $\alpha > 0$ is a constant design parameter. Then we update $\theta(t)$ and $\rho(t)$ from

$$\dot{\theta}(t) = -sign(k_p)\Gamma_\theta \zeta(t)\epsilon(t) + f_\theta(t) \tag{6.98}$$

$$\dot{\rho}(t) = -\gamma_\rho \xi(t)\epsilon(t) + f_\rho(t), \tag{6.99}$$

where

$$\Gamma_\theta = diag\{\gamma_1, \ldots, \gamma_{n_\theta}\},\ \gamma_j > 0,\ 1 \le j \le n_\theta,\ \gamma_\rho > 0$$

and $f_\theta(t)$, $f_\rho(t)$ are modifying terms for robustness with respect to $d(t)$ and for implementation of an adaptive inverse $\widehat{NI}(\cdot)$.

The projection algorithm for $f_\theta(t)$ and $f_\rho(t)$ has the same form as that given by (6.83) and (6.84) except that in this case

$$g_\theta(t) = -sign(k_p)\Gamma_\theta \zeta(t)\epsilon(t)$$

$$g_\rho(t) = -\gamma_\rho \xi(t)\epsilon(t).$$

Since $L(s)$ has order $n^* - 1$, while $P_m(s)$ has order n^*, in view of the definitions of $\zeta(t)$ and $\xi(t)$, the dynamic order of the adaptive law (6.98) - (6.99) is lower than that of the adaptive law (6.78) - (6.79).

The above Lyapunov-type adaptive scheme has the following properties.

Lemma 6.3 *The adaptive law (6.98) - (6.99) guarantees that $\theta(t)$, $\rho(t)$, $\epsilon(t) \in L_\infty$ and that*

$$\int_{t_1}^{t_2} \epsilon^2(t)m^2(t)dt \le a_1 + b_1 \int_{t_1}^{t_2} \frac{d^2(t)}{m^2(t)}dt \tag{6.100}$$

$$\int_{t_1}^{t_2} \|\dot{\theta}(t)\|_2^2 dt \le a_2 + b_2 \int_{t_1}^{t_2} \frac{d^2(t)}{m^2(t)}dt \tag{6.101}$$

$$\int_{t_1}^{t_2} |\dot{\rho}(t)|^2 dt \le a_3 + b_3 \int_{t_1}^{t_2} \frac{d^2(t)}{m^2(t)}dt \tag{6.102}$$

for some constants $a_i > 0$, $b_i > 0$, $i = 1, 2, 3$, and all $t_2 > t_1 \ge 0$, where

$$m(t) = \sqrt{1 + \zeta^T(t)\zeta(t) + \xi^2(t)}.$$

Proof: Introducing

$$\bar{d}(t) = L^{-1}(s)[d](t) = k_p L^{-1}(s)W_m(s)(1 - \theta_1^{*T}\frac{a(s)}{\Lambda(s)})[d_N](t)$$

we rewrite (6.77) as

$$e(t) = W_m(s)L(s)[\rho^*(L^{-1}(s)[\theta^T\omega] - \theta^{*T}L^{-1}(s)[\omega]) + \bar{d}](t). \tag{6.103}$$

Substituting (6.103) in (6.97), we have

$$\epsilon(t) = W_m(s)L(s)[\rho^*\tilde{\theta}^T\zeta + \tilde{\rho}\xi - \alpha\epsilon(\zeta^T\zeta + \xi^2) + \bar{d}](t). \tag{6.104}$$

Let (A_m, B_m, C_m) be a controllable realization of $W_m(s)L(s)$ and $e_\epsilon(t)$ be its state. With this notation, from (6.104) we have

$$\dot{e}_\epsilon(t) = A_m e_\epsilon(t) + B_m \nu(t), \ \epsilon(t) = C_m e_\epsilon(t), \tag{6.105}$$

where

$$\nu(t) = \rho^* \tilde{\theta}^T(t)\zeta(t) + \tilde{\rho}(t)\xi(t) - \alpha\epsilon(t)(\zeta^T(t)\zeta(t) + \xi^2(t)) + \bar{d}(t).$$

Strict positive realness of $W_m(s)L(s)$ implies [80] that there exist constant matrices $Q = Q^T > 0$, $P = P^T > 0$ such that

$$A_m^T P + P A_m = -Q, \ P B_m = C_m^T.$$

Hence the time derivative of

$$V(e_\epsilon, \phi, \psi) = e_\epsilon^T P e_\epsilon + |\rho^*| \tilde{\theta}^T \Gamma_\theta^{-1} \tilde{\theta} + \gamma_\rho^{-1} \tilde{\rho}^2$$

along the trajectories of (6.98) and (6.99) is

$$\begin{aligned} \dot{V} = & -e_\epsilon^T(t) Q e_\epsilon(t) - 2\alpha\epsilon^2(t)(\zeta^T(t)\zeta(t) + \xi^2(t)) + 2\epsilon(t)\bar{d}(t) \\ & + 2|\rho^*| \tilde{\theta}^T(t) \Gamma_\theta^{-1} f_\theta(t) + 2\tilde{\rho}(t)\gamma_\rho^{-1} f_\rho(t). \end{aligned} \tag{6.106}$$

Using (6.105) and defining $\alpha_1 = min\{\frac{\lambda_m}{\|C_m\|_2^2}, 2\alpha\} > 0$ with λ_m being the smallest eigenvalue of Q, from (6.106), we obtain

$$\begin{aligned} \dot{V} \le & -\alpha_1 \epsilon^2(t) m^2(t) + 2\epsilon(t)\bar{d}(t) \\ & + 2|\rho^*| \tilde{\theta}^T(t) \Gamma_\theta^{-1} f_\theta(t) + 2\tilde{\rho}(t)\gamma_\rho^{-1} f_\rho(t) \\ = & -\frac{\alpha_1}{2}\epsilon^2(t) m^2(t) - \frac{\alpha_1}{2}(|\epsilon(t)|m(t) - \frac{2\bar{d}(t)}{\alpha_1 m(t)})^2 + \frac{2\bar{d}^2(t)}{\alpha_1 m(t)} \\ & + 2|\rho^*| \tilde{\theta}^T(t) \Gamma_\theta^{-1} f_\theta(t) + 2\tilde{\rho}(t)\gamma_\rho^{-1} f_\rho(t). \end{aligned} \tag{6.107}$$

The parameter projection algorithm ensures that

$$\theta_j(t) \in [\theta_j^a, \theta_j^b], \ (\theta_j(t) - \theta_j^*) f_{\theta j}(t) \le 0, \ 1 \le j \le n_\theta$$
$$\rho(t) \in [\rho^a, \rho^b], \ (\rho(t) - \rho^*) f_\rho(t) \le 0.$$

Thus, $\theta(t)$ and $\rho(t)$ are bounded and

$$|\rho^*| \tilde{\theta}^T(t) \Gamma_\theta^{-1} f_\theta(t) \le 0, \ \tilde{\rho}(t)\gamma_\rho^{-1} f_\rho(t) \le 0. \tag{6.108}$$

Since $\bar{d}(t)$ is bounded, from (6.107) and (6.108) it follows that $\epsilon(t)m(t)$ is bounded. Finally we have (6.100) - (6.102) from (6.98), (6.99), (6.107), (6.108), and the fact that $L^{-1}(s)$ is a stable and strictly proper transfer function. ∇

The switching σ-modification with parameter projection for $f_\theta(t)$ and $f_\rho(t)$ has the same form as that given by (6.93) - (6.96) except that in this case

$$g(t) = -sign(k_p)\Gamma_\theta \zeta(t)\epsilon(t) - \Gamma_\theta \sigma_\theta(t)\theta(t).$$

This modification also has the desired properties stated in Lemma 6.3.

6.4 Stability and Performance

To be of interest for practical applications, the adaptive schemes developed in the preceding sections must assure that all their signals remain bounded and that with adaptation the tracking error is significantly reduced. For the most general case, analytical proofs of boundedness and performance improvement are still lacking. The proof of boundedness is available in the case of equal slopes. For all the cases, extensive simulations provide evidence of both boundedness and performance improvements. Several such examples are included in this section.

6.4.1 Boundedness

The closed-loop signal boundedness is proven in [116] for dead-zones with equal slopes $m = m_r = m_l$ and in [120] for hysteresis with equal slopes $m = m_t = m_b$, and the proofs are given in Appendix B. The signal boundedness also holds for the adaptive backlash inverse because the backlash nonlinearity $B(\cdot)$ is defined with equal slopes.

The equal slope parametrizations for dead-zone and hysteresis are

$$\begin{aligned} \theta_d^* &= (mb_r, m, mb_l)^T \\ \theta_d(t) &= (\widehat{mb_r}(t), \widehat{m}(t), \widehat{mb_l}(t))^T \\ \omega_d(t) &= (\widehat{\chi_r}(t), -v(t), 1 - \widehat{\chi_r}(t))^T \end{aligned} \tag{6.109}$$

$$\begin{aligned} \theta_h^* &= (c_t, m_t, c_b, m_r, m_r c_r, m_l, m_l c_l)^T \\ \theta_h(t) &= (\widehat{c_t}(t), \widehat{m_t}(t), \widehat{c_b}(t), \widehat{m_r}(t), \widehat{m_r c_r}(t), \widehat{m_l}(t), \widehat{m_l c_l}(t))^T \\ \omega_h(t) &= (-\widehat{\chi_t}(t), -(\widehat{\chi_t}(t) + \widehat{\chi_c}(t) + \widehat{\chi_b}(t))v(t), -\widehat{\chi_b}(t), \\ &\quad - \widehat{\chi_r}(t)v(t), \widehat{\chi_r}(t), -\widehat{\chi_l}(t)v(t), \widehat{\chi_l}(t))^T, \end{aligned} \tag{6.110}$$

where

$$\widehat{\chi_c}(t) = \widehat{\chi_w}(t) + \widehat{\chi_e}(t)$$

for $\widehat{\chi_w}(t)$, $\widehat{\chi_e}(t)$ defined in (3.50), (3.51) with $\widehat{m_t}(t) = \widehat{m_b}(t)$. For the equal slope parametrizations the properties of the continuous-time adaptive laws (6.21), (6.78), and (6.98) hold and are sufficient for the following closed-loop signal boundedness result.

Theorem 6.1 *For a dead-zone with equal slopes $m_r = m_l$, or a backlash, or a hysteresis with equal slopes $m_t = m_b$, all signals in the closed-loop system with an adaptive inverse and a fixed (adaptive) linear controller for the plant (6.1) with $N(\cdot)$ unknown and $G(s)$ known (unknown) are bounded.*

With the proof of this theorem given in Appendix B, we proceed to present examples which illustrate the achieved boundedness and tracking properties.

6.4.2 Example: Adaptive Dead-Zone Inverse

To examine the effectiveness of the adaptive dead-zone inverse, simulations were performed for the plant

$$y(t) = G(s)[u](t),\ u(t) = DZ(v(t)),\ G(s) = \frac{-2}{s^2 - s - 6}$$

with the unknown dead-zone parameters

$$b_r = 0.25,\ b_l = -0.3,\ m_r = 1.0,\ m_l = 1.25.$$

The design is performed with the reference model:

$$y_m(t) = W_m(s)[r](t),\ W_m(s) = \frac{1}{s^2 + 4s + 4}.$$

The linear controller structure (6.15) is

$$u_d(t) = \frac{\theta_{11}^* + \theta_{12}^* s}{(s+1)^2}[u_d](t) + \frac{\theta_{21}^* + \theta_{22}^* s}{(s+1)^2}[y](t) + \theta_3^* r(t). \tag{6.111}$$

The parameter values satisfying (6.14) are

$$\theta_{11}^* = -25,\ \theta_{12}^* = -5,\ \theta_{21}^* = 80,\ \theta_{22}^* = 40,\ \theta_3^* = -0.5.$$

The tracking error is $e(t) = y(t) - y_m(t)$.

We present simulation results for three controllers. The first controller uses a fixed highly detuned dead-zone inverse with

$$\widehat{b_r} = 2b_r,\ \widehat{b_l} = 1.43b_l,\ \widehat{m_r} = 1.25m_r,\ \widehat{m_l} = 1.3m_l$$

and adapts the parameters of the linear controller (6.111) for $G(s)$ unknown. This controller results in large tracking errors over a long period as shown by Figure 6.4. The adaptive control system was simulated using the scheme of Section 6.3.4 with a switching σ-modification without parameter projection because the dead-zone inverse is fixed. The adaptive task of this controller was significant, because with such a detuned dead-zone inverse, a fixed linear controller would result in an unstable system even when $G(s)$ is known. Still the adaptive linear controller resulted in bounded signals. However, for larger errors in the dead-zone parameter estimates, its response is much worse and

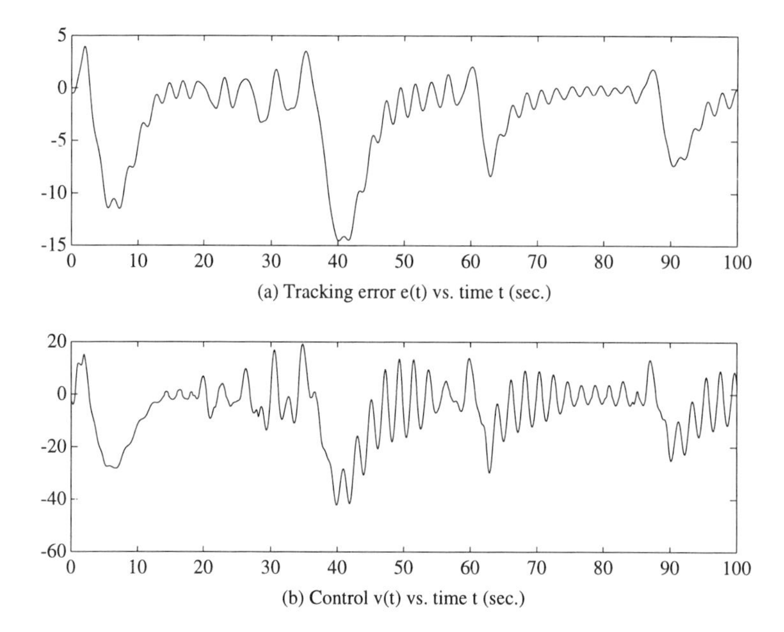

Figure 6.4: Response of a fixed dead-zone inverse.

may even be unbounded. This indicates that a controller with a fixed detuned dead-zone inverse cannot handle large uncertainties.

Dramatic improvements are achieved with the second controller in which an adaptive dead-zone inverse was introduced. When $G(s)$ is known, the adaptive controller updates only the parameters of the dead-zone inverse and uses the linear controller (6.111) with the fixed parameters θ_{11}^*, θ_{12}^*, θ_{21}^*, θ_{22}^*, θ_3^*. When $G(s)$ is unknown, in the third controller the parameters of both the dead-zone inverse and the linear controller structure are updated. This controller has the form (6.74), which in this case is

$$u_d(t) = \frac{\theta_{21} + \theta_{22}s}{(s+1)^2}[y](t) + \theta_3 r(t) + (\theta_{41}, \ldots, \theta_{48})^T \begin{pmatrix} \frac{1}{(s+1)^2}[\omega_d](t) \\ \frac{s}{(s+1)^2}[\omega_d](t) \end{pmatrix}.$$

In extensive simulations the above two adaptive dead-zone inverse controllers were able to consistently compensate for large differences in slopes m_r, m_l as well as in break-points b_r, b_l. Figure 6.5 shows a typical response for $G(s)$ known, while Figure 6.6 shows a typical response for $G(s)$ unknown. In both figures, the tracking error converges to a very small value after a short transient. The adaptive designs of Section 6.2.2 and Sections 6.3.2 and 6.3.4 with a switching σ-modification combined with parameter projection are used. Other

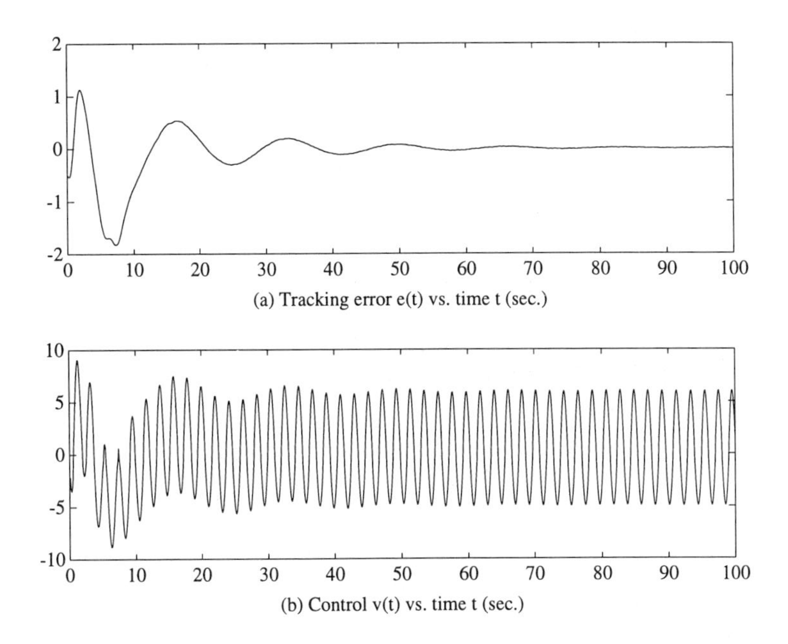

(a) Tracking error e(t) vs. time t (sec.)

(b) Control v(t) vs. time t (sec.)

Figure 6.5: Response of an adaptive dead-zone inverse: $G(s)$ known.

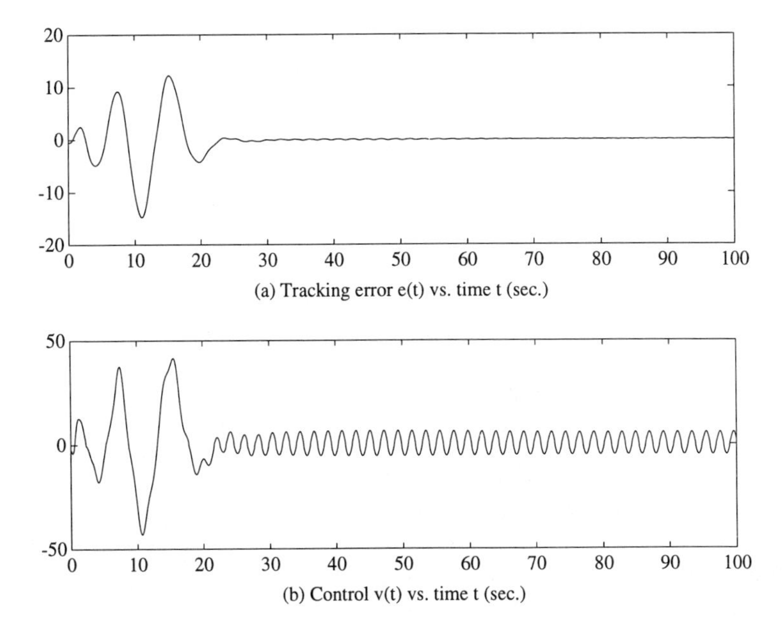

(a) Tracking error e(t) vs. time t (sec.)

(b) Control v(t) vs. time t (sec.)

Figure 6.6: Response of an adaptive dead-zone inverse: $G(s)$ unknown.

simulations also showed that for larger initial dead-zone parameter errors with which the fixed dead-zone inverse controller would result in unbounded responses, adaptive dead-zone inverse controllers achieved signals boundedness and small tracking errors. The simulation evidence is overwhelming that the adaptive dead-zone inverse significantly improves tracking performance and, at the same time, uses less control effort.

6.4.3 Example: Adaptive Backlash Inverse

The design of a continuous-time adaptive backlash inverse controller for a plant with an unknown input backlash and a known linear plant has been given in Section 6.2. It is now applied to the plant

$$y(t) = G(s)[u](t),\ u(t) = B(v(t)),\ G(s) = \frac{1}{(s-1.5)(s+1)}$$

with the unknown backlash parameters

$$m = 0.7,\ c_r = 1.3,\ c_l = -0.5.$$

The controller structure (6.15) is

$$u_d(t) = \frac{\theta_{11}^* + \theta_{12}^* s}{(s+2)^2}[u_d](t) + \frac{\theta_{21}^* + \theta_{22}^* s}{(s+2)^2}[y](t) + \theta_3^* r(t).$$

The signal at the output of the backlash inverse $v(t) = \widehat{BI}(u_d(t))$ is then applied to the plant.

With the choice of the reference model

$$y_m(t) = W_m(s)[r](t),\ W_m(s) = \frac{1}{(s+2)(s+3)}$$

we calculate the controller parameters from (6.14). They are

$$\theta_{11}^* = -\frac{129}{4},\ \theta_{12}^* = -\frac{11}{2},\ \theta_{21}^* = -\frac{627}{8},\ \theta_{22}^* = -\frac{611}{8},\ \theta_3^* = 1.$$

The knowledge of $u_d(t)$ and $\dot{u}_d(t)$ is used for implementing the adaptive backlash inverse $\widehat{BI}(\cdot)$ given by (3.15) - (3.16) and its approximation (3.17). Their parameters are adaptively updated with the adaptive law (6.21) consisting of the switching-σ and parameter projection (6.49).

The unknown backlash parameter vector is

$$\theta_N^* = \theta_b^* = (0.91, 0.7, -0.35)^T.$$

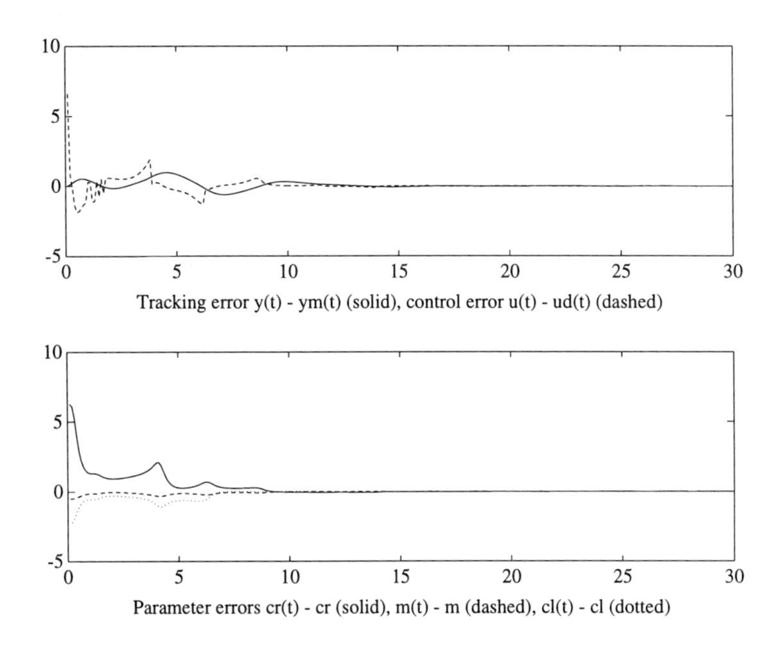

Figure 6.7: Convergence with an adaptive backlash inverse.

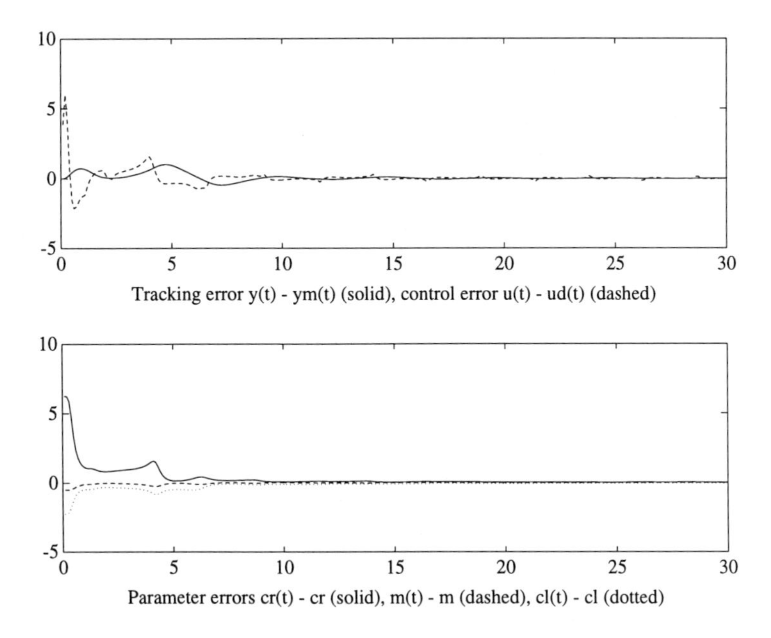

Figure 6.8: Convergence with an approximate adaptive backlash inverse.

For simulations the initial guess was

$$\theta_N(0) = \theta_b(0) = (1.51, 0.2, -0.55)^T$$

for the adaptive law (6.21) with (6.49). Figure 6.7 shows the system responses for $r(t) = 10\sin 1.3t$ with the adaptive backlash inverse (3.15) - (3.16). Figure 6.8 shows the results with the approximate continuous adaptive backlash inverse (3.17), where $\bar{c}(t)$ is generated from (3.19) with $\alpha = 6$. We see that in both cases the tracking error $e(t) = y(t) - y_m(t)$ converges to very small values, and so do the control error and the parameter error. Simulations not shown here indicate that without an adaptive backlash inverse, for example, with the fixed controller

$$v(t) = \frac{\theta_{11}^* + \theta_{12}^* s}{(s+2)^2}[v](t) + \frac{\theta_{21}^* + \theta_{22}^* s}{(s+2)^2}[y](t) + \theta_3^* r(t)$$

the tracking error remains large for large t. In contrast, adaptive inverse control schemes significantly improve the system tracking performance.

6.4.4 Example: Adaptive Hysteresis Inverse

This example compares four different designs for a plant with unknown input hysteresis. Simulation results show the effectiveness of adaptive schemes developed in Sections 6.2 and 6.3.

Consider a plant with an unknown input hysteresis $H(\cdot)$, in which the linear part $G(s)$ is unstable:

$$y(t) = G(s)[u](t), \ u(t) = H(v(t)), \ G(s) = \frac{2.5}{s^2 + 2s - 4}$$

and the unknown hysteresis parameters are

$$m_t = m_b = 1.8, \ m_r = m_l = 3$$

$$c_t = -c_b = 1.9, \ c_r = -c_l = 2.5.$$

Let the reference model be chosen as

$$y_m(t) = W_m(s)[r](t), \ W_m(s) = \frac{1}{s^2 + 3s + 2}.$$

Since the hysteresis is symmetric:

$$m_t = m_b, \ m_r = m_l, \ c_t = -c_b > 0, \ c_r = -c_l > 0$$

the reduced parametrization is

$$\theta_h(t) = (\widehat{m_t}(t), \widehat{c_t}(t), \widehat{m_r}(t), \widehat{m_r c_r}(t))^T$$

$$\omega_h(t) = (-(\widehat{\chi_t}(t) + \widehat{\chi_c}(t) + \widehat{\chi_b}(t))v(t), -(\widehat{\chi_t}(t) - \widehat{\chi_b}(t)), \\ -(\widehat{\chi_r}(t) + \widehat{\chi_l}(t))v(t), \widehat{\chi_r}(t) - \widehat{\chi_l}(t))^T.$$

We study the system performance of the following four cases.

(a) For the linear part $G(s)$ known the control scheme without hysteresis inverse is

$$v(t) = \theta_1^* \frac{1}{s+3}[v](t) + \theta_2^* \frac{1}{s+3}[y](t) + \theta_{20}^* y(t) + \theta_3^* r(t).$$

Its parameters are calculated from (6.10):

$$\theta_1^* = -1,\ \theta_2^* = -0.4,\ \theta_{20}^* = -2.8,\ \theta_3^* = 0.4.$$

(b) For $G(s)$ known the control scheme with a fixed linear controller and a fixed detuned hysteresis inverse, shown in Figure 4.5, is

$$u_d(t) = \theta_1^* \frac{1}{s+3}[u_d](t) + \theta_2^* \frac{1}{s+3}[y](t) + \theta_{20}^* y(t) + \theta_3^* r(t)$$

$$v(t) = \widehat{HI}(u_d(t)),$$

where the detuned inverse $\widehat{HI}(\cdot)$ is implemented with

$$\widehat{m_t} = \widehat{m_b} = 2.3,\ \widehat{m_r} = \widehat{m_l} = 2.5$$

$$\widehat{c_t} = -\widehat{c_b} = 0.9,\ \widehat{c_r} = -\widehat{c_l} = 0.5$$

and θ_1^*, θ_2^*, θ_{20}^*, θ_3^* are the same as in (a).

(c) For $G(s)$ known the control scheme with a fixed linear controller and an adaptive hysteresis inverse, shown in Figure 6.2, is

$$u_d(t) = \theta_1^* \frac{1}{s+3}[u_d](t) + \theta_2^* \frac{1}{s+3}[y](t) + \theta_{20}^* y(t) + \theta_3^* r(t)$$

$$v(t) = \widehat{HI}(u_d(t)),$$

where the adaptive inverse $\widehat{HI}(\cdot)$ is implemented with

$$\widehat{m_t}(t) = \widehat{m_b}(t),\ \widehat{m_r}(t) = \widehat{m_l}(t)$$

$$\widehat{c_t}(t) = -\widehat{c_b}(t),\ \widehat{c_r}(t) = -\widehat{c_l}(t),$$

which are updated by the adaptive law (6.21) with the switching-σ and parameter projection (6.49) for

$$\theta_h(t) = (\widehat{m_t}(t), \widehat{c_t}(t), \widehat{m_r}(t), \widehat{m_r c_r}(t))^T, \ \widehat{c_r} = \frac{\widehat{m_r c_r}}{\widehat{m_r}}$$

$$\widehat{m_t}(0) = 2.3, \ \widehat{m_r}(0) = 2.5, \ \widehat{c_t}(0) = 0.9, \ \widehat{c_r}(0) = 0.5.$$

(d) For $G(s)$ unknown the control scheme in Figure 6.3 with an adaptive linear controller and an adaptive hysteresis inverse is

$$u_d(t) = \theta_2(t)\frac{1}{s+3}[y](t) + \theta_{20}(t)y(t) + \theta_3(t)r(t)$$
$$+(\theta_{41}(t), \theta_{42}(t), \theta_{43}(t), \theta_{44}(t), \theta_{45}(t))\frac{1}{s+3}[\bar{\omega}_h](t)$$
$$\bar{\omega}(t) = (\omega_h^T(t), \widehat{\chi_c}(t)\widehat{c_c}(t))^T, \ \widehat{c_c}(t) = \widehat{c_w}(t) = \widehat{c_e}(t)$$
$$v(t) = \widehat{HI}(u_d(t)),$$

where the adaptive inverse $\widehat{HI}(\cdot)$, similar to that in (c), is updated by the adaptive law (6.98) - (6.99) with $L(s) = s+2$, $\alpha = 1$ and the switching-σ and parameter projection (6.93) - (6.96) for

$$\theta(t) = (\theta_h^T(t), \theta_2(t), \theta_{20}(t), \theta_3(t), \theta_4^T(t))^T$$

$$\theta(0) = (2.3, 0.9, 2.5, 1.25, -0.8, -2, 0.8, 3.45, 3.75, 1.35, 0.75, -1.5)^T.$$

Typical tracking errors $e(t) = y(t) - y_m(t)$ and control signals $v(t)$ for $r(t) = 12.7 \sin 2.3t$ are shown in Figures 6.9 - 6.12 for the four control schemes. Figure 6.9 shows that a fixed linear controller without a hysteresis inverse results in an error $e(t)$ which remains large for large t. Figure 6.10 shows that a fixed linear controller with a fixed but inaccurate hysteresis inverse can reduce the tracking error. Significant improvements of system tracking performance are achieved with the use of an adaptive hysteresis inverse. For the linear part $G(s)$ known, this is shown in Figure 6.11; and for $G(s)$ unknown, it is shown in Figure 6.12. It is remarkable that the improved tracking performance is achieved with reduced control effort.

6.5 Summary

- A continuous-time plant with an input dead-zone, backlash, or hysteresis $N(\cdot)$ parametrized by θ_N^* is described by

$$y(t) = G(s)[u](t), \ u(t) = N(v(t)).$$

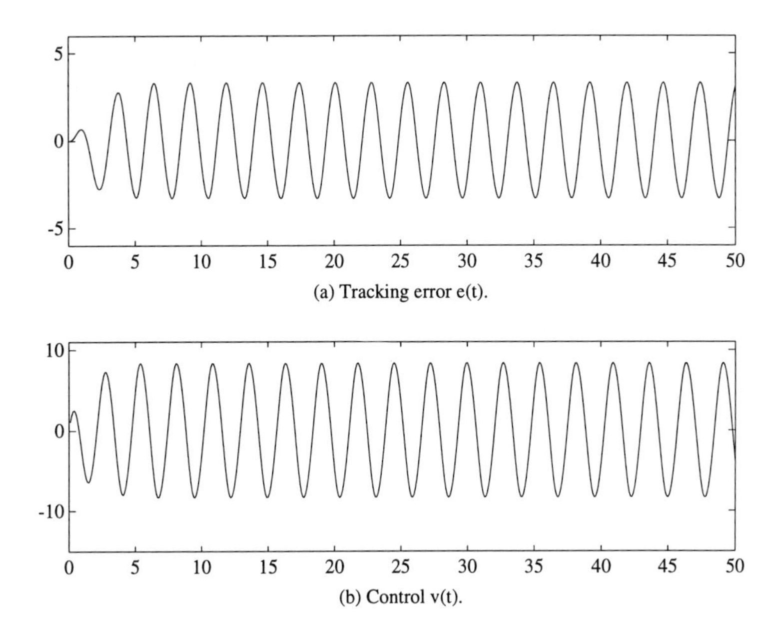

Figure 6.9: Error and control without hysteresis inverse.

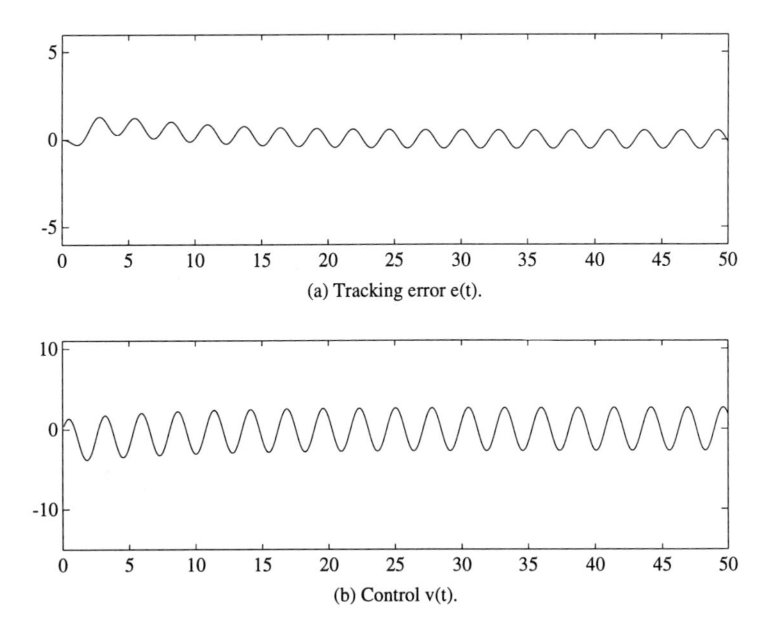

Figure 6.10: Error and control with a fixed hysteresis inverse.

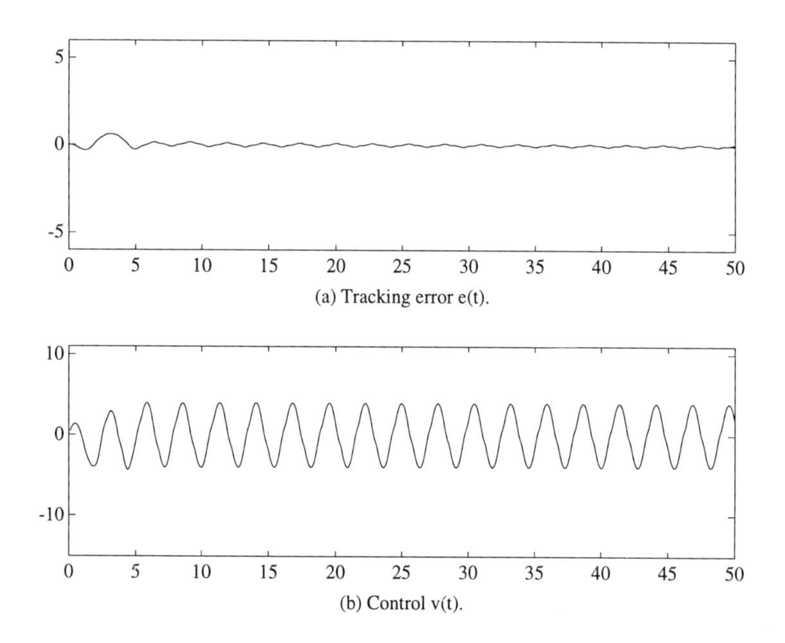

Figure 6.11: Error and control with an adaptive hysteresis inverse.

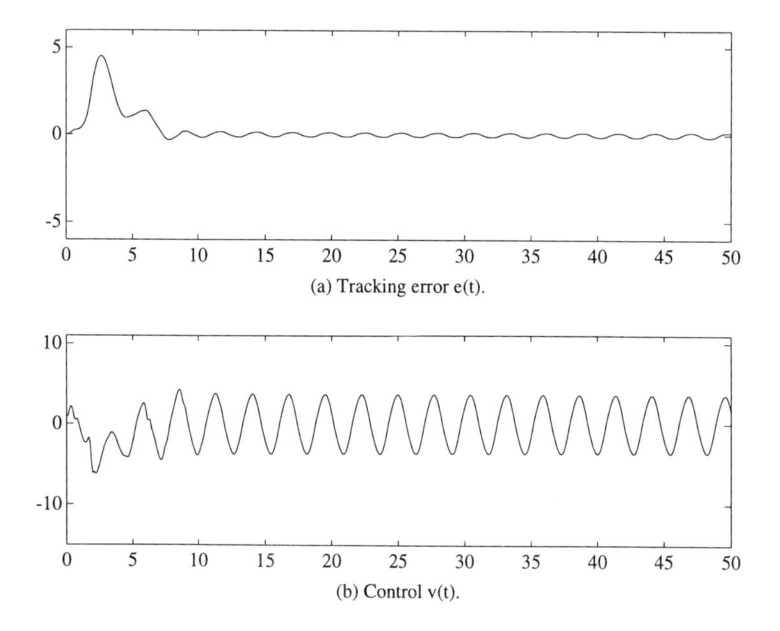

Figure 6.12: Error and control with an adaptive hysteresis inverse and adaptation for $G(s)$ unknown.

- Adaptive inverse control employs an inverse whose parameters are updated by an adaptive law to cancel the effects of the unknown $N(\cdot)$ so that a linear controller for the linear part $G(s)$ can reduce the tracking error $e(t) = y(t) - y_m(t)$ for the reference signal $y_m(t)$ generated by

$$y_m(t) = W_m(s)[r](t).$$

- When implemented with an adaptive estimate θ_N of the true parameter θ_N^*, the adaptive inverse

$$v(t) = \widehat{NI}(u_d(t))$$

results in a control error linear in parameter error:

$$u(t) - u_d(t) = (\theta_N(t) - \theta_N^*)^T \omega_N(t) + d_N(t)$$

for a known regressor $\omega_N(t)$, and an unknown but bounded $d_N(t)$. If $\theta_N = \theta_N^*$ and the inverse is properly initialized, then $d_N(t) = 0$.

- The linear controller structure for $G(s)$ known is

$$u_d(t) = \theta_1^{*T} \frac{a(s)}{\Lambda(s)}[u_d](t) + \theta_2^{*T} \frac{a(s)}{\Lambda(s)}[y](t) + \theta_{20}^* y(t) + \theta_3^* r(t).$$

- The adaptive update law for $\theta_N(t)$ is

$$\dot{\theta}_N(t) = -\frac{\Gamma_N \zeta_N(t) \epsilon_N(t)}{1 + \zeta_N^T(t)\zeta_N(t) + \xi_N^2(t)} + f_N(t),$$

where, with $W(s) = k_p W_m(s)(1 - \theta_1^{*T} \frac{a(s)}{\Lambda(s)})$ and k_p as in $G(s)$,

$$\xi_N(t) = \theta_N^T(t)\zeta_N(t) - W(s)[\theta_N^T \omega_N](t)$$

$$\zeta_N(t) = W(s)[\omega_N](t)$$

$$\epsilon_N(t) = e(t) + \xi_N(t).$$

- To achieve robustness with respect to $d(t)$ and to ensure that the components of θ_N stay in a prespecified region, two recommended modification terms $f_N(t)$ are: (1) parameter projection (6.41) - (6.42) and (2) switching σ-modification with parameter projection (6.49).

- The reparametrized controller structure for $G(s)$ unknown is

$$u_d(t) = \theta_2^T \frac{a(s)}{\Lambda(s)}[y](t) + \theta_{20}y(t) + \theta_3 r(t) + \theta_4^T \omega_4(t)$$

with a new regressor $\omega_4(t)$ and the estimate θ_4 of

$$\theta_4^* = \begin{cases} -\theta_1^* \otimes \theta_N^* & \text{for dead-zone or backlash} \\ -\theta_1^* \otimes (\theta_N^{*T}, -1)^T & \text{for hysteresis.} \end{cases}$$

- The designed gradient-type adaptive scheme is

$$\dot{\theta}(t) = -\frac{sign(k_p)\Gamma_\theta \zeta(t)\epsilon(t)}{1 + \zeta^T(t)\zeta(t) + \xi^2(t)} + f_\theta(t)$$

$$\dot{\rho}(t) = -\frac{\gamma_\rho \xi(t)\epsilon(t)}{1 + \zeta^T(t)\zeta(t) + \xi^2(t)} + f_\rho(t),$$

where $\rho(t)$ is the estimate of k_p in $G(s)$, and

$$\xi(t) = \theta^T(t)\zeta(t) - W_m(s)[\theta^T\omega](t)$$

$$\zeta(t) = W_m(s)[\omega](t)$$

$$\epsilon(t) = e(t) + \rho(t)\xi(t)$$

$$\theta(t) = (\theta_N^T(t), \theta_2^T(t), \theta_{20}(t), \theta_3(t), \theta_4^T(t))^T$$

$$\omega(t) = (\omega_N^T(t), \frac{a(s)}{\Lambda(s)}[y](t), y(t), r(t), \omega_4^T(t))^T.$$

- The designed Lyapunov-type adaptive scheme is

$$\dot{\theta}(t) = -sign(k_p)\Gamma_\theta \zeta(t)\epsilon(t) + f_\theta(t)$$

$$\dot{\rho}(t) = -\gamma_\rho \xi(t)\epsilon(t) + f_\rho(t),$$

where

$$\zeta(t) = L^{-1}(s)[\omega](t)$$

$$\xi(t) = \theta^T(t)\zeta(t) - L^{-1}(s)[\theta^T\omega](t)$$

$$\epsilon(t) = e(t) + W_m(s)L(s)[\rho\xi - \alpha\epsilon(\zeta^T\zeta + \xi^2)](t), \ \alpha > 0$$

and $W_m(s)L(s)$ is strictly positive real and strictly proper.

Chapter 7

Discrete-Time Adaptive Inverse Control

Discrete-time models with dead-zone, backlash, or hysteresis frequently appear in control applications. An example in biomedical control is the functional neuromuscular stimulation for restoring motor function by directly activating paralyzed muscles. The muscular joint systems are modeled by a dead-zone block preceding a linear block. The dead-zone is due to the difference between the agonist and antagonist muscle recruitment characteristics [31]. It is common that some parameters in such models are unknown or vary from experiment to experiment.

In this chapter, we develop a discrete-time version of our adaptive inverse control approach. Similar to the continuous-time case, we design adaptive control schemes which consist of an adaptive inverse and a linear controller structure. With true parameters, the inverse cancels the effects of the nonlinearity, while with parameter estimates it results in an error expressed in two parts. The first part admits a linear parametrization, while the second part is treated as an unknown but bounded disturbance.

We present adaptive control designs for two problems: one with only the nonlinear part unknown and the other with both parts are unknown. For the first problem, the linear part of the controller is fixed and its parameters are determined from the known parameters of the linear part of the plant. For the second problem, the linear part of the controller is modified to allow for a linear parametrization suitable for standard adaptive update laws. We use a gradient-type adaptive law with projection which in the presence of a bounded disturbance ensures that the closed-loop signals are bounded. A dead-zone example is investigated in detail to show that our adaptive inverse leads to significant improvements of tracking performance.

Discrete-time formulations of backlash and hysteresis inverses do not need the knowledge of the derivative of $u_d(t)$. Another advantage over continuous-time formulations is that a proof of the closed-loop signal boundedness can be given even for the case of unequal slopes.

7.1 Control Objective

Consider the discrete-time plant with a nonlinearity $N(\cdot)$ at the input of a linear time-invariant part $G(D)$:

$$y(t) = G(D)[u](t), \ u(t) = N(v(t)), \tag{7.1}$$

where $G(D) = k_p \frac{Z(D)}{P(D)}$ with a constant gain k_p and monic polynomials $Z(D)$, $P(D)$. Throughout this chapter the symbol D is used to denote the z-transform variable or the time advance operator $D[x](t) = x(t+1)$, as the case may be. The plant output $y(t)$ is measured, and the available control input is $v(t)$. The nonlinearity output $u(t)$ is not accessible for either measurement or control, and the nonlinear characteristic $N(\cdot)$, representing either a dead-zone or backlash or hysteresis, is unknown.

The objective is to design a feedback control $v(t)$ such that in the presence of the unknown $N(\cdot)$ the tracking of a bounded reference signal $y_m(t)$ by the plant output $y(t)$ is improved and all closed-loop signals are bounded.

With the last chapter's successes in adaptive inverse control of continuous-time plants, we now develop a discrete-time adaptive inverse approach. We will design two adaptive inverse control schemes: one for the plant with $G(D)$ known and the other for $G(D)$ unknown. The model reference control approach (see Appendix A) will be employed to design a linear controller structure, under the assumptions (A1) - (A4) in Section 4.1.

The nonlinearity $N(\cdot)$—that is, the dead-zone $DZ(\cdot)$, backlash $B(\cdot)$, or hysteresis $H(\cdot)$, shown in Figures 2.1, 2.9, or 2.16—is described in discrete time by (2.1), (2.2) or (2.11), parametrized by $\theta_N^* \in \{\theta_d^*, \theta_b^*, \theta_h^*\}$, where

$$\theta_d^* = (m_r, m_r b_r, m_l, m_l b_l)^T$$

$$\theta_b^* = (mc_r, m, mc_l)^T$$

$$\theta_h^* = (m_t, c_t, m_b, c_b, m_r, m_r c_r, m_l, m_l c_l)^T.$$

Similar to its continuous-time counterpart, our discrete-time adaptive inverse controller employs a linear controller structure and an adaptive inverse, as shown in Figure 7.1. In this controller structure, the adaptive inverse is

$$v(t) = \widehat{NI}(u_d(t)) \tag{7.2}$$

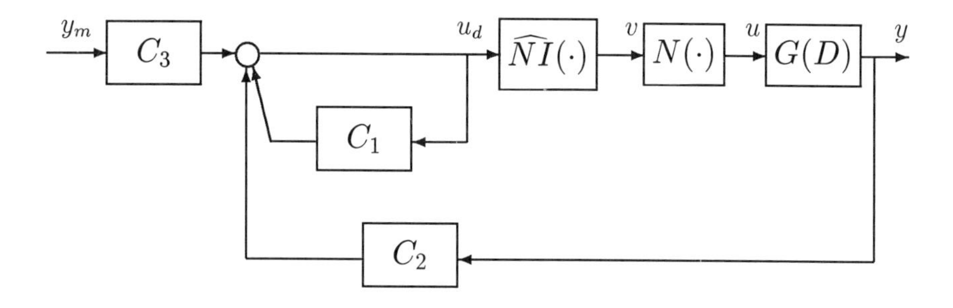

Figure 7.1: Adaptive inverse control.

as described by (3.2), (3.20), or (3.45) and shown in Figures 3.2, 3.3, or 3.5. As in the continuous-time inverse (6.5), the inverse (7.2) is parametrized by the adaptive estimate $\theta_d(t)$, $\theta_d(t)$, or $\theta_h(t)$ of θ_d^*, θ_b^*, or θ_h^*:

$$\theta_d = (\widehat{m_r}, \widehat{m_r b_r}, \widehat{m_l}, \widehat{m_l b_l})^T$$

$$\theta_b = (\widehat{mc_r}, \widehat{m}, \widehat{mc_l})^T$$

$$\theta_h = (\widehat{m_t}, \widehat{c_t}, \widehat{m_b}, \widehat{c_b}, \widehat{m_r}, \widehat{m_r c_r}, \widehat{m_l}, \widehat{m_l c_l})^T.$$

The design of $\widehat{NI}(\cdot)$ is based on the assumptions (A7a) - (A7c) in Section 6.1, and the parameter estimates are forced to meet these assumptions.

From the results of Chapter 3, the adaptive inverse (7.2) results in the control error

$$u(t) - u_d(t) = (\theta_N(t) - \theta_N^*)^T \omega_N(t) + d_N(t), \tag{7.3}$$

where $\theta_N^* \in \{\theta_d^*, \theta_b^*, \theta_h^*\}$, $\theta_N(t) \in \{\theta_b(t), \theta_b(t), \theta_h(t)\}$, and the measured signals $\omega_N(t) \in \{\omega_d(t), \omega_b(t), \omega_h(t)\}$ are defined in (3.6), (3.36), and (3.71) for the dead-zone, backlash, and hysteresis, respectively. The unparametrized part of the control error $d_N(t)$ is bounded. Moreover, $d_N(t) = 0$ for $t \geq t_0$ if $\theta_N = \theta_N^*$, and for backlash and hysteresis the inverse is initialized at t_0:

$$u_d(t_0) = B(BI(u_d(t_0))) \text{ and } u_d(t_0) = H(HI(u_d(t_0))).$$

The parameters of the linear controller consisting of the blocks C_1, C_2, and C_3 are either calculated using the knowledge of $G(D)$, if $G(D)$ is known, or updated by an adaptive law, if $G(D)$ is unknown. A modified structure for C_1 is needed in the case when $G(D)$ is unknown, to ensure a linear parametrization of the closed-loop system.

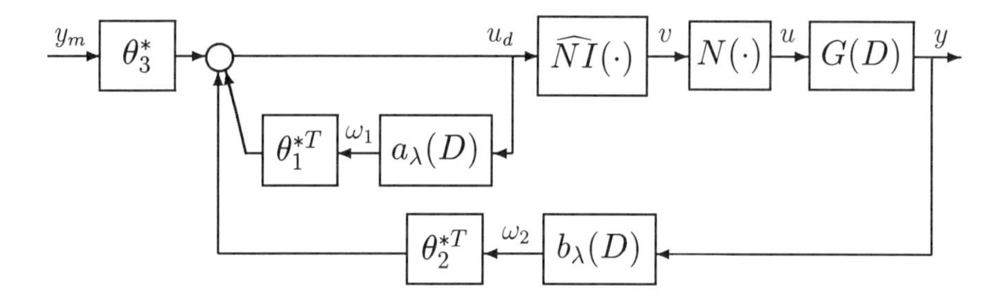

Figure 7.2: Controller structure for $G(D)$ known.

7.2 Design for $G(D)$ Known

When $G(D)$ is known, the adaptive inverse controller, shown in Figure 7.2, consists of a fixed linear controller structure and an adaptive inverse $\widehat{NI}(\cdot)$, both implemented in discrete time. The parameters of $\widehat{NI}(\cdot)$ are updated by an adaptive law.

7.2.1 Controller Structure

The linear controller structure generating $u_d(t)$ is

$$u_d(t) = \theta_1^{*T}\omega_1(t) + \theta_2^{*T}\omega_2(t) + \theta_3^* y_m(t+n^*), \tag{7.4}$$

where

$$\omega_1(t) = a_\lambda(D)[u_d](t),\ \omega_2(t) = b_\lambda(D)[y](t)$$

$$a_\lambda(D) = (D^{-n+1}, \ldots, D^{-1})^T,\ b_\lambda(D) = (D^{-n+1}, \ldots, D^{-1}, 1)^T$$

and the parameters $\theta_1^* \in R^{n-1}$, $\theta_2^* \in R^n$, $\theta_3^* \in R$ satisfy the matching equation

$$\theta_1^{*T} a_\lambda(D)P(D) + \theta_2^{*T} b_\lambda(D)k_p Z(D) = P(D) - \theta_3^* k_p Z(D)D^{n^*}. \tag{7.5}$$

Thanks to the design equation (7.5) the control $u_d(t)$ achieves the desired stability and tracking performance when applied to the plant with the nonlinearity $N(\cdot)$ cancelled by its exact inverse $NI(\cdot) = \widehat{NI}(\cdot)|_{\theta_N=\theta_N^*}$, in which case $u(t) = u_d(t)$. This is ensured because $u(t) = u_d(t)$ and (7.5) result in the the closed-loop matching $y(t+n^*) = y_m(t+n^*)$ with the closed-loop poles placed at the zeros of $Z(D)D^{n^*}$ which are stable by the assumption (A1).

Since the nonlinearity $N(\cdot)$ is unknown, an adaptive inverse $\widehat{NI}(\cdot)$ is used to generate the control signal $v(t)$: $v(t) = \widehat{NI}(u_d(t))$. To derive an adaptive update law for the inverse parameter θ_N, we operate both sides of (7.5) on $y(t)$ and use (7.1) to obtain

$$u(t) = \theta_1^{*T} a_\lambda(D)[u](t) + \theta_2^{*T} b_\lambda(D)[y](t) + \theta_3^* y(t+n^*). \tag{7.6}$$

From (7.3), (7.4), and (7.6), we obtain the tracking error equation

$$e(t) = y(t) - y_m(t) = W(D)[\tilde{\theta}_N^T \omega_N](t) + d(t), \tag{7.7}$$

where $W(D)$ is stable and strictly proper:

$$W(D) = k_p(1 - \theta_1^{*T} a_\lambda(D))D^{-n^*}.$$

This also ensures that $d(t)$ is bounded because $d_N(t)$ is bounded:

$$d(t) = W(D)[d_N](t).$$

7.2.2 Adaptive Scheme

The error equation (7.7) suggests the following adaptive update law for $\theta_N(t)$:

$$\theta_N(t+1) = \theta_N(t) - \frac{\Gamma_N \zeta_N(t) \epsilon_N(t)}{1 + \zeta_N^T(t)\zeta_N(t) + \xi_N^2(t)} + f_N(t), \tag{7.8}$$

where

$$\epsilon_N(t) = e(t) + \xi_N(t) \tag{7.9}$$

$$\xi_N(t) = \theta_N^T(t)\zeta_N(t) - W(D)[\theta_N^T \omega_N](t) \tag{7.10}$$

$$\zeta_N(t) = W(D)[\omega_N](t). \tag{7.11}$$

To ensure that the components of $\theta_N(t)$ stay in a prespecified parameter region and that the tracking error $e(t)$ reduces to zero if $d(t)$ vanishes, the term $f_N(t)$ can be designed using either parameter projection or switching σ-modification with parameter projection. The step size matrix Γ_N is to be chosen according to the choice of $f_N(t)$.

Parameter projection. For parameter projection, we first use the assumptions (A7a) - (A7c) in Section 6.1 to define a convex parameter region which contains the true parameter θ_N^*:

$$\theta_N^* = (\theta_{N1}^*, \ldots, \theta_{Nn_N}^*)^T, \ \theta_{Nj}^* \in [\theta_{Nj}^a, \theta_{Nj}^b], \ j = 1, 2, \ldots, n_N$$

for some known constants θ_{Nj}^a, θ_{Nj}^b. We choose, for *dead-zone*:

$$\theta_{N1}^a = m_{r1}, \ \theta_{N2}^a = 0, \ \theta_{N3}^a = m_{l1}, \ \theta_{N4}^a = -m_{l2} b_{l0} \tag{7.12}$$

$$\theta_{N1}^b = m_{r2},\ \theta_{N2}^b = m_{r2}b_{r0},\ \theta_{N3}^b = m_{l2},\ \theta_{N4}^b = 0 \tag{7.13}$$

for *backlash*:

$$\theta_{N1}^a = 0,\ \theta_{N2}^a = m_1,\ \theta_{N3}^a = -m_2 c_{l0} \tag{7.14}$$

$$\theta_{N1}^b = m_2 c_{r0},\ \theta_{N2}^b = m_2,\ \theta_{N3}^b = 0 \tag{7.15}$$

for *hysteresis*:

$$\theta_{N1}^a = m_{t1},\ \theta_{N2}^a = 0,\ \theta_{N3}^a = m_{b1},\ \theta_{N4}^a = -c_{b0} \tag{7.16}$$

$$\theta_{N5}^a = m_{r1},\ \theta_{N6}^a = 0,\ \theta_{N7}^a = m_{l1},\ \theta_{N8}^a = -c_{l0}m_{l2} \tag{7.17}$$

$$\theta_{N1}^b = m_{t2},\ \theta_{N2}^b = c_{t0},\ \theta_{N3}^b = m_{b2},\ \theta_{N4}^b = 0 \tag{7.18}$$

$$\theta_{N5}^b = m_{r2},\ \theta_{N6}^b = c_{r0}m_{r2},\ \theta_{N7}^b = m_{l2},\ \theta_{N8}^b = 0. \tag{7.19}$$

With the knowledge of the parameter region $[\theta_{Nj}^a, \theta_{Nj}^b]$, the projection modification term $f_N(t)$ in (7.8) is as follows.

Choosing the step size matrix Γ_N as

$$\Gamma_N = diag\{\gamma_1, \ldots, \gamma_{n_N}\},\ 0 < \gamma_j < 2,\ j = 1, 2, \ldots, n_N$$

and defining

$$g_N(t) = -\frac{\Gamma_N \zeta_N(t) \epsilon_N(t)}{1 + \zeta_N^T(t)\zeta_N(t) + \xi_N^2(t)}$$

we denote the jth component of $\theta_N(t)$, $f_N(t)$, $g_N(t)$ as $\theta_{Nj}(t)$, $f_{Nj}(t)$, $g_{Nj}(t)$, $j = 1, 2, \ldots, n_N$. Then we let

$$\bar{\theta}_{Nj}(t) = \theta_{Nj}(t) + g_{Nj}(t)$$

initialize the algorithm with

$$\theta_{Nj}(0) \in [\theta_{Nj}^a, \theta_{Nj}^b],\ j = 1, 2, \ldots, n_N$$

and set

$$f_{Nj}(t) = \begin{cases} 0 & \text{if } \bar{\theta}_{Nj}(t) \in [\theta_{Nj}^a, \theta_{Nj}^b] \\ \theta_{Nj}^b - \bar{\theta}_{Nj}(t) & \text{if } \bar{\theta}_{Nj}(t) > \theta_{Nj}^b \\ \theta_{Nj}^a - \bar{\theta}_{Nj}(t) & \text{if } \bar{\theta}_{Nj}(t) < \theta_{Nj}^a. \end{cases} \tag{7.20}$$

This design for $f_N(t)$ ensures that the parameter estimates initiated in the known intervals $[\theta_{Nj}^a, \theta_{Nj}^b]$ remain in these intervals; that is, the components of θ_N satisfy the assumptions (A7a) - (A7c). The following results show that this adaptive design has desired boundedness properties.

Lemma 7.1 *The adaptive law (7.8) with parameter projection (7.20) ensures that* $\theta_N(t)$ *and* $\frac{\epsilon_N(t)}{m_N(t)}$ *are bounded and that there exist constants* $a_i > 0$, $b_i > 0$, $i = 1, 2$, *such that*

$$\sum_{t=t_1}^{t_2} \frac{\epsilon_N^2(t)}{m_N^2(t)} \leq a_1 + b_1 \sum_{t=t_1}^{t_2} \frac{d^2(t)}{m_N^2(t)} \tag{7.21}$$

$$\sum_{t=t_1}^{t_2} \|\theta_N(t+1) - \theta_N(t)\|_2^2 \leq a_2 + b_2 \sum_{t=t_1}^{t_2} \frac{d^2(t)}{m_N^2(t)} \tag{7.22}$$

for all $t_2 > t_1 \geq 0$, *where* $m_N(t) = \sqrt{1 + \zeta_N^T(t)\zeta_N(t) + \xi_N^2(t)}$.

Proof: Substituting (7.7), (7.10), and (7.11) in (7.9), we have

$$\epsilon_N(t) = \tilde{\theta}_N^T(t)\zeta_N(t) + d(t), \ \tilde{\theta}_N(t) = \theta_N(t) - \theta_N^*. \tag{7.23}$$

Letting $\gamma_m = \max_{1 \leq i \leq n_N} \gamma_i < 2$ and using (7.23), we obtain the time increment of the positive definite function

$$V(\tilde{\theta}_N) = \tilde{\theta}_N^T \Gamma_N^{-1} \tilde{\theta}_N$$

along the trajectories of (7.8) as

$$\begin{aligned} V(\tilde{\theta}_N(t+1)) - V(\tilde{\theta}_N(t)) & \\ = -(2 - \frac{\zeta_N^T(t)\Gamma_N\zeta_N(t)}{m_N^2(t)})\frac{\epsilon_N^2(t)}{m_N^2(t)} + \frac{2\epsilon_N(t)d(t)}{m_N^2(t)} & \\ +2f_N^T(t)\Gamma_N^{-1}(\tilde{\theta}_N(t) + g_N(t) + f_N(t)) - f_N^T(t)\Gamma_N^{-1}f_N(t) & \\ \leq -\frac{2-\gamma_m}{2}\frac{\epsilon_N^2(t)}{m_N^2(t)} + \frac{2}{2-\gamma_m}\frac{d^2(t)}{m^2(t)} - f_N^T(t)\Gamma_N^{-1}f_N(t) & \\ +2f_N^T(t)\Gamma_N^{-1}(\tilde{\theta}(t) + g_N(t) + f_N(t)). & \end{aligned} \tag{7.24}$$

The projection algorithm (7.20) ensures that

$$f_{Nj}(t)(\theta_j(t) - \theta_j^* + g_{Nj}(t) + f_{Nj}(t)) \leq 0$$

for all $j = 1, \ldots, n_N$. Since Γ_N is a diagonal matrix, we have

$$\begin{aligned} & 2f_N^T(t)\Gamma_N^{-1}(\tilde{\theta}(t) + g_N(t) + f_N(t)) \\ & = 2\sum_{j=1}^{n_N} \gamma_j^{-1} f_{Nj}(t)(\theta_j(t) - \theta_j^* + g_{Nj}(t) + f_{Nj}(t)) \leq 0. \end{aligned}$$

Hence, we can rewrite (7.24) as

$$\begin{aligned} V(\tilde{\theta}_N(t+1)) - V(\tilde{\theta}_N(t)) \leq & -\frac{2-\gamma_m}{2}\frac{\epsilon_N^2(t)}{m_N^2(t)} + \frac{2}{2-\gamma_m}\frac{d^2(t)}{m^2(t)} \\ & - f_N^T(t)\Gamma_N^{-1}f_N(t). \end{aligned} \tag{7.25}$$

Since the projection (7.32) ensures that $\theta_{Nj}(t) \in [\theta_{Nj}^a, \theta_{Nj}^b]$, $j = 1, \ldots, n_N$, we have achieved the boundedness of $\theta_N(t)$. Then, from (7.23), it follows that $\frac{\epsilon_N(t)}{m_N(t)}$ is bounded. Using (7.25) and the boundedness of $d(t)$, we obtain (7.21). Finally, using (7.8) and (7.22), we obtain

$$|\theta_{Nj}(t+1) - \theta_{Nj}(t)| \leq |g_{Nj}(t)|,\ 1 \leq j \leq n_N,$$

which, together with (7.21), proves (7.22). ∇

Using Lemma 7.1, we can prove the following boundedness result.

Theorem 7.1 *The adaptive inverse controller consisting of the linear controller structure (7.4) and the adaptive inverse (7.2) updated by (7.8) and (7.32) ensures that all closed-loop signals are bounded.*

The proof is given in Appendix C.

Switching σ-modification with parameter projection. An alternative to the above projection design is to use the switching σ-modification with parameter projection:

$$f_N(t) = -\sigma_N(t)\theta_N(t) + f_p(t),$$

where

$$\sigma_N(t) = \begin{cases} \sigma_0 & \text{if } \|\theta_N(t)\|_2 > 2M_N \\ 0 & \text{otherwise} \end{cases}$$

$$0 < \sigma_0 < \frac{1}{2}(1 - \gamma_N),\ M_N \geq \|\theta_N^*\|_2$$

and γ_N is the step size of parameter adaptation:

$$\Gamma_N = \gamma_N I_{n_N},\ 0 < \gamma_N < 1.$$

The term $f_p(t)$ is for parameter projection, whose main task is to ensure the estimate $\theta_N(t)$ to meet the conditions of (A7a) - (A7c). The design steps are similar to those in Section 6.2 for the continuous-time case. For a *dead-zone inverse*, we use (A7a) to define

$$\theta_{N1}^a = m_{r1},\ \theta_{N2}^a = 0,\ \theta_{N3}^a = m_{l1},\ \theta_{N4}^b = 0. \tag{7.26}$$

For a *backlash inverse*, we use (A7b) to define

$$\theta_{N1}^a = 0,\ \theta_{N2}^a = m_1,\ \theta_{N3}^b = 0. \tag{7.27}$$

For a *hysteresis inverse*, we use (A7c) to define

$$\theta_{N1}^a = m_{t1},\ \theta_{N2}^a = 0,\ \theta_{N3}^a = m_{b1},\ \theta_{N4}^a = -c_{b0} \tag{7.28}$$

$$\theta_{N5}^a = m_{r1},\ \theta_{N6}^a = 0,\ \theta_{N7}^a = m_{l1},\ \theta_{N8}^a = -c_{l0}m_{l2} \tag{7.29}$$

$$\theta_{N1}^b = m_{t2},\ \theta_{N2}^b = c_{t0},\ \theta_{N3}^b = m_{b2} \tag{7.30}$$

$$\theta_{N4}^b = 0,\ \theta_{N6}^b = c_{r0}m_{r2},\ \theta_{N8}^b = 0. \tag{7.31}$$

In view of the fact that the switching σ-modification guarantees that $\theta_N(t) \in l^\infty$, for all other $\theta_{Nj}^a, \theta_{Nj}^b$, we let

$$\theta_{Nj}^a = -\infty,\ \theta_{Nj}^b = \infty. \tag{7.32}$$

To design a projection algorithm for $f_p(t)$, we let

$$\bar{g}_N(t) = -\frac{\Gamma_N \zeta_N(t)\epsilon_N(t)}{m_N^2(t)} - \sigma_N(t)\theta_N(t),$$

where $\Gamma_N = \gamma_N I_{n_N}$, $0 < \gamma_N < 1$, denote

$$\bar{\theta}_{Nj}(t) = \theta_{Nj}(t) + \bar{g}_{Nj}(t),$$

and set

$$f_{pj}(t) = \begin{cases} 0 & \text{if } \bar{\theta}_{Nj}(t) \in [\theta_{Nj}^a, \theta_{Nj}^b] \\ \theta_{Nj}^b - \bar{\theta}_{Nj}(t) & \text{if } \bar{\theta}_{Nj}(t) > \theta_{Nj}^b \\ \theta_{Nj}^a - \bar{\theta}_{Nj}(t) & \text{if } \bar{\theta}_{Nj}(t) < \theta_{Nj}^a \end{cases} \tag{7.33}$$

for $j = 1, 2, \ldots, n_N$. The algorithm is initialized with $\theta_{Nj}(0) \in [\theta_{Nj}^a, \theta_{Nj}^b]$.

With this adaptive design, the time increment of the positive definite function $V(\tilde{\theta}_N) = \tilde{\theta}_N^T \Gamma_N^{-1} \tilde{\theta}_N$ satisfies

$$\begin{aligned} & V(\tilde{\theta}_N(t+1)) - V(\tilde{\theta}_N(t)) \\ & \quad \le -\frac{\gamma_N}{2}\frac{\epsilon_N^2(t)}{m_N^2(t)} + \frac{2}{2-\gamma_N}\frac{d^2(t)}{m_N^2(t)} - (1-\gamma_N)\frac{\epsilon_N^2(t)}{m_N^2(t)} \\ & \quad\quad -2(\sigma_N(t)\theta_N(t) - f_p(t))^T(\gamma_N^{-1}\tilde{\theta}_N(t) - \frac{\epsilon_N(t)\zeta_N(t)}{m_N^2(t)}) \\ & \quad\quad +(\sigma_N(t)\theta_N(t) - f_p(t))^T\gamma_N^{-1}(\sigma_N(t)\theta_N(t) - f_p(t)) \\ & \quad = -\frac{\gamma_N}{2}\frac{\epsilon_N^2(t)}{m_N^2(t)} + \frac{2}{2-\gamma_N}\frac{d^2(t)}{m_N^2(t)} - (1-\gamma_N)\frac{\epsilon_N^2(t)}{m_N^2(t)} \\ & \quad\quad -2\sigma_N(t)\theta_N^T(t)\gamma_N^{-1}\tilde{\theta}_N(t) + 2\sigma_N(t)\theta_N^T(t)\frac{\epsilon_N(t)\zeta_N(t)}{m_N^2(t)} \\ & \quad\quad +\sigma_N^2(t)\theta_N^T(t)\gamma_N^{-1}\theta_N(t) - f_p^T(t)\gamma_N^{-1}f_p(t) \\ & \quad\quad +2\gamma_N^{-1}f_p^T(t)(\tilde{\theta}_N(t) + \bar{g}_N(t) + f_p(t)). \end{aligned} \tag{7.34}$$

The projection (7.33) ensures that

$$f_p^T(t)(\tilde{\theta}_N(t) + \bar{g}_N(t) + f_p(t))$$
$$= \sum_{j=1}^{n_N} f_{pj}(t)(\theta_{Nj}(t) - \theta_{Nj}^* + \bar{g}_{Nj}(t) + f_{pj}(t)) \leq 0$$

while the switching σ-modification leads to

$$-(1-\gamma_N)\frac{\epsilon_N^2(t)}{m_N^2(t)} - 2\sigma_N(t)\theta_N^T(t)\gamma_N^{-1}\tilde{\theta}_N(t) + 2\sigma_N(t)\theta_N^T(t)\frac{\epsilon_N(t)\zeta_N(t)}{m_N^2(t)}$$
$$+\sigma_N^2(t)\theta_N^T(t)\gamma_N^{-1}\theta_N(t) \leq -\frac{\sigma_N(t)}{2}\theta_N^T(t)\theta_N(t).$$

Hence, we can rewrite (7.34) as

$$V(\tilde{\theta}_N(t+1)) - V(\tilde{\theta}_N(t)) \leq -\frac{\gamma_N}{2}\frac{\epsilon_N^2(t)}{m_N^2(t)} + \frac{2}{2-\gamma_N}\frac{d^2(t)}{m_N^2(t)}$$
$$-\frac{\sigma_N(t)}{2}\theta_N^T(t)\theta_N(t) - f_p^T(t)\gamma_N^{-1}f_p(t),$$

from which we can show that the desired properties in Lemma 7.1 also hold for the switching σ-modification with parameter projection.

7.3 Design for $G(D)$ Unknown

When both the nonlinear part $N(\cdot)$ and the linear part $G(D)$ are unknown, the controller structure (7.4) would lead to a bilinear parametrization not suitable for parameter updating. We need to develop a new controller structure with linear parametrization which will allow us to adaptively update its parameters as well as the parameters of the adaptive inverse $\widehat{NI}(\cdot)$.

7.3.1 Controller Structure

The term preventing a linear parametrization is $\theta_1^T(t)\omega_1(t)$, where $\theta_1(t)$ is an estimate of θ_1^*. This term is related to each of the three inverses $\widehat{NI}(\cdot)$ derived in Chapter 3 as

$$u_d(t) = \begin{cases} -\theta_d^T\omega_d(t) & \text{for } \widehat{DI}(\cdot) \\ -\theta_b^T\omega_b(t) & \text{for } \widehat{BI}(\cdot) \\ -\theta_h^T\omega_h(t) + \widehat{\chi_w}(t)\widehat{c_w}(t) + \widehat{\chi_e}(t)\widehat{c_e}(t) & \text{for } \widehat{HI}(\cdot). \end{cases} \tag{7.35}$$

With the true values $\theta_d = \theta_d^*$, $\theta_b = \theta_b^*$, or $\theta_h = \theta_h^*$, we reparametrize the term $\theta_1^{*T}\omega_1(t)$ using an enlarged parameter θ_4^*:

$$\theta_1^{*T}\omega_1(t) = \theta_4^{*T}\omega_4(t)$$

$$\theta_4^* = \begin{cases} -\theta_1^* \otimes \theta_d^* & \text{for } \widehat{DI}(\cdot) \\ -\theta_1^* \otimes \theta_b^* & \text{for } \widehat{BI}(\cdot) \\ -\theta_1^* \otimes (\theta_h^{*T}, -1)^T & \text{for } \widehat{HI}(\cdot) \end{cases}$$

$$\omega_4(t) = \begin{cases} A_\lambda(D)[\omega_d](t) & \text{for } \widehat{DI}(\cdot) \\ A_\lambda(D)[\omega_h](t) & \text{for } \widehat{BI}(\cdot) \\ A_\lambda(D)[(\omega_h^T, a_h)^T](t) & \text{for } \widehat{HI}(\cdot), \end{cases}$$

where $a_h(t) = \widehat{\chi_w}(t)\widehat{c_w}(t) + \widehat{\chi_e}(t)\widehat{c_e}(t)$ and

$$A_\lambda(D) = \begin{cases} (D^{-n+1}I_4, \ldots, D^{-1}I_4)^T & \text{for } \widehat{DI}(\cdot) \\ (D^{-n+1}I_3, \ldots, D^{-1}I_3)^T & \text{for } \widehat{BI}(\cdot) \\ (D^{-n+1}I_9, \ldots, D^{-1}I_9)^T & \text{for } \widehat{HI}(\cdot) \end{cases}$$

with I_j the $j \times j$ identity matrix, $j = 3, 4, 9$.

With this new parametrization, we introduce the controller structure

$$u_d(t) = \theta_2^T\omega_2(t) + \theta_3 y_m(t + n^*) + \theta_4^T\omega_4(t), \tag{7.36}$$

where θ_2, θ_3, θ_4 are the estimates of θ_2^*, θ_3^*, θ_4^*, and $\omega_2(t) = b_\lambda(D)[y](t)$, $b_\lambda(D) = (D^{-n+1}, \ldots, D^{-1}, 1)^T$ as defined in (7.5).

The signal $u_d(t)$ is then applied to the adaptive inverse

$$v(t) = \widehat{NI}(u_d(t)) \tag{7.37}$$

to generate the plant input $v(t)$.

The adaptive inverse controller consists of (7.36) and (7.37), as shown in Figure 7.3 where the logic block L produces $\omega_d(t)$, $\omega_b(t)$, or $(\omega_h^T(t), a_h(t))^T$ for the dead-zone, backlash, or hysteresis, respectively. Our next task is to develop an adaptive law to update the parameters of (7.36) and (7.37).

Using (7.3) for $u(t)$ on the left side of (7.6) and using (7.3) with $u_d(t)$ from (7.35) for $u(t)$ on the right side of (7.6), we obtain

$$u_d(t) + (\theta_N(t) - \theta_N^*)^T\omega_N(t) + d_N(t) = \theta_2^{*T}\omega_2(t) + \theta_3^* r(t)$$
$$+ \theta_3^* D^{n^*}[y - y_m](t) + \theta_4^{*T}\omega_4(t) + \theta_1^{*T}a_\lambda(D)[d_N](t). \tag{7.38}$$

Defining $e(t) = y(t) - y_m(t)$ and $W_m(D) = D^{-n^*}$ and substituting (7.36) in (7.38), we derive the tracking error equation

$$e(t) = k_p W_m(D)[(\theta - \theta^*)^T\omega](t) + d(t), \tag{7.39}$$

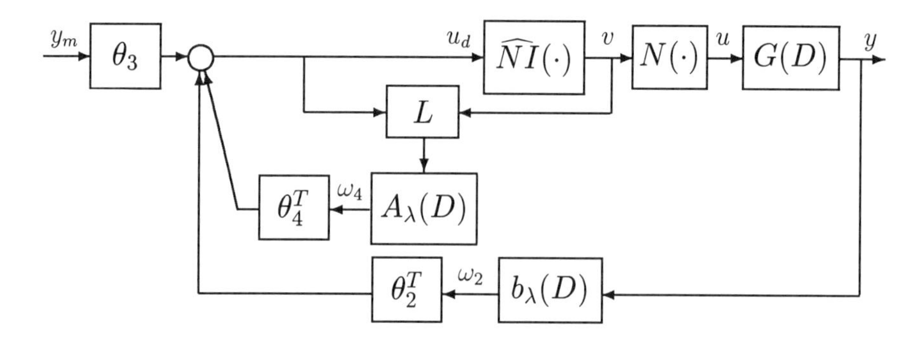

Figure 7.3: Controller structure for $G(D)$ unknown.

where

$$\omega(t) = (\omega_N^T(t), \omega_2^T(t), y_m(t+n^*), \omega_4^T(t))^T$$
$$\theta(t) = (\theta_N^T(t), \theta_2^T(t), \theta_3(t), \theta_4^T(t))^T$$
$$\theta^* = (\theta_N^{*T}, \theta_2^{*T}, \theta_3^*, \theta_4^{*T})^T$$

and $d(t) = k_p(1 - \theta_1^{*T} a_\lambda(D))D^{-n^*}[d_N](t)$ as defined in (7.7).

7.3.2 Adaptive Scheme

The equation (7.39) suggests the parameter update law

$$\theta(t+1) = \theta(t) - \frac{sign(k_p)\Gamma_\theta \epsilon(t)\omega(t-n^*)}{m^2(t)} + f_\theta(t) \tag{7.40}$$

$$\rho(t+1) = \rho(t) - \frac{\gamma_\rho \epsilon(t)\xi(t)}{m^2(t)} + f_\rho(t), \tag{7.41}$$

where $\rho(t)$ is the estimate of $\rho^* = k_p$ and

$$\epsilon(t) = e(t) + \rho(t)\xi(t)$$

$$\xi(t) = \theta^T(t)\omega(t-n^*) - \theta^T(t-n^*)\omega(t-n^*)$$
$$m(t) = \sqrt{1 + \omega^T(t-n^*)\omega(t-n^*) + \xi^2(t)}.$$

The terms $f_\theta(t)$ and $f_\rho(t)$ are to be designed for robustness with respect to $d(t)$ and for parameter projection needed for implementing the adaptive inverse $\widehat{NI}(\cdot)$. The step sizes Γ_θ, γ_ρ are to be chosen together with $f_\theta(t)$, $f_\rho(t)$.

Parameter projection. For a projection design of $f_\theta(t)$ and $f_\rho(t)$, we assume:

(**A8**) The jth component of $\theta^* \in R^{n_\theta}$ belongs to a known interval: $\theta_j^* \in [\theta_j^a, \theta_j^b]$, $j = 1, \ldots, n_\theta$, and so does ρ^*: $\rho^* \in [\rho^a, \rho^b]$.

For the nonlinearity parameter θ_N, the bounds θ_{Nj}^a and θ_{Nj}^b are characterized by (7.12) - (7.19), based on the assumptions (A7a) - (A7c).

We then define

$$g_\theta(t) = -\frac{sign(k_p)\Gamma_\theta \omega(t-n^*)\epsilon(t)}{m^2(t)} \tag{7.42}$$

$$g_\rho(t) = -\frac{\gamma_\rho \xi(t)\epsilon(t)}{m^2(t)}, \tag{7.43}$$

where $0 < \gamma_\rho < 2$ and

$$\Gamma_\theta = diag\{\gamma_1, \ldots, \gamma_{n_\theta}\},\ 0 < \gamma_j < \frac{2}{k_p^0},\ j = 1, 2, \ldots, n_\theta,$$

with k_p^0 being a known upper bound on $|k_p|$: $|k_p| \leq k_p^0$. Denoting the jth component of $\theta(t)$, $f_\theta(t)$, $g_\theta(t)$ as $\theta_j(t)$, $f_{\theta j}(t)$, $g_{\theta j}(t)$, $j = 1, 2, \ldots, n_\theta$, we choose $\theta_j(0) \in [\theta_j^a, \theta_j^b]$, $\rho(0) \in [\rho^a, \rho^b]$ and set

$$f_{\theta j}(t) = \begin{cases} 0 & \text{if } \theta_j(t) + g_{\theta j}(t) \in [\theta_j^a, \theta_j^b] \\ \theta_j^b - \theta_j(t) - g_{\theta j}(t) & \text{if } \theta_j(t) + g_{\theta j}(t) > \theta_j^b \\ \theta_j^a - \theta_j(t) - g_{\theta j}(t) & \text{if } \theta_j(t) + g_{\theta j}(t) < \theta_j^a \end{cases}$$

$$f_\rho(t) = \begin{cases} 0 & \text{if } \rho(t) + g_\rho(t) \in [\rho^a, \rho^b] \\ \rho^b - \rho(t) - g_\rho(t) & \text{if } \rho(t) + g_\rho(t) > \rho^b \\ \rho^a - \rho(t) - g_\rho(t) & \text{if } \rho(t) + g_\rho(t) < \rho^a. \end{cases}$$

This adaptive control design has the following properties.

Lemma 7.2 *The adaptive law (7.40) - (7.41) guarantees that* $\theta(t)$, $\rho(t)$, $\frac{\epsilon(t)}{m(t)} \in l^\infty$, *and* $\frac{\epsilon(t)}{m(t)}$, $\theta(t+1) - \theta(t)$, $\rho(t+1) - \rho(t)$ *satisfy*

$$\sum_{t=t_1}^{t_2} \frac{\epsilon^2(t)}{m^2(t)} \leq a_1 + b_1 \sum_{t=t_1}^{t_2} \frac{d^2(t)}{m^2(t)} \tag{7.44}$$

$$\sum_{t=t_1}^{t_2} \|\theta(t+1) - \theta(t)\|_2^2 \leq a_2 + b_2 \sum_{t=t_1}^{t_2} \frac{d^2(t)}{m^2(t)} \tag{7.45}$$

$$\sum_{t=t_1}^{t_2} (\rho(t+1) - \rho(t))^2 \leq a_3 + b_3 \sum_{t=t_1}^{t_2} \frac{d^2(t)}{m^2(t)} \tag{7.46}$$

for some constants $a_i > 0, b_i > 0$, $i = 1, 2, 3$, *and all* $t_2 > t_1 \geq 0$.

Proof: Similar to the proof of Lemma 4.4, the error equation is

$$\epsilon(t) = k_p\tilde{\theta}^T(t)\omega(t-n^*) + \tilde{\rho}(t)\xi(t) + d(t).$$

From the choice of γ_i, $i = 1, \ldots, n_\theta$, and γ_ρ, we have

$$\gamma_m = \max\{\gamma_1 k_p^0, \ldots, \gamma_{n_\theta} k_p^0, \gamma_\rho\} < 2.$$

Defining

$$V(\tilde{\theta}, \tilde{\rho}) = |k_p|\tilde{\theta}^T\Gamma_\theta^{-1}\tilde{\theta} + \gamma_\rho^{-1}\tilde{\rho}^2$$

with $\tilde{\theta}(t) = \theta(t) - \theta^*$, $\tilde{\rho}(t) = \rho(t) - \rho^*$, we obtain

$$\begin{aligned} V(\tilde{\theta}(t+1), \tilde{\rho}(t+1)) - V(\tilde{\theta}(t), \tilde{\rho}(t)) \\ \le -\frac{2-\gamma_m}{2}\frac{\epsilon^2(t)}{m^2(t)} + \frac{2}{2-\gamma_m}\frac{d^2(t)}{m^2(t)} \\ -|k_p| f_\theta^T(t)\Gamma_\theta^{-1} f_\theta(t) - f_\rho(t)\gamma_\rho^{-1} f_\rho(t). \end{aligned} \tag{7.47}$$

By parameter projection, $\theta_j(t) \in [\theta_j^a, \theta_j^b]$, $j = 1, \ldots, n_\theta$, and $\rho \in [\rho^a, \rho^b]$; that is, $\theta(t)$ and $\rho(t)$ are bounded, and so is $\frac{\epsilon(t)}{m(t)}$ from the above error equation. From the projection algorithm, we have

$$|\theta_j(t+1) - \theta_j(t)| \le |g_{\theta j}(t)|, \; 1 \le j \le n_\theta \tag{7.48}$$

$$|\rho(t+1) - \rho(t)| \le |g_\rho(t)|. \tag{7.49}$$

From (7.42), (7.43), (7.47) - (7.49), and the boundedness of $V(\tilde{\theta}(t), \tilde{\rho}(t))$, we have (7.44) - (7.46). ∇

The proof of the following boundedness result is given in Appendix C.

Theorem 7.2 *The adaptive inverse controller, consisting of the linear controller structure (7.36) and the adaptive inverse (7.37) and updated by (7.40) - (7.41), ensures that all closed-loop signals are bounded.*

Switching σ-modification with parameter projection. As an alternative, we can design the switching σ-modification with parameter projection for $f_\theta(t)$ and $f_\rho(t)$. For this design, we take

$$\Gamma_\theta = \gamma_\theta I_{n_\theta}, \; 0 < \gamma_\theta < \frac{1}{k_p^0}, \; 0 < \gamma_\rho < 1.$$

As before k_p^0 is a known upper bound on $|k_p|$. Then, we choose

$$f_\theta(t) = -\sigma_\theta(t)\theta(t) + f_p(t)$$

$$f_\rho(t) = -\sigma_\rho(t)\rho(t),$$

where $f_p(t)$ is a modification for projecting the nonlinearity parameter estimate $\theta_N(t)$ to a prespecified region,

$$\sigma_\theta(t) = \begin{cases} \sigma_0 & \text{if } \|\theta(t)\|_2 > 2M_\theta \\ 0 & \text{otherwise} \end{cases}$$

$$\sigma_\rho(t) = \begin{cases} \sigma_0 & \text{if } |\rho(t)| > 2M_\rho \\ 0 & \text{otherwise} \end{cases}$$

$$0 < \sigma_0 < \frac{1}{2}(1-\gamma_m),\ \gamma_m = \max\{\gamma_\theta k_p^0, \gamma_\rho\} < 1$$

and M_θ, M_ρ are upper bounds of $\|\theta^*\|_2$, $|\rho^*|$, respectively. Similar to the continuous-time case (see Section 6.3), we only project θ_N whose boundaries θ_{Nj}^a, θ_{Nj}^b are given in (7.26) - (7.32).

Denote the jth component of $\theta(t) \in R^{n_\theta}$ as $\theta_j(t)$ and that of $f_p(t)$ as $f_{pj}(t)$. For parameter projection, we choose

$$\theta_j(0) \in \begin{cases} [\theta_j^a, \theta_j^b] = [\theta_{Nj}^a, \theta_{Nj}^b] & \text{for } j = 1, 2, \ldots, n_N \\ \text{no constraint} & \text{otherwise} \end{cases}$$

and set

$$f_{pj}(t) = \begin{cases} f_{Nj}(t) & \text{for } j = 1, 2, \ldots, n_N \\ 0 & \text{otherwise,} \end{cases}$$

where $f_{Nj}(t)$ is the projection modification for $\theta_j(t) = \theta_{Nj}(t)$. To define $f_{Nj}(t)$, we let

$$\bar{g}(t) = -\frac{sign(k_p)\Gamma_\theta \epsilon(t)\omega(t-n^*)}{m^2(t)} - \sigma_\theta(t)\theta(t)$$

$$\bar{\theta}_j(t) = \theta_j(t) + \bar{g}_j(t),$$

where $\bar{g}_j(t)$ is the jth component of $\bar{g}(t)$, and set

$$f_{Nj}(t) = \begin{cases} 0 & \text{if } \bar{\theta}_{Nj}(t) \in [\theta_{Nj}^a, \theta_{Nj}^b] \\ \theta_{Nj}^b - \bar{\theta}_{Nj}(t) & \text{if } \bar{\theta}_{Nj}(t) > \theta_{Nj}^b \\ \theta_{Nj}^a - \bar{\theta}_{Nj}(t) & \text{if } \bar{\theta}_{Nj}(t) < \theta_{Nj}^a \end{cases}$$

for $j = 1, 2, \ldots, n_N$.

This adaptive design also has the properties established in Lemma 7.2.

7.4 Example: Adaptive Dead-Zone Inverse

We now illustrate the design and tracking performance of an adaptive inverse controller by a dead-zone example motivated by muscular stimulation studies [1], [5], [31] from which the electrically stimulated muscle can be approximately modeled as an input dead-zone followed by a second-order linear dynamics, both in discrete time, as in the form (7.1).

In our simulations, we consider the following plant:

$$y(t) = G(D)[u](t),\ u(t) = DZ(v(t))$$

$$G(D) = \frac{1}{D^2 + a_1 D + a_0},\ a_1 = -0.6,\ a_0 = -1.12$$

$$DZ(\cdot): m_r = 1.23,\ m_l = 0.87,\ b_r = 0.58,\ b_l = -0.92,$$

where the dead-zone parameters m_r, m_l, b_r, b_l are *unknown*. The reference signal is $y_m(t) = 10 \sin 0.1325t$, and the tracking error is $e(t) = y(t) - y_m(t)$.

There are two adaptive dead-zone inverse control designs: one for $G(D)$ known and the other for $G(D)$ unknown, as shown in Figures 7.2 and 7.3. In both cases, the adaptive dead-zone inverse

$$v(t) = \widehat{DI}(u_d(t)) \tag{7.50}$$

is implemented with the estimates $\widehat{m_r}$, $\widehat{m_l}$, $\widehat{b_r}$, $\widehat{b_l}$ of m_r, m_l, b_r, b_l.

When $G(D)$ is *known*, the fixed controller structure (7.4) is

$$u_d(t) = \theta_1^* u_d(t-1) + \theta_{21}^* y(t-1) + \theta_{22}^* y(t) + y_m(t+2), \tag{7.51}$$

where $\theta_1^* = -0.6$, $\theta_{21}^* = -0.672$, $\theta_{22}^* = -1.48$, as from (7.5) with $k_p = 1$.

The adaptive law (7.8) with parameter projection (7.20) is used to update the dead-zone inverse parameter estimates:

$$\theta_d(t) = (\widehat{m_r}(t), \widehat{m_r b_r}(t), \widehat{m_l}(t), \widehat{m_l b_l}(t))^T$$

with a step size 0.8 and parameter bounds:

$$\theta_d^a = (0.7, 0, 0.5, -1.5)^T,\ \theta_d^b = (2, 1.5, 1.8, 0)^T.$$

When $G(D)$ is *unknown*, the adaptive controller structure (7.36) is

$$\begin{aligned} u_d(t) = &\ \theta_{21}(t) y(t-1) + \theta_{22}(t) y(t) + y_m(t+2) \\ &-\theta_{41}(t)\widehat{\chi_r}(t)v(t-1) + \theta_{42}(t)\widehat{\chi_r}(t-1) \\ &-\theta_{43}(t)\widehat{\chi_l}(t-1)v(t-1) + \theta_{44}(t)\widehat{\chi_l}(t-1) \end{aligned} \tag{7.52}$$

with $\theta_3(t) = 1$ because $k_p = 1$ in $G(D)$.

The overall system parameter and regressor vectors are

$$\theta(t) = (\theta_d^T(t), \theta_c^T(t))^T, \; \omega(t) = (\omega_d^T(t), \omega_c^T(t))^T,$$

where

$$\theta_c(t) = (\theta_{21}(t), \theta_{22}(t), \theta_{41}(t), \theta_{42}(t), \theta_{43}(t), \theta_{44}(t))^T$$

$$\omega_c(t) = (y(t-1), y(t), -\widehat{\chi_r}(t-1)v(t-1), \widehat{\chi_r}(t-1), \\ -\widehat{\chi_l}(t-1)v(t-1), \widehat{\chi_l}(t-1)).$$

The unknown parameter vector is

$$\theta_c^* = (\theta_{21}^*, \theta_{22}^*, -\theta_1^* m_r, -\theta_1^* m_r b_r, -\theta_1^* m_l, -\theta_1^* m_l b_l)^T \\ = (-0.672, -1.48, 0.738, 0.42804, 0.522, -0.48024)^T.$$

The adaptive law (7.40) is used to update $\theta(t)$ with a step size 0.8 and parameter bounds θ_d^a, θ_d^b, and

$$\theta_c^a = (-1.3, -2.8, 0.3, 0.2, 0.25, -1)^T, \; \theta_c^b = (-0.4, -0.8, 1.4, 1, 1, -0.2)^T.$$

In the presence of dead-zone, both the adaptive dead-zone inverse control schemes, (7.51) with (7.50), (7.52) with (7.50), are good, and they result in very small tracking errors, as shown in Figure 7.4(a) (with $\theta_d(0) = \theta_d^a$) and Figure 7.4(b) (with $\theta_d(0) = \theta_d^b$) for (7.51) with (7.50), and in Figure 7.5(a) (with $\theta_d(0) = \theta_d^a$, $\theta_c(0) = \theta_c^a$) and in Figure 7.5(b) (with $\theta_d(0) = \theta_d^b$, $\theta_c(0) = \theta_c^b$) for (7.52) with (7.50).

We compare the above adaptive dead-zone inverse controllers with a fixed linear control scheme without dead-zone inverse. For $G(D)$ *known*, such a fixed controller structure is

$$v(t) = \theta_1^* v(t-1) + \theta_{21}^* y(t-1) + \theta_{22}^* y(t) + y_m(t+2) \qquad (7.53)$$

with the same θ_1^*, θ_{21}^*, θ_{22}^* in (7.52). This controller is good only for $b_r = b_l = 0$, $m_r = m_l = 1$. In the presence of dead-zone, it results in a tracking error oscillating between -0.8 and 4, in contrast to the adaptive dead-zone inverse cases where the tracking errors are rapidly reduced to less than a half of this range and then converge to negligible values. Thus, significant improvements of the system tracking performance have been achieved by the adaptive dead-zone inverse controllers.

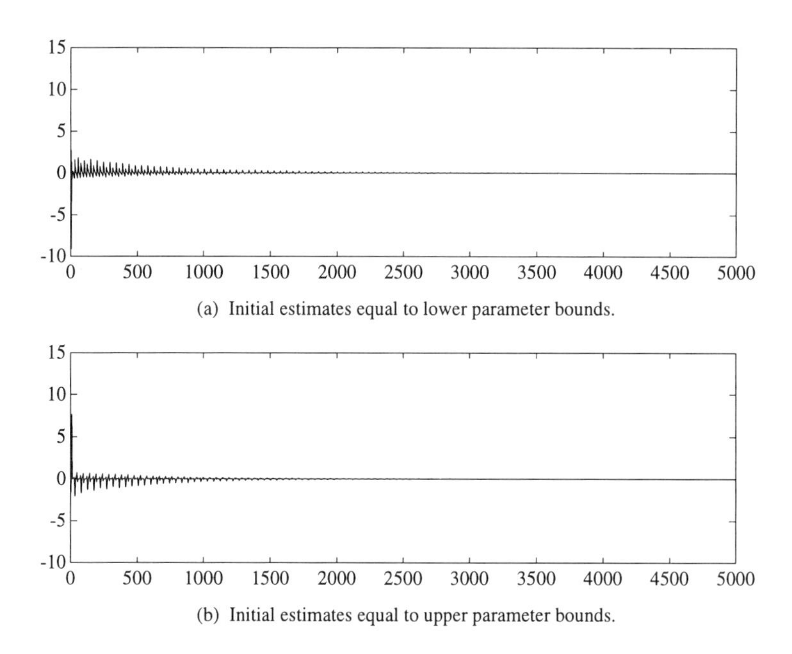

(a) Initial estimates equal to lower parameter bounds.

(b) Initial estimates equal to upper parameter bounds.

Figure 7.4: Tracking errors with adaptive inverse ($G(D)$ known).

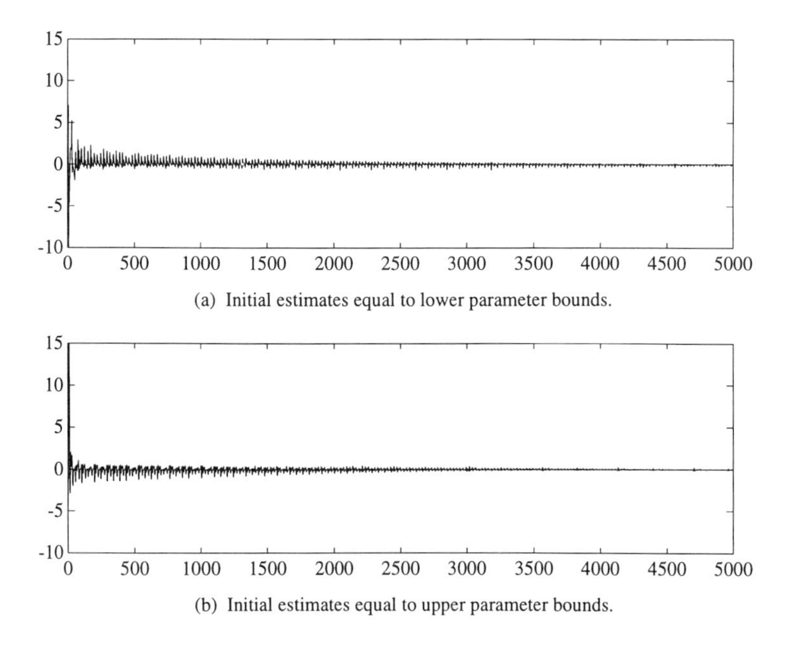

(a) Initial estimates equal to lower parameter bounds.

(b) Initial estimates equal to upper parameter bounds.

Figure 7.5: Tracking errors with adaptive inverse ($G(D)$ unknown).

Chapter 8

Fixed Inverse Control for Output Nonlinearities

So far, we have developed an adaptive inverse approach for control of plants with actuator nonlinearities located at the input of a linear part of the plant. Starting with this chapter, we will extend this approach to plants with sensor nonlinearities which appear at the output. As before, an inverse is employed to cancel the effects of the nonlinearity. The inverse can be fixed or adaptive. In this chapter, we design and unify control schemes with a fixed inverse. We first use an example to illustrate the main difficulties posed by the output nonlinearity. We then overcome these difficulties with more elaborate output inverse schemes. The fixed inverse schemes in this chapter are preliminary to the adaptive inverse designs in Chapter 9.

8.1 Difficulties with an Output Nonlinearity

Unlike the input nonlinearity whose input, when generated from its exact inverse, cancels the nonlinearity, the output nonlinearity $N(\cdot)$ cannot be cancelled by an inverse at its input. The simple reason for this is apparent from Figure 8.1: The input $z(t)$ of $N(\cdot)$ is not accessible either for control or for measurement. The measured signal is the output y of $N(\cdot)$. This is the only signal which can be used to design a feedback control u to be applied to the linear part $G(D)$. A possible cancellation of $N(\cdot)$ can only be achieved by inverting the nonlinear part $N(\cdot)$ together with the linear part $G(D)$. Therefore, the design of an output nonlinearity inverse is more complicated than that of an input nonlinearity inverse. In this section, we use an example to explain this difficulty, especially for continuous-time plants.

We start with the block diagram in Figure 8.2, where $NI(\cdot)$ is the exact

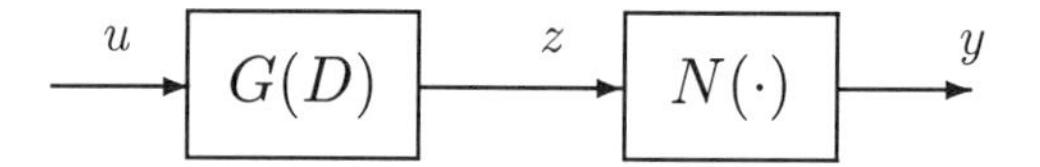

Figure 8.1: Plant with an output nonlinearity.

inverse of $N(\cdot)$ in the sense that

$$N(NI(y_m(t))) = y_m(t). \tag{8.1}$$

The desired identity $y(t) = y_m(t)$ is achieved only when a linear block $C(s)$ cancels $G(s)$.

We first study the case when $G(s)$ is stable, say $G(s) = \frac{s+2}{s+3}$. Its inverse $C(s) = \frac{s+3}{s+2}$ is a stable causal transfer function. A minimal realization of $z(t) = \frac{s+2}{s+3}[u](t)$ is

$$\dot{x} = -3x + u, \ z = -x + u$$

and that of its inverse $u(t) = \frac{s+3}{s+2}[v]$ is

$$\dot{\xi} = -2\xi + v, \ u = \xi + v,$$

where $v = NI(y_m)$. Let $\eta = \xi - x$ so that $\dot{\eta} = -3\eta$ and $y = N(NI(y_m) + e^{-3t}\eta(0))$. In this case the complete inverse is implemented and the tracking objective is achieved:

$$\lim_{t\to\infty}(y(t) - y_m(t)) = \lim_{t\to\infty}(N(NI(y_m) + e^{-3t}\eta(0)) - y_m(t)) = 0.$$

What if $G(s)$ is unstable? Consider, for example, $G(s) = \frac{s+2}{s-3}$. Its inverse $C(s) = \frac{s-3}{s+2}$ is a stable causal transfer function, and the corresponding state space realizations are

$$\dot{x} = 3x + u, \ z = 5x + u$$

$$\dot{\xi} = -2\xi + v, \ u = -5\xi + v.$$

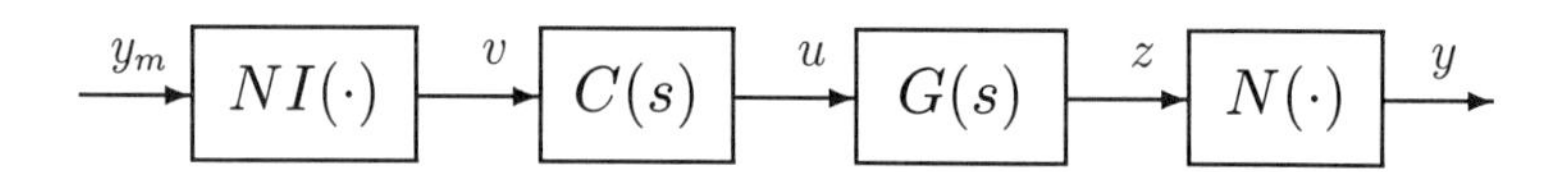

Figure 8.2: Input compensation of the output nonlinearity $N(\cdot)$.

Now $\eta = \xi - x$ results in $\dot{\eta} = 3\eta$ and $y = N(NI(y_m) + e^{3t}\eta(0))$. Because of the instability of $G(s)$, the asymptotic tracking cannot be achieved:

$$\lim_{t\to\infty}(y(t) - y_m(t)) = \lim_{t\to\infty}(N(NI(y_m) + e^{3t}\eta(0)) - y_m(t)) \neq 0.$$

If $G(s) = \frac{s-2}{s+3}$ or $G(s) = \frac{s-2}{s-3}$, then $C(s) = \frac{s+3}{s-2}$ or $C(s) = \frac{s-3}{s-2}$. Again, no stable inversion is possible.

Finally, when $G(s) = \frac{1}{s+3}$, then $C(s) = s + 3$ is improper and thus not physically implementable.

From this example we see that the input compensation scheme shown in Figure 8.2 for inverting the output nonlinearity can work only if $G(s)$ is stable and proper with a stable and proper inverse $C(s)$. In general, the same conclusion holds for the discrete-time case when $G(D) = G(z)$ is a rational function. An exception is the case when $G(D)$ is strictly proper with relative degree n^* and all zeros and poles of $G(z)$ are stable. In this case the knowledge of $y_m(t), y_m(t+1), \ldots, y_m(t+n^*)$ is sufficient for the inverse scheme analogous to that in Figure 8.2 to ensure stable asymptotic tracking.

To handle more general cases of $G(D)$, we proceed to develop a new inverse scheme more suitable than the one shown in Figure 8.2. Our development is restricted to the discrete-time case.

8.2 Control Objective

For the design of a discrete-time inverse controller, the plant in Figure 8.1 is described by

$$y(t) = N(z(t)), \; z(t) = G(D)[u](t), \tag{8.2}$$

where $u(t)$ is the accessible control input, $y(t)$ is the measured output, $G(D) = k_p\frac{Z(D)}{P(D)}$, and $P(D)$, $Z(D)$ are monic polynomials of degrees n, m, respectively. The symbol D denotes, as the case may be, the z-transform variable or the time advance operator $D[x](t) = x(t+1)$. The nonlinearity $N(\cdot)$ represents the dead-zone, backlash, or hysteresis described in Chapter 2. We also use the assumptions (A1) - (A4) stated in Section 4.1 for the linear part $G(D)$ whose relative degree is defined as $n^* = n - m$.

The objectives of this chapter are:

- to develop inverse schemes for the output dead-zone, backlash, and hysteresis characteristics,

- to design and analyze an inverse controller for the plant (8.2) with both $G(D)$ and $N(\cdot)$ known, but with the signal $z(t)$ not available for either control or measurement, and
- to design and analyze controllers employing fixed nonlinearity inverses with $G(D)$ known or unknown and with $N(\cdot)$ unknown.

It is required that all the closed-loop signals are bounded and that the plant output $y(t)$ asymptotically tracks a given reference signal $y_m(t)$.

8.3 Inverse Parametrization

The key idea in our approach is the inverse scheme shown in Figure 8.3. The block $\widehat{NI}(\cdot)$ represents an inverse for the nonlinearity $N(\cdot)$. The purpose of $\widehat{NI}(\cdot)$ is to generate an estimate $\widehat{z}(t)$ of the input $z(t)$ to $N(\cdot)$, solely based on the measurements of $y(t)$:

$$\widehat{z}(t) = \widehat{NI}(y(t)). \tag{8.3}$$

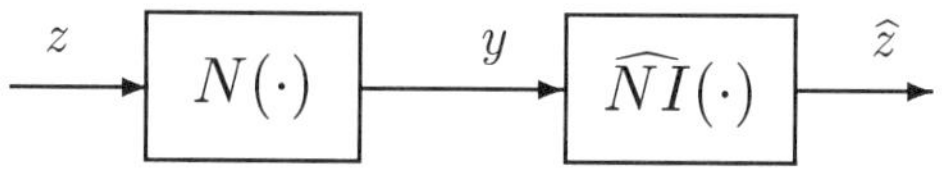

Figure 8.3: Observation of $z(t)$ with $\widehat{NI}(\cdot)$.

Treating (8.3) as an observer of $z(t)$, the observation error is $\widehat{z}(t) - z(t)$. If this error is zero: $\widehat{z}(t) = z(t)$, then a feedback controller can be designed to drive $z(t)$ to $z_m(t)$. This in turn means that $y(t)$ will track a given $y_m(t)$ which is appropriately related to $z_m(t)$. Although, in general, the observation error $\widehat{z}(t) - z(t)$ will be nonzero, we will design in Section 8.4 that a specific "output matching control" $u(t)$ which will ensure that the error $\widehat{z}(t) - z(t)$ becomes zero after a finite number of steps. In this section we proceed to develop the required inverses of the output dead-zone, backlash, and hysteresis nonlinearities.

8.3.1 Dead-Zone Inverse

The output dead-zone $DZ(\cdot)$ is described by

$$y(t) = DZ(z(t)) = \begin{cases} m_r(z(t) - b_r) & \text{if } z(t) \geq b_r \\ 0 & \text{if } b_l < z(t) < b_r \\ m_l(z(t) - b_l) & \text{if } z(t) \leq b_l. \end{cases} \tag{8.4}$$

We stress that $z(t)$ is not completely observable from $y(t)$, because of the main effect of the dead-zone, namely, $y(t) = 0$ for any $z(t) \in [b_l, b_r]$.

As an output dead-zone inverse, we use

$$\widehat{z}(t) = \widehat{DI}(y(t)) = \begin{cases} \frac{y(t)}{\widehat{m_l}} + \widehat{b_l} & \text{if } y(t) < 0 \\ 0 & \text{if } y(t) = 0 \\ \frac{y(t)}{\widehat{m_r}} + \widehat{b_r} & \text{if } y(t) > 0, \end{cases} \tag{8.5}$$

where $\widehat{m_l}$, $\widehat{m_r}$, $\widehat{b_l}$, and $\widehat{b_r}$ are the estimates of m_l, m_r, b_l, and b_r. To arrive at a compact expression, we define

$$\theta_d = (\frac{1}{\widehat{m_r}}, \widehat{b_r}, \frac{1}{\widehat{m_l}}, \widehat{b_l})^T \tag{8.6}$$

$$\begin{aligned} \omega_d(t) &= (\widehat{\chi_r}(t)y(t), \widehat{\chi_r}(t), \widehat{\chi_l}(t)y(t), \widehat{\chi_l}(t))^T \\ \widehat{\chi_r}(t) &= \chi[y(t) > 0] \\ \widehat{\chi_l}(t) &= \chi[y(t) < 0], \end{aligned} \tag{8.7}$$

where, as in Section 3.1.2, $\chi[X]$ is the indicator function of the event X:

$$\chi[X] = \begin{cases} 1 & \text{if } X \text{ is true} \\ 0 & \text{otherwise.} \end{cases}$$

In this notation, the dead-zone inverse (8.5) is

$$\widehat{z}(t) = \theta_d^T \omega_d(t). \tag{8.8}$$

From the dead-zone characteristic (8.4) rewritten as

$$y(t) = m_r(z(t) - b_r)\widehat{\chi_r}(t) + m_l(z(t) - b_l)\widehat{\chi_l}(t)$$

we get

$$\widehat{\chi_r}(t)z(t) = \widehat{\chi_r}(t)(\frac{y(t)}{m_r} + b_r)$$

$$\widehat{\chi_l}(t)z(t) = \widehat{\chi_l}(t)(\frac{y(t)}{m_l} + b_l).$$

Since $\widehat{\chi_r}(t) + \widehat{\chi_l}(t) + \chi[y(t) = 0] = 1$, it follows that

$$\begin{aligned} z(t) &= \widehat{\chi_r}(t)z(t) + \widehat{\chi_l}(t)z(t) + \chi[y(t) = 0]z(t) \\ &= \widehat{\chi_r}(t)(\frac{y(t)}{m_r} + b_r) + \widehat{\chi_l}(t)(\frac{y(t)}{m_l} + b_l) + \chi[y(t) = 0]z(t). \end{aligned}$$

Using $\omega_d(t)$ in (8.7) and defining

$$\theta_d^* = (\frac{1}{m_r}, b_r, \frac{1}{m_l}, b_l)^T \tag{8.9}$$

we express the dead-zone input $z(t)$ in the form distinguishing its parametrized and unparametrized parts:

$$z(t) = \theta_d^{*T}\omega_d(t) + d_d(t). \tag{8.10}$$

The unparametrized part

$$d_d(t) = \chi[y(t) = 0]z(t) = \chi[b_l < z(t) < b_r]z(t)$$

is bounded because $\chi[b_l < z(t) < b_r] = 0$ whenever $z(t) \notin (b_l, b_r)$.

From (8.8) and (8.10), we conclude:

- When implemented with the true parameters $m_r b_r$, m_r, $m_l b_l$, m_l—that is, with $\theta_d = \theta_d^*$—the dead-zone inverse $DI(\cdot)$ has the property

$$DI(y(t)) = DI(DZ(z(t)) = z(t) \text{ whenever } y(t) \neq 0.$$

- When $y(t) = 0$ the worst-case observation error is

$$\max|\bar{z}(t) - z(t)| = \max|d_d(t)| = \max\{b_r, |b_l|\},$$

  where

$$\bar{z}(t) = DI(y(t)) = \theta_d^{*T}\omega_d(t).$$

- When implemented with parameter estimates, the dead-zone inverse $\widehat{DI}(\cdot)$ in (8.8) results in the observation error

$$\widehat{z}(t) - z(t) = (\theta_d - \theta_d^*)^T\omega_d(t) + d_d(t), \tag{8.11}$$

  where $d_d(t)$ is a bounded unparametrizable part.

In selecting the parameter estimates, we should keep in mind their true physical meaning: $-b_{l0} \leq b_l \leq 0 \leq b_r \leq b_{r0}$, and $m_{r1} \leq \frac{1}{m_r} \leq m_{r2}$, $m_{l1} \leq \frac{1}{m_l} \leq m_{l2}$, for some positive constants b_{r0}, b_{l0}, m_{r1}, m_{r2}, m_{l1} and m_{l2}. Hence, any dead-zone parameter estimates $\frac{1}{\widehat{m_r}}$, $\frac{1}{\widehat{m_l}}$, $\widehat{c}_r$, $\widehat{c}_l$ are required to satisfy these constraints.

8.3.2 Backlash Inverse

The output backlash $B(\cdot)$ is described by

$$y(t) = B(z(t)) = \begin{cases} m(z(t) - c_l) & \text{if } z(t) \leq z_l(t-1) \\ m(z(t) - c_r) & \text{if } z(t) \geq z_r(t-1) \\ y(t-1) & \text{if } z_l(t-1) < z(t) < z_r(t-1), \end{cases} \tag{8.12}$$

where

$$z_l(t-1) = \frac{y(t-1)}{m} + c_l, \; z_r(t-1) = \frac{y(t-1)}{m} + c_r$$

are quantities similar to v_r, v_l used in (2.2). The backlash characteristic of Figure 2.9, with $v(t)$ replaced by $z(t)$ and $u(t)$ by $y(t)$, also satisfies (8.12).

As the inverse of the output backlash (8.12), we define

$$\widehat{z}(t) = \widehat{BI}(y(t)) = \begin{cases} \widehat{z}(t-1) & \text{if } y(t) = y(t-1) \\ \frac{y(t)}{\widehat{m}} + \widehat{c}_l & \text{if } y(t) < y(t-1) \\ \frac{y(t)}{\widehat{m}} + \widehat{c}_r & \text{if } y(t) > y(t-1), \end{cases} \tag{8.13}$$

where $\widehat{m}$, $\widehat{c}_l$, and $\widehat{c}_r$ are the estimates of the backlash parameters m, c_l, and c_r. Such an inverse characteristic is also represented by Figure 3.3 with $u_d(t)$ replaced by $y(t)$ and $v(t)$ by $\widehat{z}(t)$.

Similar to the output dead-zone case, the input $z(t)$ of an output backlash is not completely observable through the measurement $y(t)$. The intervals of unobservability are those in which the output $y(t)$ remains constant while the input $z(t)$ continues to change along an inner segment of the backlash characteristic $B(\cdot)$ in (8.12):

$$z_l(t-1) < z(t) < z_r(t-1), \; y(t) = y(t-1).$$

Using the parametrization

$$\theta_b(t) = (\widehat{c}_r(t), \frac{1}{\widehat{m}(t)}, \widehat{c}_l(t))^T \tag{8.14}$$

$$\omega_b(t) = (\widehat{\chi_r}(t), y(t), \widehat{\chi_l}(t))^T, \tag{8.15}$$

where

$$\widehat{\chi_r}(t) = \chi[\widehat{z}(t) = \tfrac{y(t)}{\widehat{m}} + \widehat{c}_r]$$

$$\widehat{\chi_l}(t) = \chi[\widehat{z}(t) = \tfrac{y(t)}{\widehat{m}} + \widehat{c}_l],$$

the backlash inverse (8.13) becomes

$$\widehat{z}(t) = \theta_b^T(t)\omega_b(t). \tag{8.16}$$

Introducing the backlash indicator functions

$$\chi_r(t) = \chi[y(t) > y(t-1)]$$
$$\chi_l(t) = \chi[y(t) < y(t-1)]$$
$$\chi_s(t) = \chi[y(t) = y(t-1)]$$

the backlash characteristic (8.12) is rewritten as

$$y(t) = \chi_r(t)m(z(t) - c_r) + \chi_l(t)m(z(t) - c_l) + \chi_s(t)y_s, \tag{8.17}$$

where y_s is a generic constant corresponding to the value of $y(t)$ on any active inner segment of $B(\cdot)$ characterized by

$$\frac{y_s}{m} + c_l \leq z(t) \leq \frac{y_s}{m} + c_r.$$

From the expression (8.17) of $y(t)$ we get

$$\chi_r(t)z(t) = \chi_r(t)(\frac{y(t)}{m} + c_r)$$

$$\chi_l(t)z(t) = \chi_l(t)(\frac{y(t)}{m} + c_l).$$

Since $\chi_r(t) + \chi_l(t) + \chi_s(t) = 1$, it follows that

$$\begin{aligned} z(t) &= \chi_r(t)z(t) + \chi_l(t)z(t) + \chi_s(t)z(t) \\ &= \widehat{\chi_r}(t)(\frac{y(t)}{m} + c_r) + \widehat{\chi_l}(t)(\frac{y(t)}{m} + c_l) + d_b(t). \end{aligned} \tag{8.18}$$

To show that

$$\begin{aligned} d_b(t) = (\chi_r - \widehat{\chi_r}(t))(\frac{y(t)}{m} + c_r) \\ +(\chi_l(t) - \widehat{\chi_l}(t))(\frac{y(t)}{m} + c_l) + \chi_s(t)z(t) \end{aligned}$$

is bounded, we examine the following three different cases:

(i) If $\chi_l(t) = 1$, $\chi_r(t) = \chi_s(t) = 0$, then

$$d_b(t) = \begin{cases} 0 & \text{if } \widehat{\chi_l}(t) = 1,\ \widehat{\chi_r}(t) = 0 \\ c_l - c_r & \text{if } \widehat{\chi_l}(t) = 0,\ \widehat{\chi_r}(t) = 1. \end{cases}$$

(ii) If $\chi_r(t) = 1$, $\chi_l(t) = \chi_s(t) = 0$, then

$$d_b(t) = \begin{cases} 0 & \text{if } \widehat{\chi_l}(t) = 0,\ \widehat{\chi_r}(t) = 1 \\ c_r - c_l & \text{if } \widehat{\chi_l}(t) = 1,\ \widehat{\chi_r}(t) = 0. \end{cases}$$

(iii) If $\chi_s(t) = 1$, $\chi_l(t) = \chi_r(t) = 0$, then $y(t) = y_s$, and

$$d_b(t) = \begin{cases} c_{ls} - c_l & \text{if } \widehat{\chi_l}(t) = 1,\ \widehat{\chi_r}(t) = 0 \\ c_{rs} - c_r & \text{if } \widehat{\chi_l}(t) = 0,\ \widehat{\chi_r}(t) = 1. \end{cases}$$

In the last case, c_{ls} and c_{rs} depend on the motion of $z(t)$ and $y(t)$ on an inner segment when $z_l(t-1) < z(t) < z_r(t-1)$, $y(t) = y(t-1)$—that is, when $\chi_s(t) = 1$. The values of c_{ls} and c_{rs} are in the interval (c_l, c_r); that is, $z(t)$ can be expressed as $z(t) = \frac{y_s}{m} + c_{ls}$ or $z(t) = \frac{y_s}{m} + c_{rs}$. From these expressions, it is clear that $d_b(t)$ is bounded.

To arrive at a compact form of the backlash (8.12), we define

$$\theta_b^* = (c_r, \frac{1}{m}, c_l)^T \tag{8.19}$$

and use (8.15) to express $z(t)$ in (8.18) as

$$z(t) = \theta_b^{*T}(t)\omega_b(t) + d_b(t), \tag{8.20}$$

where the unparametrized $d_b(t)$ is bounded by $\max|b_b(t)| = c_r + |c_l|$.

From (8.16) and (8.20), we conclude:

- If $z(\tau+1) - z(\tau)$ does not change sign for $\tau \in \{t_0, t_0+1, \ldots, t-1\}$, the backlash inverse $BI(\cdot)$ is implemented with the true parameters m, c_r, c_l; that is, when $\theta_b = \theta_b^*$, it has the property

$$BI(B(z(t_0))) = z(t_0) \Rightarrow BI(B(z(t))) = z(t).$$

  Otherwise, the worst-case observation error is

$$\max|\bar{z}(t) - z(t)| = \max|d_b(t)| = c_r + |c_l|,$$

  where

$$\bar{z}(t) = BI(y(t)) = \theta_b^{*T}\omega_b(t).$$

- When implemented with parameter estimates, the backlash inverse $\widehat{BI}(\cdot)$ in (8.16) results in the observation error

$$\hat{z}(t) - z(t) = (\theta_b - \theta_b^*)^T\omega_b(t) + d_b(t) \tag{8.21}$$

  for some bounded unparametrizable error $d_b(t)$.

To implement a backlash inverse, we need to use the properties of m, c_r and c_l: $m_1 \leq \frac{1}{m} \leq m_2$, $-c_{l0} \leq c_l \leq 0 \leq c_r \leq c_{r0}$, for some positive constants m_1, m_2, c_{l0} and c_{r0}, and select the parameter estimates such that $m_1 \leq \frac{1}{\widehat{m}} \leq m_2$, $-c_{l0} \leq \widehat{c}_l \leq 0 \leq \widehat{c}_r \leq c_{r0}$. In the adaptive law to be developed in Chapter 9, such a property will be employed for parameter projection which will ensure that the parameter estimates remain within the above constraints.

8.3.3 Hysteresis Inverse

The hysteresis characteristic shown in Figure 2.16 can be used to describe an output hysteresis $H(\cdot)$, when the variables $v(t)$, $u(t)$ are replaced by $z(t)$, $y(t)$, respectively. Analogous to (2.12), the output hysteresis $H(\cdot)$ is

$$y(t) = \begin{cases} y(t-1) & \text{if } z(t) = z(t-1) \\ m_t z(t) + c_t & \text{if } z(t) \geq z_r, \text{ or} \\ & \text{if } m_t < m_b, \\ & y(t-1) = m_t z(t-1) + c_t \text{ and} \\ & z(t-1) < z(t) < z_r \\ m_b z(t) + c_b & \text{if } z(t) \leq z_l, \text{ or} \\ & \text{if } m_t > m_b, \\ & y(t-1) = m_b z(t-1) + c_b \text{ and} \\ & z_l < z(t) < z(t-1) \\ m_t z(t) + c_d & \text{if } z_d < z(t) < z(t-1) \\ m_b z(t) + c_u & \text{if } z_u > z(t) > z(t-1) \\ m_l(z(t) - c_l) & \text{if } z_d \geq z(t) \geq z_l \\ m_r(z(t) - c_r) & \text{if } z_u \leq z(t) \leq z_r, \end{cases} \tag{8.22}$$

where

$$z_r = \frac{c_t + m_r c_r}{m_r - m_t}, \quad z_l = \frac{c_b + m_l c_l}{m_l - m_b}$$

$$c_d = y(t-1) - m_t z(t-1), \quad c_u = y(t-1) - m_b z(t-1)$$

$$z_d = \frac{m_l c_l + c_d}{m_l - m_t}, \quad z_u = \frac{m_r c_r + c_u}{m_r - m_b}$$

such that $z_l \leq z_d \leq z_u \leq z_r$. Note that the quantities z_d, z_u, z_r, z_l are analogous to the quantities v_d, v_u, v_3, v_4 defined in Section 2.3.

To construct an inverse for (8.22), we use the characteristic in Figure 3.5 with $u_d(t)$ replaced by $y(t)$ and $v(t)$ by $\widehat{z}(t)$. The signal motion of such an inverse is confined to two half-lines, two line segments, and the quadrilateral formed by these half-lines and segments. More precisely, we let $\widehat{m}_t$, $\widehat{c}_t$, $\widehat{m}_b$, $\widehat{c}_b$,

$\widehat{m_r}$, $\widehat{c_r}$, $\widehat{m_l}$, $\widehat{c_l}$ be the estimates of the true hysteresis parameters m_t, c_t, m_b, c_b, m_r, c_r, m_l, c_l, respectively, and introduce the quantities

$$\widehat{y_r} = \frac{\widehat{m_r}(\widehat{m_t}\widehat{c_r} + \widehat{c_t})}{\widehat{m_r} - \widehat{m_t}}, \quad \widehat{y_l} = \frac{\widehat{m_l}(\widehat{m_b}\widehat{c_l} + \widehat{c_b})}{\widehat{m_l} - \widehat{m_b}}$$

$$\widehat{y_e} = \frac{\widehat{m_r}(\widehat{m_t}\widehat{c_r} + \widehat{c_w})}{\widehat{m_r} - \widehat{m_t}}, \quad \widehat{y_w} = \frac{\widehat{m_l}(\widehat{m_b}\widehat{c_l} + \widehat{c_e})}{\widehat{m_l} - \widehat{m_b}}$$

$$\widehat{c_w} = y(t-1) - \widehat{m_t}\widehat{z}(t-1), \ \widehat{c_e} = y(t-1) - \widehat{m_b}\widehat{z}(t-1)$$

such that $\widehat{y_l} \leq \widehat{y_w} \leq \widehat{y_r} \leq \widehat{y_e}$. Then we define the output hysteresis inverse as

$$\widehat{z}(t) = \widehat{HI}(y(t)) = \begin{cases} \widehat{z}(t-1) & \text{if } y(t) = y(t-1) \\ \frac{1}{\widehat{m_t}}(y(t) - \widehat{c_t}) & \text{if } y(t) \geq \widehat{y_r} \\ \frac{1}{\widehat{m_b}}(y(t) - \widehat{c_b}) & \text{if } y(t) \leq \widehat{y_l} \\ \frac{1}{\widehat{m_r}}y(t) + \widehat{c_r} & \text{if } \widehat{y_r} \geq y(t) \geq \widehat{y_r} \\ \frac{1}{\widehat{m_l}}y(t) + \widehat{c_l} & \text{if } \widehat{y_l} \leq y(t) \leq \widehat{y_w} \\ \frac{1}{\widehat{m_t}}(y(t) - \widehat{c_w}) & \text{if } \widehat{y_w} < y(t) < y(t-1) \\ \frac{1}{\widehat{m_b}}(y(t) - \widehat{c_e}) & \text{if } y(t-1) < y(t) < \widehat{y_e}. \end{cases} \tag{8.23}$$

This hysteresis inverse has the following desired property.

Proposition 8.1 *When implemented with the true parameters m_t, c_t, m_b, c_b, m_r, c_r, m_l, c_l, the hysteresis inverse (8.23), denoted by $HI(\cdot)$, has the inversion property:*

$$HI(H(z(t_0))) = z(t_0) \Rightarrow HI(H(z(t))) = z(t), \ \forall t \geq t_0$$

for any $z(t)$ and any $t_0 \geq 0$.

This property implies that, in contrast with the dead-zone and backlash inverses, the hysteresis inverse $HI(\cdot)$ is a true inverse. With inexact initialization, the worst-case error $\widehat{z}(t) - z(t)$ is $\max\{z_i - z_j$: (z_i, y_i), (z_j, y_j) are inside the hysteresis loop.$\}$. As an illustration, in the equal-slope case $m_r = m_l$, the worst-case error is $c_r + |c_l|$.

For a compact expression, we first define the hysteresis inverse indicator functions

$$\widehat{\chi_t}(t) = \chi[\widehat{z}(t) = \tfrac{1}{\widehat{m_t}}(y(t) - \widehat{c_t})] \tag{8.24}$$

$$\widehat{\chi_b}(t) = \chi[\widehat{z}(t) = \tfrac{1}{\widehat{m_b}}(y(t) - \widehat{c_b})] \tag{8.25}$$

$$\widehat{\chi_r}(t) = \chi[\widehat{z}(t) = \tfrac{1}{\widehat{m_r}} y(t) + \widehat{c_r}] \tag{8.26}$$

$$\widehat{\chi_l}(t) = \chi[\widehat{z}(t) = \tfrac{1}{\widehat{m_l}} y(t) + \widehat{c_l}] \tag{8.27}$$

$$\widehat{\chi_w}(t) = \chi[\widehat{z}(t) = \tfrac{1}{\widehat{m_t}} (y(t) - \widehat{c_w})] \tag{8.28}$$

$$\widehat{\chi_e}(t) = \chi[\widehat{z}(t) = \tfrac{1}{\widehat{m_b}} (y(t) - \widehat{c_e})], \tag{8.29}$$

the parameter vector

$$\theta_h = (\frac{1}{\widehat{m_t}}, \frac{\widehat{c_t}}{\widehat{m_t}}, \frac{1}{\widehat{m_b}}, \frac{\widehat{c_b}}{\widehat{m_b}}, \frac{1}{\widehat{m_r}}, \widehat{c_r}, \frac{1}{\widehat{m_l}}, \widehat{c_l})^T, \tag{8.30}$$

and the regressor

$$\begin{aligned}\omega_h(t) = ((\widehat{\chi_t}(t) + \widehat{\chi_w}(t))y(t) - \widehat{\chi_w}(t)\widehat{c_w}, -\widehat{\chi_t}(t), (\widehat{\chi_b}(t) + \widehat{\chi_e}(t))y(t) \\ -\widehat{\chi_e}(t)\widehat{c_e}, -\widehat{\chi_b}(t), \widehat{\chi_r}(t)y(t), \widehat{\chi_r}(t), \widehat{\chi_l}(t)y(t), \widehat{\chi_l}(t))^T.\end{aligned} \tag{8.31}$$

In this notation, the hysteresis inverse (8.23) is

$$\widehat{z}(t) = \theta_h^T(t)\omega_h(t). \tag{8.32}$$

We further define the indicator functions

$$\chi_t(t) = \chi[y(t) = m_t z(t) + c_t]$$

$$\chi_b(t) = \chi[y(t) = m_b z(t) + c_b]$$

$$\chi_r(t) = \chi[y(t) = m_r(z(t) - c_r)]$$

$$\chi_l(t) = \chi[y(t) = m_l(z(t) - c_l)z(t)]$$

$$\chi_d(t) = \chi[y(t) = m_t z(t) + c_d]$$

$$\chi_u(t) = \chi[y(t) = m_b z(t) + c_u]$$

which have the properties:

$$\chi_k^2(t) = \chi_k(t),\ \chi_i(t)\chi_j(t) = 0,\ i \neq j,\ i, j, k \in \{t, b, r, l, d, u\} \tag{8.33}$$

$$\chi_t(t) + \chi_b(t) + \chi_r(t) + \chi_l(t) + \chi_d(t) + \chi_u(t) = 1. \tag{8.34}$$

Then the hysteresis characteristic (8.20) is more compactly expressed as

$$\begin{aligned}y(t) = \chi_t(t)(m_t z(t) + c_t) + \chi_b(t)(m_b z(t) + c_b) \\ +\chi_r(t)m_r(z(t) - c_r) + \chi_l(t)m_l(z(t) - c_l) \\ +\chi_d(t)(m_t z(t) + c_d) + \chi_u(t)(m_b z(t) + c_u).\end{aligned} \tag{8.35}$$

From (8.33) - (8.35), we obtain the identities

$$\chi_t(t)z(t) = \chi_t(t)(\frac{y(t)}{m_t} - \frac{c_t}{m_t})$$

$$\chi_b(t)z(t) = \chi_b(t)(\frac{y(t)}{m_b} - \frac{c_b}{m_b})$$

$$\chi_r(t)z(t) = \chi_r(t)(\frac{y(t)}{m_r} + c_r)$$

$$\chi_l(t)z(t) = \chi_l(t)(\frac{y(t)}{m_l} + c_l)$$

$$\chi_d(t)z(t) = \chi_d(t)(\frac{y(t)}{m_t} - \frac{c_d}{m_t})$$

$$\chi_u(t)z(t) = \chi_u(t)(\frac{y(t)}{m_b} - \frac{c_u}{m_b}),$$

which help us to parametrize the signal $z(t)$:

$$\begin{aligned} z(t) &= \chi_t(t)z(t) + \chi_b(t)z(t) + \chi_r(t)z(t) \\ &\quad +\chi_l(t)z(t) + \chi_d(t)z(t) + \chi_u(t)z(t) \\ &= \widehat{\chi_t}(t)(\frac{y(t)}{m_t} - \frac{c_t}{m_t}) + \widehat{\chi_b}(t)(\frac{y(t)}{m_b} - \frac{c_b}{m_b}) \\ &\quad +\widehat{\chi_r}(t)(\frac{y(t)}{m_r} + c_r) + \widehat{\chi_l}(t)(\frac{y(t)}{m_l} + c_l) \\ &\quad +\widehat{\chi_w}(t)\frac{y(t)}{m_t} - \widehat{\chi_w}(t)\frac{\widehat{c_w}}{m_t} + \widehat{\chi_e}(t)\frac{y(t)}{m_b} - \widehat{\chi_e}(t)\frac{\widehat{c_e}}{m_b} + d_h(t). \end{aligned} \quad (8.36)$$

The unparametrized part is

$$\begin{aligned} d_h(t) &= (\chi_t(t) - \widehat{\chi_t}(t))(\frac{y(t)}{m_t} - \frac{c_t}{m_t}) + (\chi_b(t) - \widehat{\chi_b}(t))(\frac{y(t)}{m_b} - \frac{c_b}{m_b}) \\ &\quad +(\chi_r(t) - \widehat{\chi_r}(t))(\frac{y(t)}{m_r} + c_r) + (\chi_l(t) - \widehat{\chi_l})(\frac{y(t)}{m_l} + c_l) \\ &\quad +(\chi_d(t) - \widehat{\chi_w}(t))(\frac{y(t)}{m_t} - \frac{c_d}{m_t}) + (\chi_u(t) - \widehat{\chi_e}(t))(\frac{y(t)}{m_b} - \frac{c_u}{m_b}) \\ &\quad +\widehat{\chi_w}(t)\frac{\widehat{c_w}}{m_t} - \widehat{\chi_w}(t)\frac{c_d}{m_t} + \widehat{\chi_e}(t)\frac{\widehat{c_e}}{m_b} - \widehat{\chi_e}(t)\frac{c_u}{m_b}. \end{aligned} \quad (8.37)$$

From the hysteresis (8.22), we see that there are positive constants m_{t1}, m_{t2}, m_{b1}, m_{b2}, m_{r1}, m_{r2}, m_{l1}, m_{l2} and c_{t0}, c_{b0}, c_{r0}, c_{l0} such that $\max\{m_{r2}, m_{l2}\} < m_{t1} \le \frac{1}{m_t} \le m_{t2}$, $\max\{m_{r2}, m_{l2}\} < m_{b1} \le \frac{1}{m_b} \le m_{b2}$, $-c_{b0} \le c_b \le 0 \le c_t \le c_{t0}$, $-c_{l0} \le c_l \le 0 \le c_r \le c_{r0}$. Choices of the parameter estimates $\widehat{m_t}$, $\widehat{c_t}$, $\widehat{m_b}$,

$\widehat{c_b}$, $\widehat{m_r}$, $\widehat{c_r}$, $\widehat{m_l}$, $\widehat{c_l}$ should also satisfy this condition, which in turn means that $0 \leq \frac{c_t}{m_t} \leq c_{t0}m_{t2}$, $-c_{b0}m_{b2} \leq \frac{c_b}{m_b} \leq 0$, $0 \leq \frac{\widehat{c_t}}{\widehat{m_t}} \leq c_{t0}m_{t2}$, and $-c_{b0}m_{b2} \leq \frac{\widehat{c_b}}{\widehat{m_b}} \leq 0$. Under this condition, the unparametrized part $d_h(t)$ in (8.37) is bounded. This is so because for any $t \geq 0$ and large $y(t)$, either $\chi_t(t) = \widehat{\chi_t}(t)$ or $\chi_b(t) = \widehat{\chi_b}(t)$, while all other indicator functions are zero, so that $d_h(t) = 0$.

Defining the true parameter vector

$$\theta_h^* = (\frac{1}{m_t}, \frac{c_t}{m_t}, \frac{1}{m_b}, \frac{c_b}{m_b}, \frac{1}{m_r}, c_r, \frac{1}{m_l}, c_l)^T \tag{8.38}$$

and using $\omega_h(t)$ in (8.31), we compactly express $z(t)$ in (8.36) as

$$z(t) = \theta_h^{*T}\omega_h(t) + d_h(t). \tag{8.39}$$

From (8.32) and (8.39), we obtain the observation error

$$\widehat{z}(t) - z(t) = (\theta_h - \theta_h^*)^T\omega_h(t) + d_h(t), \tag{8.40}$$

where $d_h(t)$ is bounded for any finite parameter estimates $\widehat{c_t}$, $\widehat{c_b}$, $\widehat{c_r}$, $\widehat{c_l}$. Furthermore, with a correctly initialized inverse and $\theta_h = \theta_h^*$, the error $d_h(t)$ becomes zero because in this case $\widehat{\chi_k}(t) = \chi_k(t)$, $k = t, b, r, l, d, u$, and $\widehat{c_w} = c_d$, $\widehat{c_e} = c_u$. Note that with $\theta_h(t)$ defined in (8.30), the estimates of c_t and c_b are

$$\widehat{c_t} = \frac{\theta_{h2}}{\theta_{h1}}, \quad \widehat{c_t} = \frac{\theta_{h4}}{\theta_{h3}},$$

where

$$\theta_{h1} = \frac{1}{\widehat{m_t}}, \quad \theta_{h2} = \frac{\widehat{c_t}}{\widehat{m_t}}, \quad \theta_{h3} = \frac{1}{\widehat{m_b}}, \quad \theta_{h4} = \frac{\widehat{c_b}}{\widehat{m_b}}.$$

Similar to the unified expression in Section 3.4, using the subscript "N" to denote "d", "b", or "h" for the dead-zone, backlash, or hysteresis inverse, we express the inverses (8.8), (8.16), and (8.32) and the observation errors (8.11), (8.21), and (8.40) in the unified form:

$$\widehat{z}(t) = \widehat{NI}(y(t)) = \theta_N^T\omega_N(t) \tag{8.41}$$

$$\widehat{z}(t) - z(t) = (\theta_N - \theta_N^*)^T\omega_N(t), +d_N(t), \tag{8.42}$$

where $d_N(t)$ is a bounded error. When implemented with the true inverse parameter $\theta_N = \theta_N^*$, the inverse $\widehat{z}(t) = \widehat{NI}(y(t))$ is denoted as

$$\bar{z}(t) = NI(y(t)) = \theta_N^{*T}\omega_N(t).$$

Unlike the input nonlinearity case, where $d_N(t)$ becomes zero whenever the estimate θ_N is equal to the true parameter θ_N^*, in the output nonlinearity case $d_N(t)$ becomes zero only for the hysteresis inverse with $\theta_h = \theta_h^*$. For the output dead-zone or backlash, the signal $z(t)$ is not completely observable from $y(t)$, which introduces a bounded observation error even if the parameter estimate is equal to the true parameter.

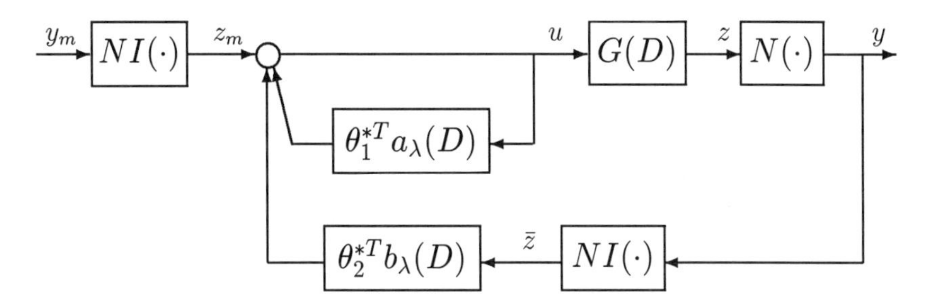

Figure 8.4: Output matching controller structure.

8.4 Output Matching Control

In spite of the difficulties with output nonlinearities, we will be able to employ the nonlinearity inverses constructed in Section 8.3 in the design of a feedback controller structure which guarantees that the closed-loop system is stable and the output tracking $y(t) = y_m(t)$ is achieved. In other words, we will design an inverse control scheme which leads to the exact cancellation of the sensor nonlinearities $DZ(\cdot)$, $B(\cdot)$, or $H(\cdot)$.

8.4.1 Design for $G(D)$ Known and $z(t)$ Unknown

For the discrete-time plant (8.2) in Figure 8.1, we assume that both the output nonlinearity $N(\cdot)$ and the linear part $G(D) = k_p \frac{Z(D)}{P(D)}$ are known and that $G(D)$ is minimum phase. Without loss of generality we let $k_p = 1$.

For a given reference output $y_m(t)$ to be tracked by $y(t)$, our inverse controller is shown in Figure 8.4. It has two identical inverses $NI(\cdot)$ and a linear controller structure consisting of two blocks $\theta_1^{*T} a_\lambda(D)$ and $\theta_2^{*T} b_\lambda(D)$. We require that their parameters $\theta_1^* \in R^{n-1}$ and $\theta_2^* \in R^n$ satisfy the matching equation

$$\theta_1^{*T} a_\lambda(D) P(D) + \theta_2^{*T} b_\lambda(D) Z(D) = P(D) - Z(D) D^{n^*}, \tag{8.43}$$

where

$$a_\lambda(D) = (D^{-n+1}, \ldots, D^{-1})^T \tag{8.44}$$

$$b_\lambda(D) = (D^{-n+1}, \ldots, D^{-1}, 1)^T. \tag{8.45}$$

The regressor $\omega_m(t)$ which parametrizes the exact inverse for the reference signal $y_m(t)$ is defined by

$$z_m(t) = NI(y_m(t)) = \theta_N^{*T} \omega_m(t). \tag{8.46}$$

With the same regressor $\omega_m(t)$ we parametrize the estimated inverse

$$\widehat{z_m}(t) = \widehat{NI}(y_m(t)) = \theta_N^T \omega_m(t), \tag{8.47}$$

where $\theta_N = \theta_d$, θ_b, or θ_h is an estimate of θ_N^*. For specific inverses the following definitions and relationships are used:

For a *dead-zone inverse*:

$$\widehat{z_m}(t) = \widehat{DI}(y_m(t)) = \theta_d^T \omega_{dm}(t)$$

$$\omega_{dm}(t) = (\widehat{\chi_{rm}}(t) y_m(t), \widehat{\chi_{rm}}(t), \widehat{\chi_{lm}}(t) y_m(t), \widehat{\chi_{lm}}(t))^T \tag{8.48}$$

$$\widehat{\chi_{rm}}(t) = \chi[y_m(t) > 0] = \chi[\widehat{z_m}(t) = \tfrac{y_m(t)}{\widehat{m_r}} + \widehat{b_r}]$$

$$\widehat{\chi_{lm}}(t) = \chi[y_m(t) < 0] = \chi[\widehat{z_m}(t) = \tfrac{y_m(t)}{\widehat{m_l}} + \widehat{b_l}].$$

For a *backlash inverse*:

$$\widehat{z_m}(t) = \widehat{BI}(y_m(t)) = \theta_b^T \omega_{bm}(t)$$

$$\omega_{bm}(t) = (\widehat{\chi_{rm}}(t), y_m(t), \widehat{\chi_{lm}}(t))^T \tag{8.49}$$

$$\begin{aligned}\widehat{\chi_{rm}}(t) &= \chi[\widehat{z_m}(t) = \tfrac{y_m(t)}{\widehat{m}} + \widehat{c_r}] \\ &= \chi[y_m(t) > y_m(t-1) \text{ or } y_m(t) = y_m(t-1), \chi_{rm}(t-1) = 1]\end{aligned}$$

$$\begin{aligned}\widehat{\chi_{lm}}(t) &= \chi[\widehat{z_m}(t) = \tfrac{y_m(t)}{\widehat{m}} + \widehat{c_l}] \\ &= \chi[y_m(t) < y_m(t-1) \text{ or } y_m(t) = y_m(t-1), \chi_{lm}(t-1) = 1]\end{aligned}$$

$$\chi_{rm}(t_0) + \chi_{lm}(t_0) = 1, \; \chi_{rm}(t_0)\chi_{lm}(t_0) = 0.$$

For a *hysteresis inverse*:

$$\widehat{z_m}(t) = \widehat{HI}(y_m(t)) = \theta_h^T \omega_{hm}(t)$$

$$\begin{aligned}\omega_{hm}(t) = (&(\widehat{\chi_{tm}}(t) + \widehat{\chi_{wm}}(t)) y_m(t) - \widehat{\chi_{wm}}(t+n^*)\widehat{c_{wm}}, -\widehat{\chi_{tm}}(t), \\ &-\widehat{\chi_{em}}(t)\widehat{c_{em}} + (\widehat{\chi_{bm}}(t) + \widehat{\chi_{em}}(t)) y(t), -\widehat{\chi_{bm}}(t), \\ &\widehat{\chi_{rm}}(t) y_m(t), \widehat{\chi_{rm}}(t), \widehat{\chi_{lm}}(t) y_m(t), \widehat{\chi_{lm}}(t))^T\end{aligned} \tag{8.50}$$

$$\widehat{\chi_{tm}}(t) = \chi[\widehat{z_m}(t) = \tfrac{1}{\widehat{m_t}}(y_m(t) - \widehat{c_t})]$$

$$\widehat{\chi_{bm}}(t) = \chi[\widehat{z_m}(t) = \tfrac{1}{\widehat{m_b}}(y_m(t) - \widehat{c_b})]$$

$$\widehat{\chi_{rm}}(t) = \chi[\widehat{z_m}(t) = \frac{1}{m_r} y_m(t) + \widehat{c_r}]$$

$$\widehat{\chi_{lm}}(t) = \chi[\widehat{z_m}(t) = \frac{1}{m_l} y_m(t) + \widehat{c_l}]$$

$$\widehat{\chi_{wm}}(t) = \chi[\widehat{z_m}(t) = \frac{1}{m_t} (y_m(t) - \widehat{c_w})]$$

$$\widehat{\chi_{em}}(t) = \chi[\widehat{z_m}(t) = \frac{1}{m_b} (y_m(t) - \widehat{c_e})]$$

with $\widehat{c_{wm}} \in (\widehat{c_t}, \widehat{c_1})$, $\widehat{c_{em}} \in (\widehat{c_2}, \widehat{c_b})$, for $\widehat{c_1}$, $\widehat{c_2}$ defined in Section 3.3.

In the unified notation, each regressor $\omega_{dm}(t)$ in (8.48), $\omega_{bm}(t)$ in (8.49), or $\omega_{hm}(t)$ in (8.50) is denoted by $\omega_m(t)$ for (8.46) and (8.47).

We select the controller structure

$$u(t) = \theta_1^{*T} \omega_1(t) + \theta_2^{*T} \omega_2(t) + z_m(t + n^*), \tag{8.51}$$

where

$$\omega_1(t) = a_\lambda(D)[u](t), \ \omega_2(t) = b_\lambda(D)[\bar{z}](t). \tag{8.52}$$

The signal $\bar{z}(t)$ is the observed value of $z(t)$:

$$\bar{z}(t) = NI(y(t)) = \theta_N^{*T} \omega_N(t) \tag{8.53}$$

and the signal $z_m(t)$ is generated from

$$z_m(t) = NI(y_m(t)) = \theta_N^{*T} \omega_m(t). \tag{8.54}$$

Recall that $NI(\cdot)$ represents $DI(\cdot)$, $BI(\cdot)$, or $HI(\cdot)$, which are the dead-zone inverse $\widehat{DI}(\cdot)$, backlash inverse $\widehat{BI}(\cdot)$, or hysteresis inverse $\widehat{HI}(\cdot)$ implemented with the true parameter θ_d^*, θ_b^*, or θ_h^*.

The signals $z_m(t)$, $z(t)$, $y_m(t)$, and $y(t)$ satisfy a key relationship.

Proposition 8.2 *The inverse scheme (8.53) - (8.54) has the following properties:*

- *For dead-zone,*

$$z(t) = z_m(t) \Rightarrow y(t) = y_m(t).$$

- *For backlash or hysteresis,*

$$z(t-1) = z_m(t-1), \ z(t) = z_m(t), \ y(t-1) = y_m(t-1)$$
$$\Rightarrow y(t) = y_m(t).$$

The two relationships differ because dead-zone is a static nonlinearity, while backlash and hysteresis are dynamic nonlinearities and for output matching their inverses need to be initialized.

It is crucial that, for output matching $y(t) = y_m(t)$ to be achieved for all t after a finite interval, the conditions in Proposition 8.2 are required to hold only during that interval.

Theorem 8.1 *The control scheme (8.51) - (8.54) ensures that all closed-loop signals are bounded. Moreover, the output tracking*

$$y(t) = y_m(t),\ t \geq t_0 + n^* + n - 1 \tag{8.55}$$

is achieved, under the finite transient interval conditions:

$$\begin{aligned}\bar{z}(\tau) &= NI(N(z(\tau))) \\ &= z(\tau) \begin{cases} \tau = t_0, \ldots, t_0 + n^* + n - 2 & \text{for } DI(\cdot) \\ \tau = t_0, \ldots, t_0 + n^* + n - 1 & \text{for } BI(\cdot) \\ \tau = t_0 & \text{for } HI(\cdot) \end{cases}\end{aligned} \tag{8.56}$$

$$y(\tau) = y_m(\tau) \begin{cases} \text{not needed} & \text{for } DI(\cdot) \\ \tau = t_0 + n^* + n - 1 & \text{for } BI(\cdot) \\ \tau = t_0 + n^* + n - 1 & \text{for } HI(\cdot). \end{cases} \tag{8.57}$$

Proof: Using the plant equation $P(D)[z](t) = Z(D)[u](t)$ in (8.2), operating both sides of (8.43) on $z(t)$, and ignoring the effect of the initial conditions, we obtain

$$u(t) = \theta_1^{*T}(t)\omega_1(t) + \theta_2^{*T} b_\lambda(D)[z](t) + z(t+n^*). \tag{8.58}$$

From (8.41), (8.42), and (8.53), it follows that $\bar{z}(t) - z(t) = d_N(t)$, where $d_N(t)$ is bounded. Using this fact and (8.51), (8.58), we have

$$z(t+n^*) = z_m(t+n^*) + \theta_2^{*T} b_\lambda(D)[d_N](t).$$

Thus $z(t)$ is bounded, and so is $u(t)$ because $G(D)$ is minimum phase.

To prove the output tracking, we consider the dead-zone, backlash, and hysteresis separately.

Dead-zone: If $\bar{z}(\tau) = z(\tau)$ for $\tau = t_0, t_0 + 1, \ldots, t_0 + n - 1$, then from (8.51) and (8.58) it follows that $z(t_0 + n^* + n - 1) = z_m(t_0 + n^* + n - 1)$, which implies the exact tracking and the exact observation of $z(t)$ by $\bar{z}(t)$ at $t = t_0 + n^* + n - 1$. In other words, $y(t_0 + n^* + n - 1) = y_m(t_0 + n^* + n - 1)$,

$\bar{z}(t_0 + n^* + n - 1) = z(t_0 + n^* + n - 1)$. Note that for the dead-zone inverse, $z(t) = z_m(t)$ implies that $y(t) = y_m(t)$ and $\bar{z}(t) = z(t)$. Now we have that $\bar{z}(\tau) = z(\tau)$ for $\tau = t_0+1, t_0+2, \ldots, t_0+n, t_0+n+1, \ldots, t_0+n^*+n-1$, which is sufficient to ensure that $y(t) = y_m(t)$ and $\bar{z}(t) = z(t)$ for $t = t_0 + n^* + n$. By induction with (8.51) and (8.58), it follows that $y(t) = y_m(t)$, $\bar{z}(t) = z(t)$, $t = t_0 + t_1, t_0 + t_1 + 1, \ldots, t_0 + t_1 + n - 1, t_0 + t_1 + n, \ldots, t_0 + t_1 + n^* + n - 2$ for any $t_1 > 0$. This shows that (8.55) holds.

Backlash: If $\bar{z}(\tau) = z(\tau)$ for $\tau = t_0, t_0 + 1, \ldots, t_0 + n - 1, t_0 + n$, then from (8.51) and (8.58) we have $z(t_0 + n^* + n - 1) = z_m(t_0 + n^* + n - 1)$ and $z(t_0+n^*+n) = z_m(t_0+n^*+n)$. This property, together with $y(t_0+n^*+n-1) = y_m(t_0 + n^* + n - 1)$, implies that $y(t_0 + n^* + n) = y_m(t_0 + n^* + n)$.

To show that $\bar{z}(t_0+n^*+n) = z(t_0+n^*+n)$, we first consider the case when $\widehat{\chi_{rm}}(t_0 + n^* + n) = 1$. Then $y(t_0 + n^* + n) \geq y(t_0 + n^* + n - 1)$. Furthermore, $z(t_0+n^*+n) = z_m(t_0+n^*+n)$ and $y(t_0+n^*+n) = y_m(t_0+n^*+n)$ imply that $y(t_0+n^*+n) = m(z(t_0+n^*+n)-c_r)$. If $y(t_0+n^*+n) > y(t_0+n^*+n-1)$, then $\bar{z}(t_0 + n^* + n) = \frac{1}{m}y(t_0 + n^* + n) + c_r$ and so $\bar{z}(t_0 + n^* + n) = z(t_0 + n^* + n)$. If $y(t_0+n^*+n) = y(t_0+n^*+n-1)$, that is, if $y_m(t_0+n^*+n) = y_m(t_0+n^*+n-1)$, then $z_m(t_0+n^*+n) = z_m(t_0+n^*+n-1)$ and so $z(t_0+n^*+n) = z(t_0+n^*+n-1)$. But $y(t_0 + n^* + n) = y(t_0 + n^* + n - 1)$ also implies that $\bar{z}(t_0 + n^* + n) = \bar{z}(t_0 + n^* + n - 1)$. Therefore we have $\bar{z}(t_0 + n^* + n) = z(t_0 + n^* + n)$ because $\bar{z}(t_0+n^*+n-1) = z(t_0+n^*+n-1)$. The case when $\widehat{\chi_{lm}}(t_0+n^*+n) = 1$ can be similarly analyzed to reach the same conclusion. By induction, we prove that (8.55) holds.

Hysteresis: If $\bar{z}(t_0) = z(t_0)$, then, with $NI(\cdot) = HI(\cdot)$ and $N(\cdot) = H(\cdot)$, from Proposition 8.1, we have $\bar{z}(t) = HI(H(z(t)) = z(t)$, $t \geq t_0$, and from (8.51) and (8.58), we have $z(t + n^* + n - 1) = z_m(t + n^* + n - 1)$, $t \geq t_0$. Given that $z(t_0 + n^* + n - 1) = z_m(t_0 + n^* + n - 1)$ and $y(t_0 + n^* + n - 1) = y_m(t_0 + n^* + n - 1)$, the property $z(t_0 + n^* + n) = z_m(t_0 + n^* + n)$—that is, $\bar{z}(t_0 + n^* + n) = z_m(t_0 + n^* + n)$—implies: $y(t_0 + n^* + n) = y_m(t_0 + n^* + n)$. Hence, we can use induction to conclude that (8.55) holds. ∇

8.4.2 Design for Both $G(D)$ and $z(t)$ Known

A special case of the previous design is when $z(t)$ is available for measurement. Then we can design a feedback controller using $z(t)$ to generate the plant input $u(t)$ such that the output $z(t)$ of $G(D)$ tracks any bounded signal $z_m(t)$. If $z_m(t)$ is generated from an inverse with $y_m(t)$ as its input, then the output tracking $y(t) = y_m(t)$ is achieved. In view of Theorem 8.1 and the inverse control scheme (8.51) - (8.54), the controller using $z(t)$ is

$$u(t) = \theta_1^{*T}(t)\omega_1(t) + \theta_2^{*T} b_\lambda(D)[z](t) + z_m(t+n^*), \tag{8.59}$$

where $z_m(t)$ is defined in (8.54).

The control scheme (8.59) ensures the output tracking $y(t) = y_m(t)$, $t \geq t_0 + n^* + n - 1$, under the condition (8.57):

$$y(t_0 + n^* + n - 1) = y_m(t_0 + n^* + n - 1),$$

which is needed only for the backlash or hysteresis inverse.

8.4.3 Design for Both $G(D)$ and $z(t)$ Unknown

The linear controller structure for $G(D)$ unknown is

$$u(t) = \theta_1^T(t)\omega_1(t) + \theta_2^T\omega_2(t) + \theta_3 z_m(t+n^*), \tag{8.60}$$

where $\omega_1(t)$, $\omega_2(t)$ and $z_m(t)$ are as in (8.52) - (8.54). The new aspect of (8.60), compared with (8.51), is that now θ_1, θ_2 and θ_3 are the estimates of the unknown parameters $\theta_1^* \in R^{n-1}$, $\theta_2^* \in R^n$ and $\theta_3^* = \frac{1}{k_p}$ which satisfy

$$\theta_1^{*T} a_\lambda(D)P(D) + \theta_2^{*T} b_\lambda(D)k_p Z(D) = P(D) - \theta_3^* k_p Z(D) D^{n^*}. \tag{8.61}$$

Using (8.61) and (8.2), we obtain

$$u(t) = \theta_1^{*T}(t)\omega_1(t) + \theta_2^{*T} b_\lambda(D)[z](t) + \theta_3^* z(t+n^*). \tag{8.62}$$

Since $\bar{z}(t) - z(t) = d_N(t)$, it follows from (8.60) and (8.62) that

$$u(t) = \theta^{*T}\omega(t) - d(t+n^*), \tag{8.63}$$

where

$$\theta^* = (\theta_1^{*T}, \theta_2^{*T}, \theta_3^*)^T \tag{8.64}$$

$$\omega(t) = (\omega_1^T(t), \omega_2^T(t), \bar{z}(t+n^*))^T \tag{8.65}$$

$$d(t+n^*) = \theta_3^* d_N(t+n^*) + \theta_2^{*T} b_\lambda(D)[d_N](t). \tag{8.66}$$

Let $\theta(t)$ be the estimate of θ^* and define the estimation error

$$\epsilon(t) = \theta^T(t-1)\omega(t-n^*) - u(t-n^*). \tag{8.67}$$

From (8.63) and (8.67), an equivalent expression for $\epsilon(t)$ is

$$\epsilon(t) = \tilde{\theta}(t-1)\omega(t-n^*) + d(t), \ \tilde{\theta}(t-1) = \theta(t-1) - \theta^*, \tag{8.68}$$

which suggests the following update law for $\theta(t)$:

$$\theta(t) = \theta(t-1) - \frac{\gamma\omega(t-n^*)\epsilon(t)}{1+\omega^T(t-n^*)\omega(t-n^*)} + f(t), \tag{8.69}$$

where $f(t)$ is either the parameter projection modification with $0 < \gamma < 2$ or the switching σ-modification with or without projection and with $0 < \gamma < 1$. These modifications are the same as in Chapter 4. They achieve robustness with respect to the bounded $d(t)$ in (8.66). A parameter projection is also needed to ensure

$$\text{sign}(\theta_3(t)) = \text{sign}(k_p),\ |\theta_3(t)| \geq \theta_3^0 > 0$$

for some constant $\theta_3^0 \leq \frac{1}{|k_p|}$.

Since $d(t)$ is bounded, as in robust adaptive control, the adaptive controller (8.60) updated by (8.69) ensures that all closed-loop signals are bounded and that the tracking error $e(t) = y(t) - y_m(t)$ satisfies

$$\sum_{t=t_1}^{t_2} e^2(t) \leq a_1 + b_1 \sum_{t=t_1}^{t_2} d^2(t),\ t_2 \geq t_1$$

for some constants $a_1 > 0$, $b_1 > 0$. As in (8.66), $d(t)$ depends on the observation error $d_N(t) = NI(N(z(t))) - z(t)$ which reduces to zero over the intervals when $z(t)$ is observable from $y(t)$ (see Section 8.3).

8.4.4 Design for $G(D)$ Unknown and $z(t)$ Known

Finally, if the signal $z(t)$ is available for measurement, then the controller (8.60) is replaced by

$$u(t) = \theta_1^T(t)\omega_1(t) + \theta_2^T b_\lambda(D)[z](t) + \theta_3 z_m(t+n^*) \tag{8.70}$$

and the parametrization (8.63) becomes

$$u(t) = \theta^{*T}\omega(t),$$

where θ^* is defined in (8.64), and $\omega(t)$ in (8.65) is replaced by

$$\omega(t) = (\omega_1^T(t), (b_\lambda(D)[z](t))^T, z(t+n^*))^T.$$

The estimation error $\epsilon(t)$ defined in (8.67) now has the form

$$\epsilon(t) = \tilde{\theta}(t-1)\omega(t-n^*)$$

which suggests the following update law for $\theta(t)$:

$$\theta(t) = \theta(t-1) - \frac{\gamma\omega(t-n^*)\epsilon(t)}{1+\omega^T(t-n^*)\omega(t-n^*)}, \quad 0 < \gamma < 2. \tag{8.71}$$

A parameter projection is also needed to ensure $\mathrm{sign}(\theta_3(t)) = \mathrm{sign}(k_p)$ and $|\theta_3(t)| \geq \theta_3^0 > 0$, for some constant $\theta_3^0 \leq \frac{1}{|k_p|}$.

Since the signal $z(t)$ is directly used for feedback control, the resulting adaptive system is a standard model reference adaptive control system (see Appendix A), with $z_m(t)$ as the reference and $z(t)$ as the output. Therefore, the adaptive controller (8.70) updated by (8.71) ensures that all closed-loop signals are bounded and that the asymptotic tracking of $z_m(t)$ by $z(t)$ is achieved:

$$\lim_{t\to\infty}(z(t) - z_m(t)) = 0.$$

Following this result and that in Proposition 8.2, we now analyze the output tracking performance of the controller (8.70). For *dead-zone*, $z(t) = z_m(t)$ implies that $y(t) = y_m(t)$. Hence, the convergence of $z(t) - z_m(t)$ to zero implies the convergence of $y(t) - y_m(t)$ to zero—that is, asymptotic output tracking. For *backlash* or *hysteresis*, from Proposition 8.2, the situation is different: $z(t) = z_m(t)$, $z(t-1) = z_m(t-1)$, and $y(t-1) = y_m(t-1)$ are needed for $y(t) = y_m(t)$. The fact that $z(t) - z_m(t)$ converges to zero may not mean that $y(t) - y_m(t)$ converges to zero. However, if there exists a $T_0 > 0$ such that if $|y(T_0) - y_m(T_0)|$ is small, then $|y(t) - y_m(t)|$, $t \geq T_0$, is also small.

8.4.5 Example: Output Backlash Compensation

To illustrate the inverse control designs presented in Sections 8.4.1 - 8.4.4 for $N(\cdot)$ known, we consider the plant (8.2) with a known or unknown second-order linear part $G(D)$ and a known output backlash $B(\cdot)$:

$$y(t) = B(z(t)), \; z(t) = G(D)[u](t)$$

$$G(D) = \frac{D + b_1}{D^2 + a_2 D + a_1}$$

$$B(\cdot): \; m = 0.014, \; c_r = 22.5, \; c_l = -24.7.$$

Different control schemes are now designed for the situations when the plant parameters $a_2 = 0.5$, $a_1 = 3$, and $b_1 = 0.7$ are known or unknown.

We examine the four control schemes in Sections 8.4.1, 8.4.2, 8.4.3, and 8.4.4, respectively, with $n = 2$, $n^* = 1$ given, and θ_1^*, $\theta_2^* = (\theta_{21}^*, \theta_{22}^*)^T$ calculated from (8.43) or (8.61) with $k_p = \theta_3^* = \theta_3 = 1$:

$$\theta_1^* = -b_1, \; \theta_{21}^* = a_1, \; \theta_{22}^* = a_2.$$

Since $B(\cdot)$ is known, we can construct the backlash inverse $BI(\cdot)$ from (8.13) with the exact parameters $\widehat{m} = m$, $\widehat{c_r} = c_r$, and $\widehat{c_l} = c_l$. For a reference output $y_m(t)$ to be tracked by $y(t)$, we generate the signals

$$z_m(t) = BI(y_m(t)$$

$$\bar{z}(t) = BI(y(t)),$$

where $\bar{z}(t)$ is an observation of $z(t)$, used for control when $z(t)$ is unknown.

(a) The controller for $G(D)$ known and $z(t)$ unknown is

$$u(t) = \theta_1^* u(t-1) + \theta_{21}^* \bar{z}(t-1) + \theta_{22}^* \bar{z}(t) + z_m(t+1).$$

(b) The controller for both $G(D)$ and $z(t)$ known is

$$u(t) = \theta_1^* u(t-1) + \theta_{21}^* z(t-1) + \theta_{22}^* z(t) + z_m(t+1).$$

(c) The controller for both $G(D)$ and $z(t)$ unknown is

$$u(t) = \theta_1(t) u(t-1) + \theta_{21}(t) \bar{z}(t-1) + \theta_{22}(t) \bar{z}(t) + z_m(t+1).$$

(d) The controller for $G(D)$ unknown and $z(t)$ known is

$$u(t) = \theta_1(t) u(t-1) + \theta_{21}(t) z(t-1) + \theta_{22}(t) z(t) + z_m(t+1).$$

In the controllers (c) and (d), $\theta_1(t)$, $\theta_{21}(t)$, and $\theta_{22}(t)$ are the adaptive estimates of θ_1^*, θ_{21}^*, and θ_{22}^*, respectively, updated with (8.69) for the controller (c) and with (8.71) for the controller (d). Typical responses of these control systems, with $y_m(t) = 15 \sin 0.05024t$, $z(-1) = 1.0$, $z(0) = 0.5$, $y(0) = 2.0$, $\bar{z}(-1) = \bar{z}(0) = 0$, $u(-1) = u(0) = 0$, are summarized as follows:

- With the controller (a), $y(t) = y_m(t)$, $t \geq 3$, and with the controller (b), $y(t) = y_m(t)$, $t \geq 2$.
- With the controllers (c) and (d), the tracking error $y(t) - y_m(t)$ converges to small values, and so does the parameter error. The transient response with the controller (d) is better, as shown in Figures 8.5 and 8.6.

This simulation study shows that with the controllers (a) and (b), exact output tracking $y(t) = y_m(t))$ can be achieved even without the initialization conditions (8.56) and (8.57). With the controllers (c) and (d), the tracking error $y(t) - y_m(t)$ becomes small for large t.

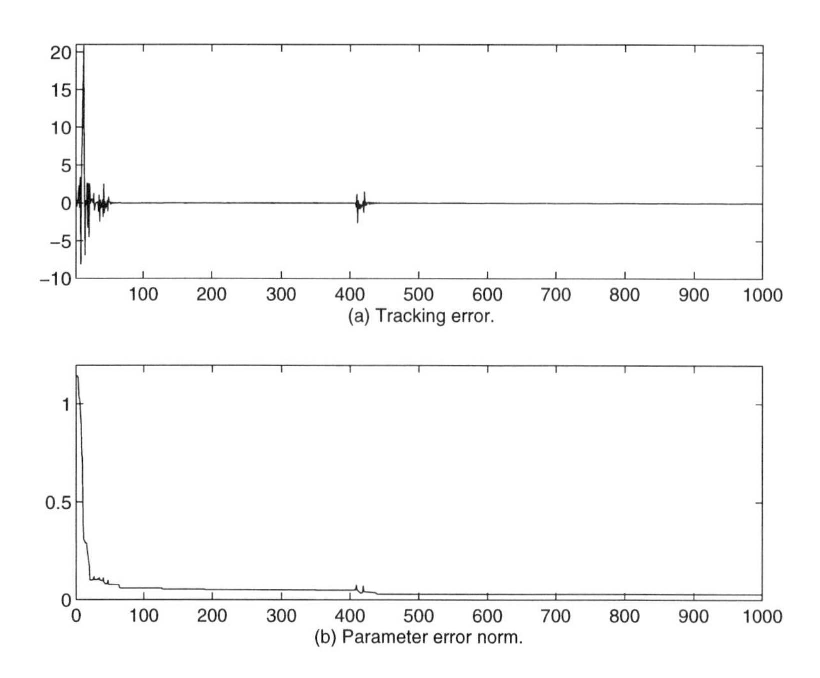

Figure 8.5: Tracking error and parameter error with controller (c).

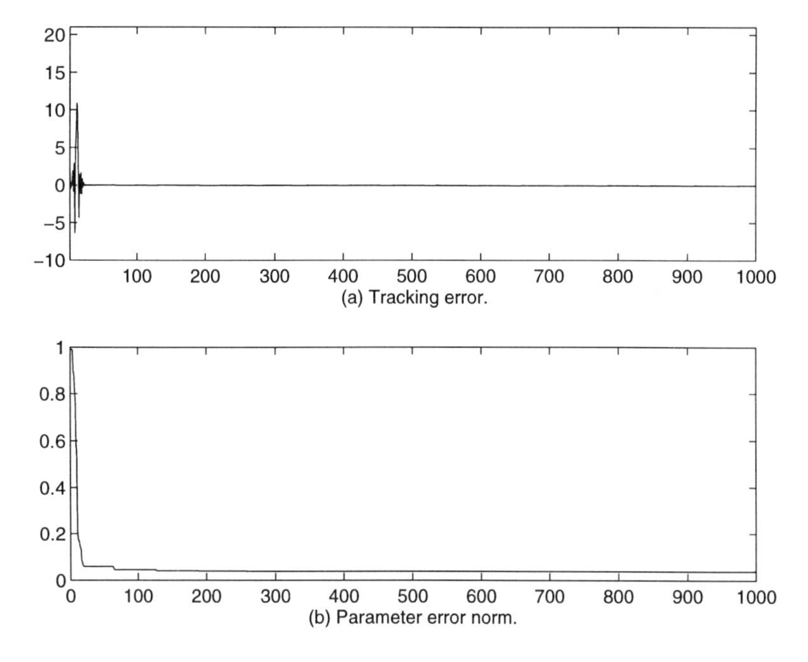

Figure 8.6: Tracking error and parameter error with controller (d).

8.5 Detuned Inverse Control

We now examine the properties of the controllers (8.51), (8.59), (8.60), and (8.70) with the dead-zone inverse $\widehat{DI}(\cdot)$, backlash inverse $\widehat{BI}(\cdot)$, or hysteresis inverse $\widehat{HI}(\cdot)$ implemented with a fixed "detuned" parameter estimate $\theta_d \neq \theta_d^*$, $\theta_b \neq \theta_b^*$, or $\theta_h \neq \theta_h^*$.

We first derive an expression for the observation error $\widehat{z}(t) - z(t)$ which will be used for the closed-loop system analysis.

For *dead-zone*, using (8.4), (8.5) and (8.11), we express

$$\widehat{z}(t) - z(t) = \tilde{k}_d(t)z(t) + \bar{d}_d, \tag{8.72}$$

where

$$\tilde{k}_d(t) = -(\frac{\widehat{m_r} - m_r}{\widehat{m_r}}\widehat{\chi_r}(t) + \frac{\widehat{m_l} - m_l}{\widehat{m_l}}\widehat{\chi_l}(t))$$

$$\begin{aligned}\bar{d}_d(t) = & \frac{\widehat{m_r} - m_r}{\widehat{m_r}}\widehat{\chi_r}(t)b_r + (\widehat{b_r} - b_r)\widehat{\chi_r}(t) \\ & + \frac{\widehat{m_l} - m_l}{\widehat{m_l}}\widehat{\chi_l}(t)b_l + (\widehat{b_l} - b_l)\widehat{\chi_l}(t) + d_d(t).\end{aligned}$$

For *backlash*, we first rewrite (8.12) as

$$y(t) = B(z(t)) = m(z(t) - c_s(t)),\ c_s(t) \in [c_l, c_r]. \tag{8.73}$$

Then, using (8.21) and (8.73), we express

$$\widehat{z}(t) - z(t) = \tilde{k}_b(t)z(t) + \bar{d}_b, \tag{8.74}$$

where

$$\tilde{k}_b(t) = -\frac{\widehat{m} - m}{\widehat{m}}$$

$$\bar{d}_b(t) = \frac{\widehat{m} - m}{\widehat{m}}c_s(t) + (\widehat{c_r} - c_r)\widehat{\chi_r}(t) + (\widehat{c_l} - c_l)\widehat{\chi_l}(t) + d_b(t).$$

For *hysteresis*, using (8.23) - (8.25), we have

$$\widehat{\chi_t}(t)y(t) = \widehat{\chi_t}(t)(m_t z(t) + c_t) + \widehat{\chi_t}(t)b_t(t) \tag{8.75}$$

$$\widehat{\chi_b}(t)y(t) = \widehat{\chi_b}(t)(m_b z(t) + c_b) + \widehat{\chi_b}(t)b_b(t) \tag{8.76}$$

for some bounded $b_t(t)$ and $b_b(t)$ whose maximum values are proportional to the width of the hysteresis loop. Then, from (8.23), (8.40), (8.75), and (8.76), we obtain

$$\widehat{z}(t) - z(t) = \tilde{k}_h(t)z(t) + \bar{d}_h, \tag{8.77}$$

where

$$\tilde{k}_h(t) = -(\frac{\widehat{m_t} - m_t}{\widehat{m_t}}\widehat{\chi_t}(t) + \frac{\widehat{m_b} - m_b}{\widehat{m_b}}\widehat{\chi_b}(t))$$

$$\begin{aligned}\bar{d}_h(t) = & (\frac{1}{\widehat{m_t}} - \frac{1}{m_t})\widehat{\chi_t}(t)(c_t + d_t(t)) + (\frac{1}{\widehat{m_t}} - \frac{1}{m_t})\widehat{\chi_w}(t)(y(t) - \widehat{c_w}) \\ & -(\frac{\widehat{c_t}}{\widehat{m_t}} - \frac{c_t}{m_t})\widehat{\chi_t}(t) + (\frac{1}{\widehat{m_b}} - \frac{1}{m_b})\widehat{\chi_b}(t)(c_b + d_b(t)) \\ & +(\frac{1}{\widehat{m_b}} - \frac{1}{m_b})\widehat{\chi_e}(t)(y(t) - \widehat{c_e}) - (\frac{\widehat{c_b}}{\widehat{m_b}} - \frac{c_b}{m_b})\widehat{\chi_b}(t) \\ & +(\frac{1}{\widehat{m_r}} - \frac{1}{m_r})\widehat{\chi_r}(t)y(t) + (\widehat{c_r} - c_r)\widehat{\chi_r}(t) \\ & +(\frac{1}{\widehat{m_l}} - \frac{1}{m_l})\widehat{\chi_l}(t)y(t) + (\widehat{c_l} - c_l)\widehat{\chi_l}(t) + d_h(t).\end{aligned}$$

With "N" denoting "d", "b", or "h" for dead-zone, backlash, or hysteresis, we unify (8.72), (8.73), and (8.77) as

$$\widehat{z}(t) - z(t) = \tilde{k}_N(t)z(t) + \bar{d}_N(t), \tag{8.78}$$

where $\bar{d}_N(t) \in \{\bar{d}_d(t), \bar{d}_b(t), \bar{d}_h(t)\}$ is bounded for all $t \geq 0$, and $\tilde{k}_N(t)$ depends on the dead-zone slope errors $\widehat{m_r} - m_r$, $\widehat{m_l} - m_l$, or the backlash slope error $\widehat{m} - m$, or the hysteresis slope errors $\widehat{m_t} - m_t$, $\widehat{m_b} - m_b$. The size of $\tilde{k}_N(t)$ is proportional to the quantity

$$\mu_N = \begin{cases} \max\{|\widehat{m_r} - m_r|, |\widehat{m_l} - m_l|\} & \text{for } \widehat{DI}(\cdot) \\ |\widehat{m} - m| & \text{for } \widehat{BI}(\cdot) \\ \max\{|\widehat{m_t} - m_t|, |\widehat{m_b} - m_b|\} & \text{for } \widehat{HI}(\cdot), \end{cases} \tag{8.79}$$

which is important in characterizing the closed-loop stability.

8.5.1 Design for $G(D)$ Known

With the linear part $G(D)$ known ($k_p = 1$) and the nonlinear part $N(\cdot)$ unknown, the controller is as in Figure 8.4 with $\bar{z}(t) = NI(y(t))$ replaced by $\widehat{z}(t) = \widehat{NI}(y(t))$. It consists of a fixed linear controller structure

$$u(t) = \theta_1^{*T}(t)\omega_1(t) + \theta_2^{*T}\omega_2(t) + \widehat{z_m}(t + n^*), \tag{8.80}$$

where

$$\omega_1(t) = a_\lambda(D)[u](t),\ \omega_2(t) = b_\lambda(D)[\widehat{z}](t) \tag{8.81}$$

and $\theta_1^* \in R^{n-1}$, $\theta_2^* \in R^n$ satisfy (8.43), as well as two identical inverses with a fixed detuned estimate θ_N of θ_N^*:

$$\widehat{z}(t) = \widehat{NI}(y(t)) = \theta_N^T \omega_N(t) \tag{8.82}$$

$$\widehat{z_m}(t) = \widehat{NI}(y_m(t)) = \theta_N^T \omega_m(t). \tag{8.83}$$

To describe the closed-loop system, we substitute (8.78) and (8.80) into (8.85), and obtain

$$z(t+n^*) = \theta_2^{*T} b_\lambda(D)[\tilde{k}_N z](t) + \theta_2^{*T} b_\lambda(D)[\bar{d}_N](t) + \widehat{z_m}(t+n^*). \tag{8.84}$$

Since $\widehat{k}_N(t)$ is proportional to μ_N defined in (8.79), $\theta_2^{*T} b_\lambda(D)$ is a stable filter, and both $\bar{d}_N(t)$ and $\widehat{z_m}(t+n^*)$ are bounded, from (8.83) and the assumption (A1), we derive the following result.

Theorem 8.2 *There exists a constant $\mu^* > 0$ such that for any $N(\cdot) \in \{DZ(\cdot), B(\cdot), H(\cdot)\}$ with its detuned inverse $\widehat{NI}(\cdot) \in \{\widehat{DI}(\cdot), \widehat{BI}(\cdot), \widehat{HI}(\cdot)\}$ satisfying $\mu_N \in [0, \mu^*]$, the inverse controller consisting of the fixed linear controller structure (8.80) and the detuned inverses (8.82) and (8.83), applied to the plant (8.2), ensures that all closed-loop signals are bounded.*

8.5.2 Design for $G(D)$ Unknown

When both the linear part $G(D)$ and the nonlinear part $N(\cdot)$ are unknown, an inverse controller employs an adaptive linear controller structure and a fixed detuned inverse.

The adaptive linear controller structure is

$$u(t) = \theta_1^T(t)\omega_1(t) + \theta_2^T \omega_2(t) + \theta_3 \widehat{z_m}(t+n^*), \tag{8.85}$$

where $\omega_1(t)$, $\omega_2(t)$, and $\widehat{z_m}(t)$ are as in (8.81) - (8.83), and θ_1, θ_2, and θ_3 are the estimates of the unknown parameters $\theta_1^* \in R^{n-1}$, $\theta_2^* \in R^n$, and $\theta_3^* = \frac{1}{k_p}$ which satisfy (8.61).

To derive an error equation, we rewrite (8.78) as

$$\widehat{z}(t) - z(t) = \frac{\tilde{k}_N(t)}{1+\tilde{k}_N(t)}\widehat{z}(t) + \frac{1}{1+\tilde{k}_N(t)}\bar{d}_N(t) \tag{8.86}$$

and, using (8.62), we obtain

$$u(t) = \theta^{*T}\omega(t) - d(t+n^*) - \eta(t+n^*), \tag{8.87}$$

where

$$\theta^* = (\theta_1^{*T}, \theta_2^{*T}, \theta_3^*)^T$$

$$\omega(t) = (\omega_1^T(t), \omega_2^T(t), \widehat{z}(t+n^*))^T$$

$$d(t+n^*) = \theta_3^*(D^{n^*} + \theta_2^{*T} b_\lambda(D))[\frac{1}{1+\tilde{k}_N}\bar{d}_N](t)$$

$$\eta(t+n^*) = \theta_3^*(D^{n^*} + \theta_2^{*T} b_\lambda(D))[\frac{\tilde{k}_N}{1+\tilde{k}_N}\widehat{z}](t)$$

Let $\theta(t)$ be the estimate of θ^* and define the estimation error

$$\epsilon(t) = \theta^T(t-1)\omega(t-n^*) - u(t-n^*), \tag{8.88}$$

which, using (8.87) and (8.88), can be expressed as

$$\epsilon(t) = \tilde{\theta}(t-1)\omega(t-n^*) + d(t) + \eta(t), \tag{8.89}$$

where $\tilde{\theta}(t-1) = \theta(t-1) - \theta^*$.

Based on the equation (8.89), we select the update law

$$\theta(t) = \theta(t-1) - \frac{\gamma\omega(t-n^*)\epsilon(t)}{1+\omega^T(t-n^*)\omega(t-n^*)} + f(t), \tag{8.90}$$

where $f(t)$ is either the parameter projection with $0 < \gamma < 2$ or the switching σ-modification with or without projection, $0 < \gamma < 1$. A parameter projection is also used for (8.90) to ensure $\text{sign}(\theta_3(t)) = \text{sign}(k_p)$ and $|\theta_3(t)| \geq \theta_3^0 > 0$, for some constant $\theta_3^0 \leq \frac{1}{|k_p|}$. Since $\eta(t)$ is proportional to $\frac{\tilde{k}_N}{1+\tilde{k}_N}\widehat{z}(t)$ and the components of $b_\lambda(D)[\frac{\tilde{k}_N}{1+\tilde{k}_N}\widehat{z}](t)$, the desired normalization for robustness:

$$\frac{|\eta(t)|}{\sqrt{1+\omega^T(t-n^*)\omega(t-n^*)}} \leq k_\eta \mu_N,$$

for some constant $k_\eta > 0$, is achieved.

This control scheme has the following property.

Theorem 8.3 *There exists a constant $\mu^* > 0$ such that for any $N(\cdot) \in \{DZ(\cdot), B(\cdot), H(\cdot)\}$ and its detuned inverse $\widehat{NI}(\cdot) \in \{\widehat{DI}(\cdot), \widehat{BI}(\cdot), \widehat{HI}(\cdot)\}$ with $\mu_N \in [0, \mu^*]$, the controller, consisting of the linear controller (8.85), updated by the adaptive law (8.90), and the detuned inverses (8.82) and (8.83), applied to the plant (8.2), ensures that all closed-loop signals are bounded.*

This is a robustness property similar to that in Theorem 4.6 for the actuator nonlinearities.

8.6 Summary

- A plant with an output (sensor) nonlinearity is

$$y(t) = N(z(t)),\ z(t) = G(D)[u](t),$$

where $N(\cdot)$ is a dead-zone, backlash, or hysteresis parametrized by θ_N^*, and $G(D)$ is a known or unknown linear part.

- A control scheme employs an inverse $NI(\cdot)$ to construct an estimate $\bar{z}(t)$ of $z(t)$ from the output $y(t)$:

$$\bar{z}(t) = NI(y(t)) = \theta_N^{*T}\omega_N(t).$$

- With such an inverse scheme, the observation error

$$\bar{z}(t) - z(t) = d_N(t)$$

is bounded. Moreover, because of the unobservability of $z(t)$ from $y(t)$ in a dead-zone or backlash, the error $d_N(t)$ may be nonzero even when the inverse is exact.

- In spite of the unobservability of $z(t)$, when $G(D)$ and $N(\cdot)$ are known, the inverse control scheme

$$u(t) = \theta_1^{*T}(t)a_\lambda(D)[u](t) + \theta_2^{*T}b_\lambda(D)[\bar{z}](t) + z_m(t+n^*)$$

$$\bar{z}(t) = NI(y(t)) = \theta_N^{*T}\omega_N(t)$$

$$z_m(t) = NI(y_m(t)) = \theta_N^{*T}\omega_m(t)$$

ensures closed-loop signal boundedness. Moreover, under the initialization conditions (8.56) and (8.57), the output tracking is achieved:

$$y(t) = y_m(t),\ t \geq t_0 + n^* + n - 1.$$

- When $z(t)$ is measured and $G(D)$ and $N(\cdot)$ are known, the output tracking is achieved by the inverse controller

$$u(t) = \theta_1^{*T}(t)a_\lambda(D)[u](t) + \theta_2^{*T}b_\lambda(D)[z](t) + z_m(t+n^*)$$

under the condition $y(t_0 + n^* + n - 1) = y_m(t_0 + n^* + n - 1)$, which is needed only for backlash and hysteresis but not for dead-zone.

- Two more adaptive control schemes with exact inverse are, for $G(D)$ and $z(t)$ unknown:

$$u(t) = \theta_1^T(t)a_\lambda(D)[u](t) + \theta_2^T a_\lambda(D)[\bar{z}](t) + \theta_3 z_m(t+n^*)$$

and for $G(D)$ unknown but $z(t)$ known:

$$u(t) = \theta_1^T(t)a_\lambda(D)[u](t) + \theta_2^T b_\lambda(D)[z](t) + \theta_3 z_m(t+n^*).$$

- When $N(\cdot)$ is unknown, a detuned inverse implemented with an estimate θ_N of θ_N^*:

$$\widehat{z}(t) = \widehat{NI}(y(t)) = \theta_N^T \omega_N(t)$$

results in the observation error

$$\widehat{z}(t) - z(t) = \tilde{k}_N(t)z(t) + \bar{d}_N(t)$$

with both $\tilde{k}_N(t)$ and $\bar{d}_N(t)$ bounded.

- With detuned inverses

$$\widehat{z}(t) = \widehat{NI}(y(t)) = \theta_N^T \omega_N(t)$$

$$\widehat{z_m}(t) = \widehat{NI}(y_m(t)) = \theta_N^T \omega_m(t)$$

the fixed linear controller for $G(D)$ known:

$$u(t) = \theta_1^{*T}(t)a_\lambda(D)[u](t) + \theta_2^{*T} b_\lambda(D)[\widehat{z}](t) + \widehat{z_m}(t+n^*)$$

and the adaptive linear controller for $G(D)$ unknown:

$$u(t) = \theta_1^T(t)a_\lambda(D)[u](t) + \theta_2^T b_\lambda(D)[\widehat{z}](t) + \theta_3 \widehat{z_m}(t+n^*)$$

ensure closed-loop signal boundedness when $\tilde{k}_N$ is small.

Chapter 9

Adaptive Inverse Control for Output Nonlinearities

When the output nonlinearity is unknown, the control structure designed in Chapter 8 is augmented with an adaptive nonlinearity inverse. Such adaptive inverse controllers are developed in this chapter. As in Chapters 6 and 7, we will design two types of adaptive inverse controllers: one for a known linear part and the other for an unknown linear part. In both cases the output nonlinearity is unknown. For the linear part known, the linear control law is fixed, and for the linear part unknown, the linear controller structure is adaptive, with a new linear parametrization. Simulation results to show that the adaptive inverse controllers significantly improve tracking performance.

9.1 Control Objective

As in Section 8.2, we consider the discrete-time plant with a nonlinearity $N(\cdot)$ at the output of a linear part $G(D)$:

$$y(t) = N(z(t)), \ z(t) = G(D)[u](t), \tag{9.1}$$

where $N(\cdot)$ represents the dead-zone, backlash, or hysteresis characteristic parametrized by θ_d^*, θ_b^*, θ_h^* as in (8.9), (8.19), (8.33), and the linear part $G(D) = k_p \frac{Z(D)}{P(D)}$ satisfies the assumptions (A1) - (A4) in Section 4.1. Without loss of generality, throughout this chapter we assume that $k_p = 1$, considering that a $k_p \neq 1$ is represented by the unknown slopes of $N(\cdot)$.

As before, the adaptive control problems addressed in this chapter have two key features:

- The output $z(t)$ of the linear part $G(D)$ is not accessible for either measurement or control.

- The parameters of the nonlinear part $N(\cdot)$ are unknown.

Our approach is to develop adaptive versions of the inverses

$$\bar{z}(t) = NI(y(t)), \; z_m(t+n^*) = NI(y_m(t+n^*)).$$

From the discussion in Section 8.3, expressions (8.46) - (8.50), and Figure 8.4, with an estimate θ_N of θ_N^*, the adaptive inverses have the forms

$$\widehat{z}(t) = \widehat{NI}(y(t)) = \theta_N^T \omega_N(t) \tag{9.2}$$

$$\widehat{z_m}(t+n^*) = \widehat{NI}(y_m(t+n^*)) = \theta_N^T \omega_m(t+n^*), \tag{9.3}$$

where

$$\omega_N(t) \in \{\omega_d(t), \omega_b(t), \omega_h(t)\}$$

defined in (8.7), (8.15), (8.31), and

$$\omega_m(t) \in \{\omega_{dm}(t), \omega_{bm}(t), \omega_{hm}(t)\}$$

defined in (8.46) - (8.50). Furthermore, the observation error with the adaptive inverse (9.2) is

$$\widehat{z}(t) - z(t) = (\theta_N - \theta_N^*)^T \omega_N(t) + d_N(t), \tag{9.4}$$

where $d_N(t)$ is a bounded error.

For implementing the adaptive inverses (9.2) and (9.3) and for closed-loop stability, we need the components of θ_N to satisfy some conditions such as $m_0 < \widehat{m_r}(t) < \infty$, $m_0 < \widehat{m_l}(t) < \infty$, $-\infty < \widehat{c_l}(t) \leq \widehat{c_r}(t) < \infty$ for the dead-zone inverse. Such conditions are characterized by the following assumptions which are similar to those stated in Section 6.1:

> (**A8a**) For *dead-zone*, positive constants b_{r0}, b_{l0}, m_{r1}, m_{r2}, m_{l1}, m_{l2} are known such that $-b_{l0} \leq b_l \leq 0 \leq b_r \leq b_{r0}$, and $m_{r1} \leq \frac{1}{m_r} \leq m_{r2}$, $m_{l1} \leq \frac{1}{m_l} \leq m_{l2}$.
>
> (**A8b**) For *backlash*, positive constants m_1, m_2, c_{l0}, c_{r0} are known such that $m_1 \leq \frac{1}{m} \leq m_2$, $-c_{l0} \leq c_l \leq 0 \leq c_r \leq c_{r0}$.
>
> (**A8c**) For *hysteresis*, positive constants m_{t1}, m_{t2}, m_{b1}, m_{b2}, m_{r1}, m_{r2}, m_{l1}, m_{l2}, c_{t0}, c_{b0}, c_{r0}, c_{l0} are known such that $\max\{m_{r2}, m_{l2}\} < m_{t1} \leq \frac{1}{m_t} \leq m_{t2}$, $\max\{m_{r2}, m_{l2}\} < m_{b1} \leq \frac{1}{m_b} \leq m_{b2}$, $-c_{b0} \leq c_b \leq 0 \leq c_t \leq c_{t0}$, $-c_{l0} \leq c_l \leq 0 \leq c_r \leq c_{r0}$.

Assumption (A8c) implies that $0 \leq \frac{c_t}{m_t} \leq c_{t0} m_{t2}$ and $-c_{b0} m_{b2} \leq \frac{c_b}{m_b} \leq 0$. These assumptions will be used for parameter projection to ensure that the components of the inverse parameter θ_N remain in the sets defined by (A8a) - (A8c).

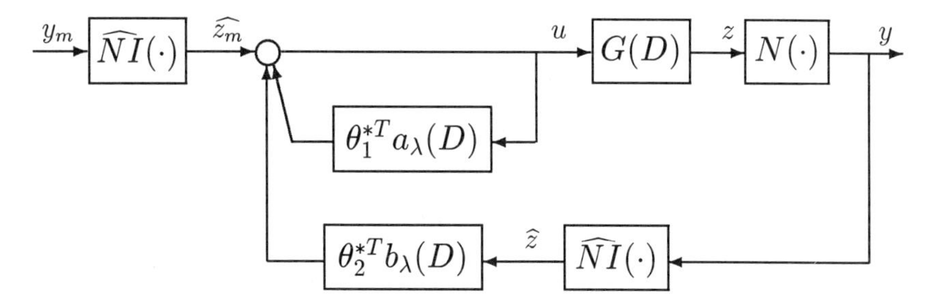

Figure 9.1: Explicit inverse controller for $G(D)$ known.

9.2 Designs for $G(D)$ Known

To choose a controller structure for the plant (9.1) with the linear part $G(D)$ known, we use an adaptive inverse $\widehat{NI}(\cdot)$ to replace the fixed inverse $NI(\cdot)$ in (8.51) and Figure 8.4. We present two designs: one with an explicit adaptive version $\widehat{z}(t) = \widehat{NI}(y(t))$ of the fixed inverse $\bar{z} = NI(y(t))$, and the other with an implicit adaptive version of $\bar{z} = NI(y(t))$.

9.2.1 Explicit Inverse Scheme

The first controller structure, shown in Figure 9.1, employs an explicit adaptive version $\widehat{z}(t) = \widehat{NI}(y(t))$ of the fixed inverse $\bar{z} = NI(y(t))$ in (8.51), namely,

$$u(t) = \theta_1^{*T} a_\lambda(D)[u](t) + \theta_2^{*T} b_\lambda(D)[\widehat{z}](t) + \widehat{z_m}(t+n^*), \tag{9.5}$$

where $\theta_1^* \in R^{n-1}$, $\theta_2^* \in R^n$, $a_\lambda(D)$, and $b_\lambda(D)$ are defined in (8.43) - (8.45). The signals $\widehat{z}(t)$ and $\widehat{z_m}(t)$ are obtained from the adaptive inverses

$$\widehat{z}(t) = \widehat{NI}(y(t)) = \theta_N^T(t)\omega_N(t) \tag{9.6}$$

$$\widehat{z_m}(t+n^*) = \widehat{NI}(y_m(t+n^*)) = \theta_N^T(t)\omega_m(t+n^*). \tag{9.7}$$

The parameter estimate $\theta_N(t) \in R^{n_N}$ is to be updated by an adaptive law. We first derive an error equation for $\theta_N(t) - \theta_N^*$. From (9.3) and (9.4) we have

$$z(t) = \theta_N^{*T}\omega_N(t) - d_N(t). \tag{9.8}$$

Substituting (9.8) into (8.68) and introducing

$$\omega_{NN}(t) = \theta_2^{*T} b_\lambda(D) I_{n_N}[\omega_N](t) + \omega_N(t+n^*) \tag{9.9}$$

$$d(t+n^*) = d_N(t+n^*) + \theta_2^{*T} b_\lambda(D)[d_N](t), \tag{9.10}$$

where I_{n_N} is the $n_N \times n_N$ identity matrix, we obtain

$$u(t) = \theta_1^{*T}\omega_1(t) + \theta_N^{*T}\omega_{NN}(t) - d(t+n^*), \tag{9.11}$$

where

$$\omega_1(t) = a_\lambda(D)[u](t).$$

We know that $d(t)$ is bounded for any $t \geq 0$, which in the hysteresis case is ensured by bounding the parameter estimates $\widehat{c_t}$, $\widehat{c_b}$, $\widehat{c_r}$, and $\widehat{c_l}$.

Defining the estimation error

$$\epsilon_N(t) = \theta_1^{*T}\omega_1(t-n^*) + \theta_N^T(t-1)\omega_{NN}(t-n^*) - u(t-n^*) \tag{9.12}$$

and using (9.11), we obtain

$$\epsilon_N(t) = (\theta_N(t-1) - \theta_N^*)^T \omega_{NN}(t-n^*) + d(t). \tag{9.13}$$

Based on this error equation we choose the adaptive update law for $\theta_N(t)$:

$$\theta_N(t) = \theta_N(t-1) - \frac{\Gamma_N \omega_{NN}(t-n^*)\epsilon_N(t)}{m_N^2(t)} + f_N(t), \tag{9.14}$$

where

$$m_N(t) = \sqrt{1 + \omega_{NN}^T(t-n^*)\omega_{NN}(t-n^*) + \xi_N^2(t)} \tag{9.15}$$

$$\begin{aligned} \xi_N(t) = {} & \theta_N^T(t-1)\theta_2^{*T} b_\lambda(D)[\omega_N](t-n^*) + \theta_N^T(t-1)\omega_N(t) \\ & -\theta_2^{*T} b_\lambda(D)[\theta_N^T\omega_N](t-n^*) - \theta_N^T(t-n^*)\omega_N(t). \end{aligned} \tag{9.16}$$

Again, $f_N(t)$ is a modification for robustness with respect to $d(t)$, and Γ_N is a positive diagonal step-size matrix. Both $f_N(t)$ and Γ_N are yet to be specified.

9.2.2 Implicit Inverse Scheme

The second controller structure is shown in Figure 9.2, where the logic block L implements (8.7), (8.15), and (8.31) to generate $\omega_N(t)$. For this design, we employ an implicit adaptive version of $\bar{z} = NI(y(t))$:

$$\begin{aligned} u(t) &= \theta_1^{*T} a_\lambda(D)[u](t) + \theta_N^T(t)\theta_2^{*T} b_\lambda(D) I_{n_N}[\omega_N](t) + \widehat{z_m}(t+n^*) \\ &= \theta_1^{*T} a_\lambda(D)[u](t) + \theta_N^T(t)\omega_{Nm}(t), \end{aligned} \tag{9.17}$$

where

$$\omega_{Nm}(t) = \theta_2^{*T} b_\lambda(D) I_{n_N}[\omega_N](t) + \omega_m(t+n^*)$$

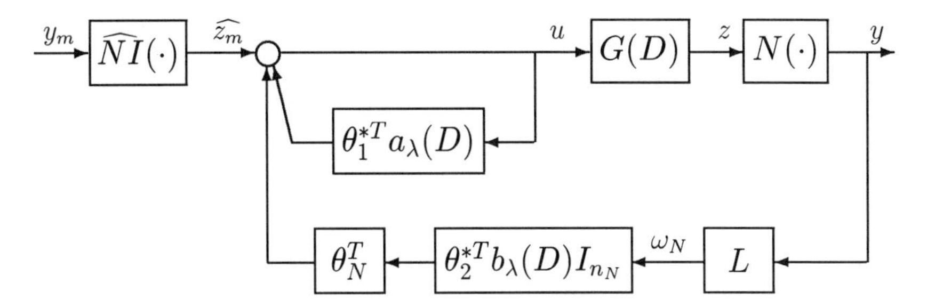

Figure 9.2: Implicit inverse controller for $G(D)$ known.

and θ_1^*, θ_2^*, $a_\lambda(D)$, $b_\lambda(D)$, $\omega_N(t)$, $\theta_N(t)$, $\widehat{z_m}(t)$ are as in Section 9.2.1.

In this design, the adaptive version $\widehat{z}(t) = \widehat{NI}(y(t))$ of the inverse $\bar{z}(t) = NI(y(t)) = \theta_N^{*T}\omega_N(t)$ is implicit because $\theta_N^T(t)\theta_2^{*T}b_\lambda(D)I_{n_N}[\omega_N](t)$ is the adaptive version of $\theta_2^{*T}b_\lambda(D)[\bar{z}](t)$; that is, for $\theta_N = \theta_N^*$, it follows that

$$\theta_N^T\theta_2^{*T}b_\lambda(D)I_{n_N}[\omega_N](t) = \theta_N^{*T}\theta_2^{*T}b_\lambda(D)I_{n_N}[\omega_N](t) = \theta_2^{*T}b_\lambda(D)[\theta_N^{*T}\omega_N](t).$$

The same estimation error expressions (9.12) and (9.13) are valid and suggest the following adaptive update law for $\theta_N(t)$:

$$\theta_N(t) = \theta_N(t-1) - \frac{\Gamma_N\omega_{NN}(t-n^*)\epsilon_N(t)}{m_N^2(t)} + f_N(t), \tag{9.18}$$

where $m_N(t)$ is as in (9.15), $\omega_{NN}(t)$ is as in (9.9), and

$$\begin{aligned}\xi_N(t) = {} & \theta_N^T(t-1)(\theta_2^{*T}b_\lambda(D)I_{n_N}[\omega_N](t-n^*) + \omega_N(t)) \\ & -\theta_N^T(t-n^*)(\theta_2^{*T}b_\lambda(D)I_{n_N}[\omega_N](t-n^*) + \omega_N(t)) \\ & (\theta_N(t-1) - \theta_N(t-n^*))^T\omega_{NN}(t-n^*).\end{aligned} \tag{9.19}$$

The quantities $f_N(t)$, a modification for robustness with respect to $d(t)$ and for parameter projection, and Γ_N, a positive diagonal matrix, are to be specified.

To compare the two designs, we use (9.5), (9.6), and (9.16) to express the estimation error $\epsilon_N(t)$ in (9.12) for the explicit inverse design as

$$\begin{aligned}\epsilon_N(t) &= \widehat{z}(t) - \widehat{z_m}(t) + \theta_N^T(t-1)\theta_2^{*T}b_\lambda(D)I_{n_N}[\omega_N](t-n^*) \\ &\quad -\theta_2^{*T}b_\lambda(D)[\theta_N^T\omega_N](t-n^*) \\ &= \theta_N^T(t-n^*)(\omega_N(t) - \omega_m(t)) + \xi_N(t)\end{aligned}$$

and (9.17) and (9.19) to express $\epsilon_N(t)$ for the implicit inverse design as

$$\begin{aligned}\epsilon_N(t) &= \widehat{z}(t) - \widehat{z_m}(t) + \theta_N^T(t-1)\theta_2^{*T}b_\lambda(D)I_{n_N}\omega_N](t-n^*) \\ &\quad -\theta_N^T(t-n^*)\theta_2^{*T}b_\lambda(D)I_{n_N}\omega_N](t-n^*) \\ &= \theta_N^T(t-n^*)(\omega_N(t) - \omega_m(t)) + \xi_N(t).\end{aligned}$$

Although the estimation error expressions are the same for both designs, they have different $\xi_N(t)$ in (9.16) and (9.19).

9.2.3 Adaptive Law Modifications

As in our earlier designs, there are two versions of the modification term $f_N(t)$: parameter projection and switching σ-modification with parameter projection.

Parameter projection. For parameter projection, we use assumptions (A8a) - (A8c) to specify a convex parameter region containing the true parameter θ_N^*.

For *dead-zone inverse*, assumption (A8a) implies that

$$\theta_N^* = \theta_d^* = (\frac{1}{m_r}, b_r, \frac{1}{m_l}, b_l)^T = (\theta_{N1}^*, \theta_{N2}^*, \theta_{N3}^*, \theta_{N4}^*)^T \tag{9.20}$$

$$\theta_{Nj}^* \in [\theta_{Nj}^a, \theta_{Nj}^b],\ j = 1, 2, 3, 4 \tag{9.21}$$

$$\theta_{N1}^a = m_{r1},\ \theta_{N2}^a = 0,\ \theta_{N3}^a = m_{l1},\ \theta_{N4}^a = -b_{l0} \tag{9.22}$$

$$\theta_{N1}^b = m_{r2},\ \theta_{N2}^b = b_{r0},\ \theta_{N3}^b = m_{l2},\ \theta_{N4}^b = 0. \tag{9.23}$$

For *backlash inverse*, assumption (A8b) implies that

$$\theta_N^* = \theta_b^* = (c_r, \frac{1}{m}, c_l)^T = (\theta_{N1}^*, \theta_{N2}^*, \theta_{N3}^*)^T \tag{9.24}$$

$$\theta_{Nj}^* \in [\theta_{Nj}^a, \theta_{Nj}^b],\ j = 1, 2, 3 \tag{9.25}$$

$$\theta_{N1}^a = 0,\ \theta_{N2}^a = m_1,\ \theta_{N3}^a = -c_{l0} \tag{9.26}$$

$$\theta_{N1}^b = c_{r0},\ \theta_{N2}^b = m_2,\ \theta_{N3}^b = 0. \tag{9.27}$$

For *hysteresis inverse*, assumption (A8c) implies that

$$\begin{aligned} \theta_N^* = \theta_h^* &= (\frac{1}{m_t}, \frac{c_t}{m_t}, \frac{1}{m_b}, \frac{c_b}{m_b}, \frac{1}{m_r}, c_r, \frac{1}{m_l}, c_l)^T \\ &= (\theta_{N1}^*, \theta_{N2}^*, \theta_{N3}^*, \theta_{N4}^*, \theta_{N5}^*, \theta_{N6}^*, \theta_{N7}^*, \theta_{N8}^*)^T \end{aligned} \tag{9.28}$$

$$\theta_{Nj}^* \in [\theta_{Nj}^a, \theta_{Nj}^b],\ j = 1, \ldots, 8 \tag{9.29}$$

$$\theta_{N1}^a = m_{t1},\ \theta_{N2}^a = 0,\ \theta_{N3}^a = m_{b1},\ \theta_{N4}^a = -c_{b0} \tag{9.30}$$

$$\theta_{N5}^a = m_{r1},\ \theta_{N6}^a = 0,\ \theta_{N7}^a = m_{l1},\ \theta_{N8}^a = -c_{l0} \tag{9.31}$$

$$\theta_{N1}^b = m_{t2},\ \theta_{N2}^b = c_{t0},\ \theta_{N3}^b = m_{b2},\ \theta_{N4}^b = 0 \tag{9.32}$$

$$\theta_{N5}^b = m_{r2},\ \theta_{N6}^b = c_{r0},\ \theta_{N7}^b = m_{l2},\ \theta_{N8}^b = 0. \tag{9.33}$$

To project the parameter estimates in $\theta_N(t) \in R^{n_N}$ to the parameter region $[\theta_{Nj}^a, \theta_{Nj}^b]$, $j = 1, 2, \ldots, n_N$, we first define

$$g_N(t) = -\frac{\Gamma_N \omega_{NN}(t - n^*)\epsilon_N(t)}{m_N^2(t)}$$

and denote the jth component of $\theta_N(t)$, $g_N(t)$, $f_N(t) \in R^{n_N}$ as $\theta_{Nj}(t)$, $g_{Nj}(t)$, $f_{Nj}(t)$, respectively, then choose

$$\Gamma_N = \{\gamma_1, \ldots, \gamma_{n_N}\},\ 0 < \gamma_j < 2$$

$$\theta_{Nj}(0) \in [\theta_{Nj}^a, \theta_{Nj}^b]$$

for $j = 1, 2, \ldots, n_N$. Finally, we let

$$\bar{\theta}_{Nj}(t) = \theta_{Nj}(t-1) + g_{Nj}(t)$$

and set

$$f_{Nj}(t) = \begin{cases} 0 & \text{if } \bar{\theta}_{Nj}(t) \in [\theta_{Nj}^a, \theta_{Nj}^b] \\ \theta_{Nj}^b - \bar{\theta}_{Nj}(t) & \text{if } \bar{\theta}_{Nj}(t) > \theta_{Nj}^b \\ \theta_{Nj}^a - \bar{\theta}_{Nj}(t) & \text{if } \bar{\theta}_{Nj}(t) < \theta_{Nj}^a. \end{cases} \tag{9.34}$$

Switching σ-modification with parameter projection. In this case,

$$f_N(t) = -\sigma_N(t)\theta_N(t) + f_p(t), \tag{9.35}$$

where

$$\sigma_N(t) = \begin{cases} \sigma_0 & \text{if } \|\theta_N(t)\|_2 > 2M_N \\ 0 & \text{otherwise} \end{cases}$$

$$0 < \sigma_0 < \frac{1}{2}(1 - \gamma_N),\ M_N \geq \|\theta_N^*\|_2$$

γ_N is the scalar adaptation step-size:

$$\Gamma_N = \gamma_N I_{n_N},\ 0 < \gamma_N < 1,$$

and $f_p(t)$ is for parameter projection which ensures that the estimate $\theta_N(t)$ remains in the region specified by (A8a) - (A8c). To implement parameter projection, for *dead-zone inverse*, we use (A8a) with

$$\theta_{N1}^a = m_{r1},\ \theta_{N2}^a = 0,\ \theta_{N3}^a = m_{l1},\ \theta_{N4}^b = 0 \tag{9.36}$$

for *backlash inverse*, we use (A8b) with

$$\theta_{N1}^a = 0,\ \theta_{N2}^a = m_1,\ \theta_{N3}^b = 0, \tag{9.37}$$

and, for *hysteresis inverse*, we use (A8c) with

$$\theta^a_{N1} = m_{t1},\ \theta^a_{N2} = 0,\ \theta^a_{N3} = m_{b1},\ \theta^a_{N4} = -c_{b0} \tag{9.38}$$

$$\theta^a_{N5} = m_{r1},\ \theta^a_{N6} = 0,\ \theta^a_{N7} = m_{l1},\ \theta^a_{N8} = -c_{l0} \tag{9.39}$$

$$\theta^b_{N1} = m_{t2},\ \theta^b_{N2} = c_{t0},\ \theta^b_{N3} = m_{b2} \tag{9.40}$$

$$\theta^b_{N4} = 0,\ \theta^b_{N6} = c_{r0},\ \theta^b_{N8} = 0. \tag{9.41}$$

For all other θ^a_{Nj}, θ^b_{Nj}, we let

$$\theta^a_{Nj} = -\infty,\ \theta^b_{Nj} = \infty. \tag{9.42}$$

For these parameters, projection is not needed because the switching-σ guarantees that $\theta_N(t) \in l^\infty$.

To construct an algorithm for $f_p(t)$, we let

$$\bar{g}_N(t) = -\frac{\Gamma_N \zeta_N(t)\epsilon_N(t)}{m_N^2(t)} - \sigma_N(t)\theta_N(t),$$

where $\Gamma_N = \gamma_N I_{n_N}$, $0 < \gamma_N < 1$, denote the jth component of $\theta_N(t)$, $f_p(t)$, $\bar{g}_N(t)$ as $\theta_{Nj}(t)$, $f_{pj}(t)$, $\bar{g}_{Nj}(t)$, respectively, choose $\theta_{Nj}(0) \in [\theta^a_{Nj}, \theta^b_{Nj}]$, and set

$$f_{pj}(t) = \begin{cases} 0 & \text{if } \theta_{Nj}(t-1) + \bar{g}_{Nj}(t) \in [\theta^a_{Nj}, \theta^b_{Nj}] \\ \theta^b_{Nj} - \theta_{Nj}(t-1) - \bar{g}_{Nj}(t) & \text{if } \theta_{Nj}(t-1) + \bar{g}_{Nj}(t) > \theta^b_{Nj} \\ \theta^a_{Nj} - \theta_{Nj}(t-1) - \bar{g}_{Nj}(t) & \text{if } \theta_{Nj}(t-1) + \bar{g}_{Nj}(t) < \theta^a_{Nj} \end{cases}$$

for $j = 1, 2, \ldots, n_N$.

The main properties of the explicit and implicit inverse designs are established, as in Lemma 9.1, by evaluating the time increment $V(\tilde{\theta}_N(t)) - V(\tilde{\theta}_N(t-1))$ of the positive definite function

$$V(\tilde{\theta}_N) = \tilde{\theta}_N^T \Gamma_N^{-1} \tilde{\theta}_N,\ \tilde{\theta}_N = \theta_N - \theta_N^*$$

along the trajectories of (9.14) or (9.18) with either of the two algorithms (9.34) or (9.35).

Lemma 9.1 *The adaptive law (9.14) or (9.18) with parameter projection (9.34) or switching σ-modification with parameter projection (9.35) ensures that $\theta_N(t)$ and $\frac{\epsilon_N(t)}{m_N(t)}$ are bounded and that*

$$\sum_{t=t_1}^{t_2} \frac{\epsilon_N^2(t)}{m_N^2(t)} \leq a_1 + b_1 \sum_{t=t_1}^{t_2} \frac{d^2(t)}{m_N^2(t)}$$

$$\sum_{t=t_1}^{t_2} \|\theta_N(t) - \theta_N(t-1)\|_2^2 \leq a_2 + b_2 \sum_{t=t_1}^{t_2} \frac{d^2(t)}{m_N^2(t)}$$

for some constants $a_i > 0$, $b_i > 0$, $i = 1, 2$, and all $t_2 > t_1 \geq 0$.

Using Lemma 9.1, in Appendix D, we prove the following boundedness result.

Theorem 9.1 *The adaptive inverse controller consisting of the linear controller structure (9.5) or (9.17) and the adaptive inverses (9.6) and (9.7) updated by (9.14) or (9.18) ensures that all closed-loop signals are bounded.*

9.3 Design for $G(D)$ Unknown

When both $G(D)$ and $N(\cdot)$ in (9.1) are unknown, we need an adaptive design which leads to a linear parametrization suitable for adaptation of an inverse and a linear controller structure.

9.3.1 Controller Structure

A bilinear parametrization would result when a parameter estimate $\theta_2(t)$ is used to replace θ_2^* in the explicit inverse controller (9.5). Instead, replacing θ_1^* and θ_2^* by their estimates in the implicit inverse controller (9.17), our new controller structure for $G(D)$ unknown, shown in Figure 9.3, employs

$$\theta_5(t) = \theta_2(t) \otimes \theta_N(t),$$

which is the estimate of the new parameter vector

$$\theta_5^* = \theta_2^* \otimes \theta_N^*.$$

The new filter $B_\lambda(D)$ is

$$B_\lambda(D) = \begin{cases} (D^{-n+1}I_4, \ldots, D^{-1}I_4, I_4)^T & \text{for } \widehat{DI}(\cdot) \\ (D^{-n+1}I_3, \ldots, D^{-1}I_3, I_3)^T & \text{for } \widehat{BI}(\cdot) \\ (D^{-n+1}I_8, \ldots, D^{-1}I_8, I_8)^T & \text{for } \widehat{HI}(\cdot) \end{cases}$$

while the logic block L generates the signal $\omega_N(t)$. The adaptive version of $\theta_N^T(t)\theta_2^{*T}b_\lambda(D)I_{n_N}[\omega_N](t)$ in (9.17) becomes

$$\theta_N^T(t)\theta_2^T(t)b_\lambda(D)I_{n_N}[\omega_N](t) = (\theta_2(t) \otimes \theta_N(t))^T B_\lambda(D)[\omega_N](t)$$

and, hence, the adaptive controller structure is

$$u(t) = \theta_1^T(t)a_\lambda(D)u(t) + \theta_5^T(t)B_\lambda(D)[\omega_N](t) + \widehat{z_m}(t+n^*). \tag{9.43}$$

As in Section 9.2, the signal $\widehat{z_m}(t+n^*)$ is obtained from the adaptive inverse

$$\widehat{z_m}(t+n^*) = \widehat{NI}(y_m(t+n^*) = \theta_N^T(t)\omega_m(t+n^*). \tag{9.44}$$

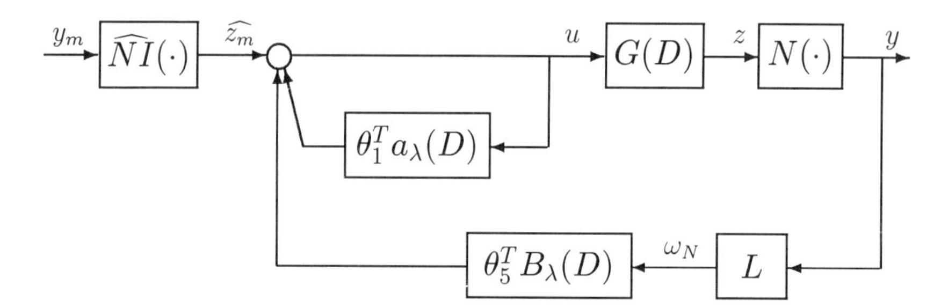

Figure 9.3: Adaptive inverse controller for $G(D)$ unknown.

9.3.2 Adaptive Update Law

To develop an adaptive update law for $\theta_1(t)$, $\theta_5(t)$, and $\theta_N(t)$ in (9.43), we introduce the new regressor

$$\omega_5(t) = B_\lambda(D)[\omega_N](t),$$

substitute (9.8) in (8.68), and use (9.10) to obtain

$$u(t) = \theta_1^{*T}\omega_1(t) + \theta_5^{*T}\omega_5(t) + \theta_N^{*T}\omega_N(t+n^*) - d(t). \tag{9.45}$$

In view of (9.45), we define the estimation error as

$$\begin{aligned}\epsilon(t) = {} & \theta_1^T(t-1)\omega_1(t-n^*) + \theta_5^T(t-1)\omega_5(t-n^*) \\ & + \theta_N^T(t-1)\omega_N(t) - u(t-n^*).\end{aligned} \tag{9.46}$$

Denoting the overall system parameter vectors by

$$\theta(t) = (\theta_N^T(t), \theta_1^T(t), \theta_5^T(t))^T \tag{9.47}$$

$$\theta^* = (\theta_N^*, \theta_1^*, \theta_5^*)^T \tag{9.48}$$

and the corresponding regressor by

$$\omega(t) = (\omega_N^T(t+n^*), \omega_1^T(t), \omega_5^T(t))^T \tag{9.49}$$

and using (9.46) - (9.49), we obtain the error equation

$$\epsilon(t) = (\theta(t-1) - \theta^*)^T\omega(t-n^*) + d(t).$$

Based on this error equation, we update $\theta(t)$ with

$$\theta(t) = \theta(t-1) - \frac{\Gamma\omega(t-n^*)\epsilon(t)}{1 + \omega^T(t-n^*)\omega(t-n^*) + \xi^2(t)} + f(t), \tag{9.50}$$

where

$$\xi(t) = (\theta(t-1) - \theta(t-n^*))^T \omega(t-n^*).$$

As before, there are two designs for $f(t)$: parameter projection and switching σ-modification with parameter projection.

Parameter projection. For parameter projection, we make the following assumption:

(**A9**) Each component θ_j^* of $\theta^* \in R^{n_\theta}$, $j = 1, \ldots, n_\theta$, belongs to a known interval: $\theta_j^* \in [\theta_j^a, \theta_j^b]$.

The bounds θ_{Nj}^a and θ_{Nj}^b for θ_N are characterized by (9.20) - (9.33), based on the assumptions (A8a) - (A8c). For Γ, we take

$$\Gamma = diag\{\gamma_1, \ldots, \gamma_{n_\theta}\},\ 0 < \gamma_j < 2,\ 1 \le j \le n_\theta.$$

Then we define

$$g(t) = -\frac{\Gamma\omega(t-n^*)\epsilon(t)}{1 + \omega^T(t-n^*)\omega(t-n^*) + \xi^2(t)},$$

denote the jth components of $\theta(t)$, $f_\theta(t)$, $g(t)$ as $\theta_j(t)$, $f_{\theta j}(t)$, $g_j(t)$, respectively, and set

$$f_j(t) = \begin{cases} 0 & \text{if } \theta_j(t-1) + g_j(t) \in [\theta_j^a, \theta_j^b] \\ \theta_j^b - \theta_j(t-1) - g_j(t) & \text{if } \theta_j(t-1) + g_j(t) > \theta_j^b \\ \theta_j^a - \theta_j(t-1) - g_j(t) & \text{if } \theta_j(t-1) + g_j(t) < \theta_j^a. \end{cases} \tag{9.51}$$

This algorithm is initialized with $\theta_j(0) \in [\theta_j^a, \theta_j^b]$.

Switching σ-modification with parameter projection. We choose

$$f(t) = -\sigma(t)\theta(t) + f_p(t) \tag{9.52}$$

$$\Gamma = \gamma I_{n_\theta},\ 0 < \gamma < 1$$

$$\sigma(t) = \begin{cases} \sigma_0 & \text{if } \|\theta(t)\|_2 > 2M_\theta \\ 0 & \text{otherwise} \end{cases}$$

$$0 < \sigma_0 < \frac{1}{2}(1-\gamma),\ \|\theta^*\|_2 < M_\theta$$

and use $f_p(t)$ to project θ_N within the boundaries (9.36) - (9.42). To design $f_p(t)$, we denote the jth components of $\theta(t)$, $f_p(t) \in R^{n_\theta}$ as $\theta_j(t)$, $f_{pj}(t)$, respectively, and set

$$f_{pj}(t) = \begin{cases} f_{Nj}(t) & \text{for } j = 1, 2, \ldots, n_N \\ 0 & \text{otherwise} \end{cases}$$

$$f_{Nj}(t) = \begin{cases} 0 & \text{if } \theta_{Nj}(t-1) + \bar{g}_j(t) \in [\theta_{Nj}^a, \theta_{Nj}^b] \\ \theta_{Nj}^b - \theta_{Nj}(t-1) - \bar{g}_j(t) & \text{if } \theta_{Nj}(t-1) + \bar{g}_j(t) > \theta_{Nj}^b \\ \theta_{Nj}^a - \theta_{Nj}(t-1) - \bar{g}_j(t) & \text{if } \theta_{Nj}(t-1) + \bar{g}_j(t) < \theta_{Nj}^a, \end{cases}$$

where $\bar{g}_j(t)$ is the jth component of

$$\bar{g}(t) = -\frac{\Gamma\omega(t-n^*)\epsilon(t)}{1+\omega^T(t-n^*)\omega(t-n^*)+\xi^2(t)} - \sigma(t)\theta(t)$$

for $j = 1, 2, \ldots, n_N$.

When initialized with

$$\theta_j(0) \in \begin{cases} [\theta_j^a, \theta_j^b] = [\theta_{Nj}^a, \theta_{Nj}^b] & \text{for } j = 1, 2, \ldots, n_N \\ \text{no constraint} & \text{otherwise} \end{cases}$$

the above adaptive scheme has the following properties.

Lemma 9.2 *The adaptive law (9.50) with (9.51) or (9.52) guarantees that* $\theta(t)$, $\frac{\epsilon(t)}{m(t)} \in l^\infty$ *and that*

$$\sum_{t=t_1}^{t_2} \frac{\epsilon^2(t)}{m^2(t)} \leq a_1 + b_1 \sum_{t=t_1}^{t_2} \frac{d^2(t)}{m^2(t)}$$

$$\sum_{t=t_1}^{t_2} \|\theta(t) - \theta(t-1)\|_2^2 \leq a_2 + b_2 \sum_{t=t_1}^{t_2} \frac{d^2(t)}{m^2(t)}$$

for some constants $a_i > 0, b_i > 0$, $i = 1, 2$, *and all* $t_2 > t_1 \geq 0$, *where*

$$m(t) = \sqrt{1+\omega^T(t-n^*)\omega(t-n^*)+\xi^2(t)}.$$

Proof: As in Lemmas 4.4 and 7.1, we let $\gamma_m = \max\{\gamma_1, \ldots, \gamma_{n_\theta}\} < 2$ and define $V(\tilde{\theta}) = \tilde{\theta}^T\Gamma^{-1}\tilde{\theta}$, $\tilde{\theta}(t) = \theta(t) - \theta^*$. For the parameter projection algorithm, we have

$$V(\tilde{\theta}(t)) - V(\tilde{\theta}(t-1)) \leq -\frac{2-\gamma_m}{2}\frac{\epsilon^2(t)}{m^2(t)} + \frac{2}{2-\gamma_m}\frac{d^2(t)}{m^2(t)} - f^T(t)\Gamma f(t)$$

$$|\theta_j(t+1) - \theta_j(t)| \leq |g_j(t)|,\ 1 \leq j \leq n_\theta$$

and $\theta_j(t) \in [\theta_j^a, \theta_j^b]$, $j = 1, \ldots, n_\theta$. Hence, $\frac{\epsilon(t)}{m(t)}$ is also bounded and the two inequalities in the lemma hold.

For the switching σ-modification with parameter projection, we have

$$\begin{aligned} V(\tilde{\theta}(t)) - V(\tilde{\theta}(t-1)) \leq & -\frac{\gamma}{2}\frac{\epsilon^2(t)}{m^2(t)} + \frac{2}{2-\gamma}\frac{d^2(t)}{m^2(t)} \\ & -\frac{\sigma(t)}{2}\theta^T(t)\theta(t) - f_p^T(t)\gamma^{-1}f_p(t), \end{aligned}$$

which also proves the lemma. ∇

In Appendix D we prove the following boundedness result.

Theorem 9.2 *The adaptive inverse controller, consisting of the linear controller structure (9.43) and the adaptive inverse (9.44) and updated by (9.50), ensures that all closed-loop signals are bounded.*

9.4 Examples: Adaptive Dead-Zone Inverse

We present two output dead-zone inverse examples: one to illustrate various control schemes applied to a first-order plant, and the other as an illustration of the scheme in Section 9.2.2 applied to a second-order plant.

9.4.1 First-Order Plant

Four different controllers were simulated for the plant:

$$y(t) = DZ(z(t)),\ z(t) = G(D)[u](t),\ G(D) = \frac{1}{D + a_1},\ a_1 = 1.7$$

with the unknown dead-zone parameters

$$m_r = 0.01,\ m_l = 0.013,\ b_r = 21,\ b_l = -27.$$

From (8.43), we have $\theta_1^* = 0$ and $\theta_2^* = a_1 = 1.7$.

(a) The fixed linear controller without dead-zone inverse:

$$u(t) = \theta_2^* \theta_m^* y(t) + \theta_m^* y_m(t+1),\ \theta_m^* = 100$$

would achieve output tracking $y(t) = y_m(t),\ t \geq 1$, if the dead-zone were absent: $m_l = m_r = 0.01,\ b_r = b_l = 0$.

(b) The adaptive linear controller without dead-zone inverse:

$$u(t) = \theta_y(t)y(t) + \theta_m(t)y_m(t+1)$$

would achieve asymptotic tracking $\lim_{t\to\infty}(y(t) - y_m(t)) = 0$ if the dead-zone were absent. The initial estimates $\theta_y(0) = 294.5$ of $\theta_y^* = 170$ and $\theta_m(0) = 95$ of $\theta_m^* = 100$ were used for simulation.

(c) The fixed linear controller with an implicit adaptive dead-zone inverse as in (9.17) is

$$u(t) = \theta_d^T(t)\omega_{dm}(t+1)$$

$$\omega_{dm}(t+1) = (\theta_2^* \widehat{\chi_r}(t) y(t) + \widehat{\chi_{rm}}(t+1) y_m(t+1), \theta_2^* \widehat{\chi_r}(t) + \widehat{\chi_{rm}}(t+1),$$
$$\theta_2^* \widehat{\chi_l}(t) y(t) + \widehat{\chi_{lm}}(t+1) y_m(t+1), \theta_2^* \widehat{\chi_l}(t) + \widehat{\chi_{lm}}(t+1))^T$$

with the initial estimate $\theta_d(0) = (95, 23.5, 80, -22.3)^T$ of the true dead-zone parameter vector $\theta_d^* = (100, 21, 76.92, -27)^T$.

(d) The adaptive linear controller with the adaptive dead-zone inverse as in (9.43) is

$$u(t) = \theta_5^T(t)\omega_5(t) + \theta_d^T(t)\omega_m(t+1)$$

$$\omega_5(t) = (\widehat{\chi_r}(t)y(t), \widehat{\chi_r}(t), \widehat{\chi_l}(t)y(t), \widehat{\chi_l}(t))^T$$

$$\omega_m(t+1) = (\widehat{\chi_{rm}}(t+1)y_m(t+1), \widehat{\chi_{rm}}(t+1),$$
$$\widehat{\chi_{lm}}(t+1)y_m(t+1), \widehat{\chi_{lm}}(t+1))^T$$

with the same initial estimate $\theta_d(0)$ as in the controller (c), and the initial estimate $\theta_5(0) = (294.5, 70.5, 248, 69.13)^T$ of the true parameter vector $\theta_5^* = (170, 35.7, 130.78, 45.9)^T$.

Typical responses in Figure 9.4 for $y_m(t) = 10 \sin 0.0942t$ show that the control schemes (a) and (b) which ignore the presence of the dead-zone result in large tracking errors, while the adaptive dead-zone inverse control schemes (c) and (d) lead to asymptotic tracking. For the control schemes (c) and (d), the estimates $\widehat{m_r}$, $\widehat{m_l}$, $\widehat{b_r}$, $\widehat{b_l}$ of m_r, m_l, b_r, b_l, shown in Figure 9.5, converge to their true values.

9.4.2 Second-Order Plant

Consider the plant (9.1) with a known second-order linear part

$$G(D) = \frac{1}{D^2 - 3D + 5}$$

and an unknown dead-zone $DZ(\cdot)$:

$$m_r = 0.01,\ m_l = 0.013,\ b_r = 21,\ b_l = -27.$$

From (8.43) we obtain $\theta_1^* = -3$, $\theta_2^* = (15, -4)^T$. The controller structure is (9.17) with $\theta_d(0) = (95, 23.5, 80, -22.3)^T$ as the initial estimate of $\theta_d^* = (100, 21, 76.92, -27)^T$.

The adaptive controller reduces the tracking error to a very small value, as shown in Figure 9.6. The estimates $\widehat{m_r}$, $\widehat{m_l}$, $\widehat{b_r}$, $\widehat{b_l}$ of m_r, m_l, b_r, b_l are shown in Figure 9.7 for different $y_m(t)$. It is significant that for some $y_m(t)$ the

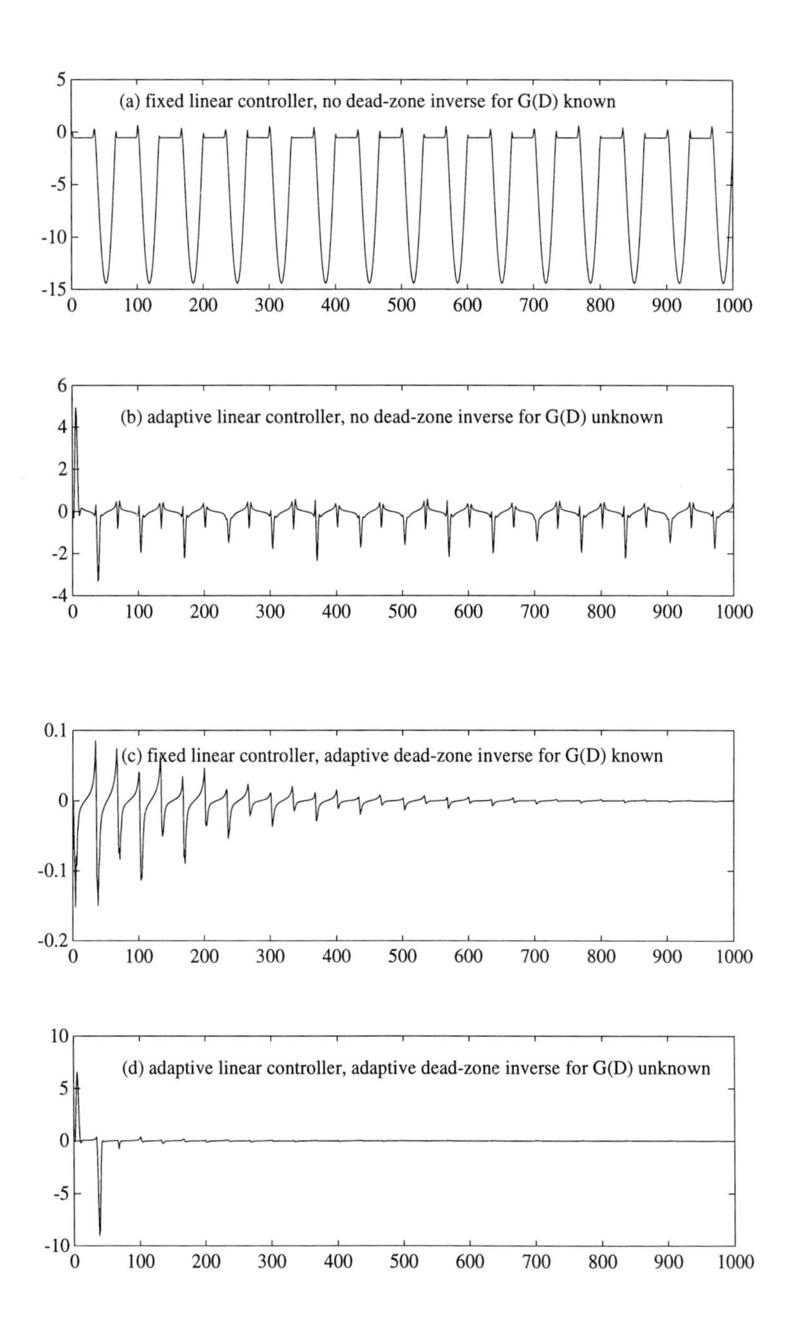

Figure 9.4: Tracking errors with controller (a) - (d) for the first-order example.

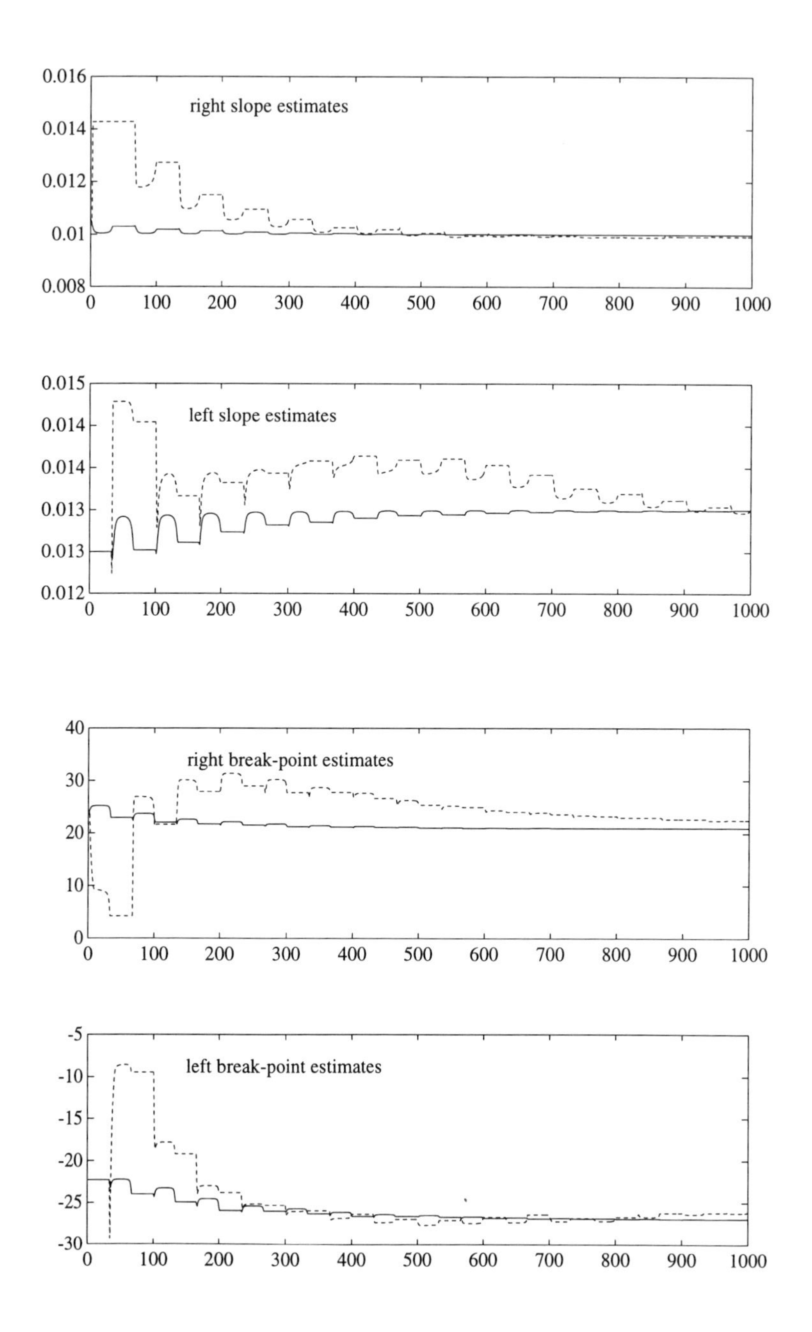

Figure 9.5: Estimates of the slopes m_r, m_l and break-points b_r, b_l for the first-order example (solid: controller (c), dashed: controller (d)).

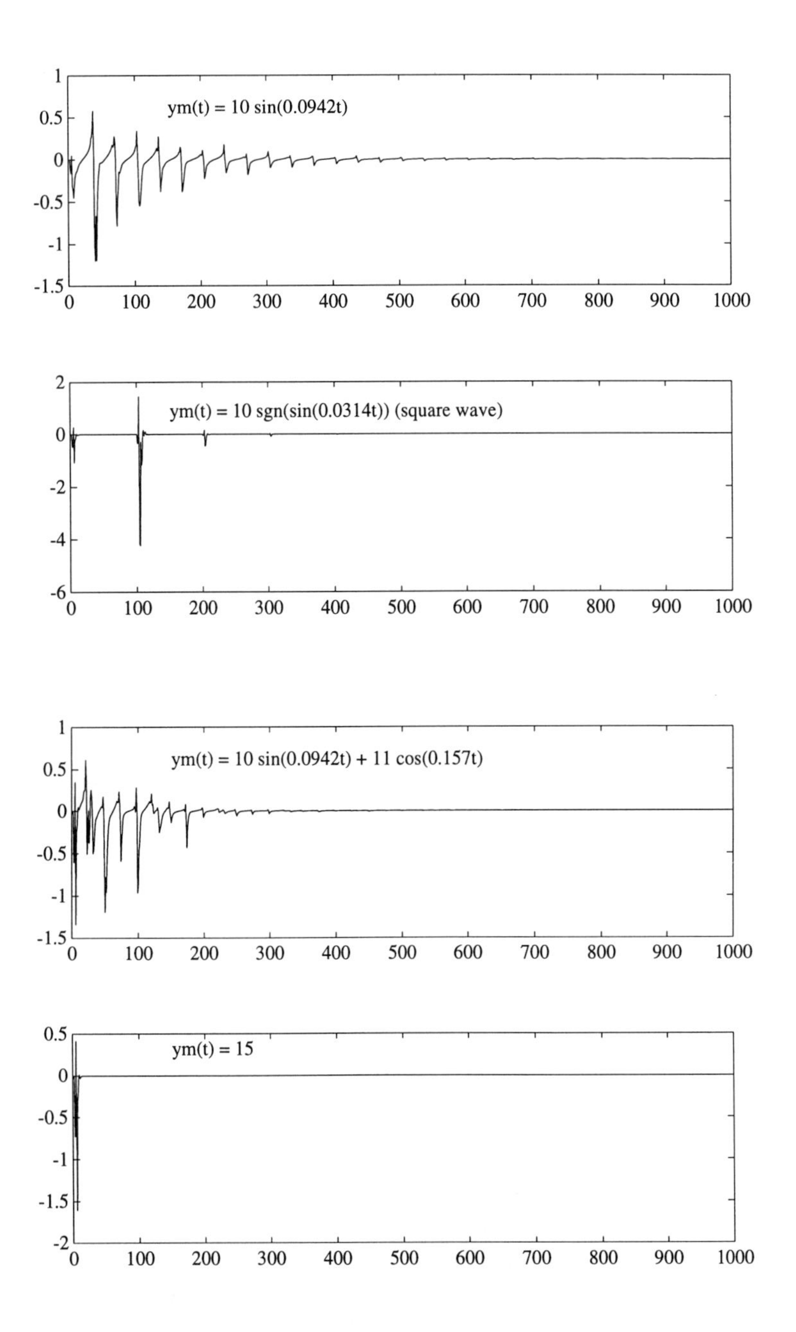

Figure 9.6: Tracking errors for different $y_m(t)$ for the second-order example.

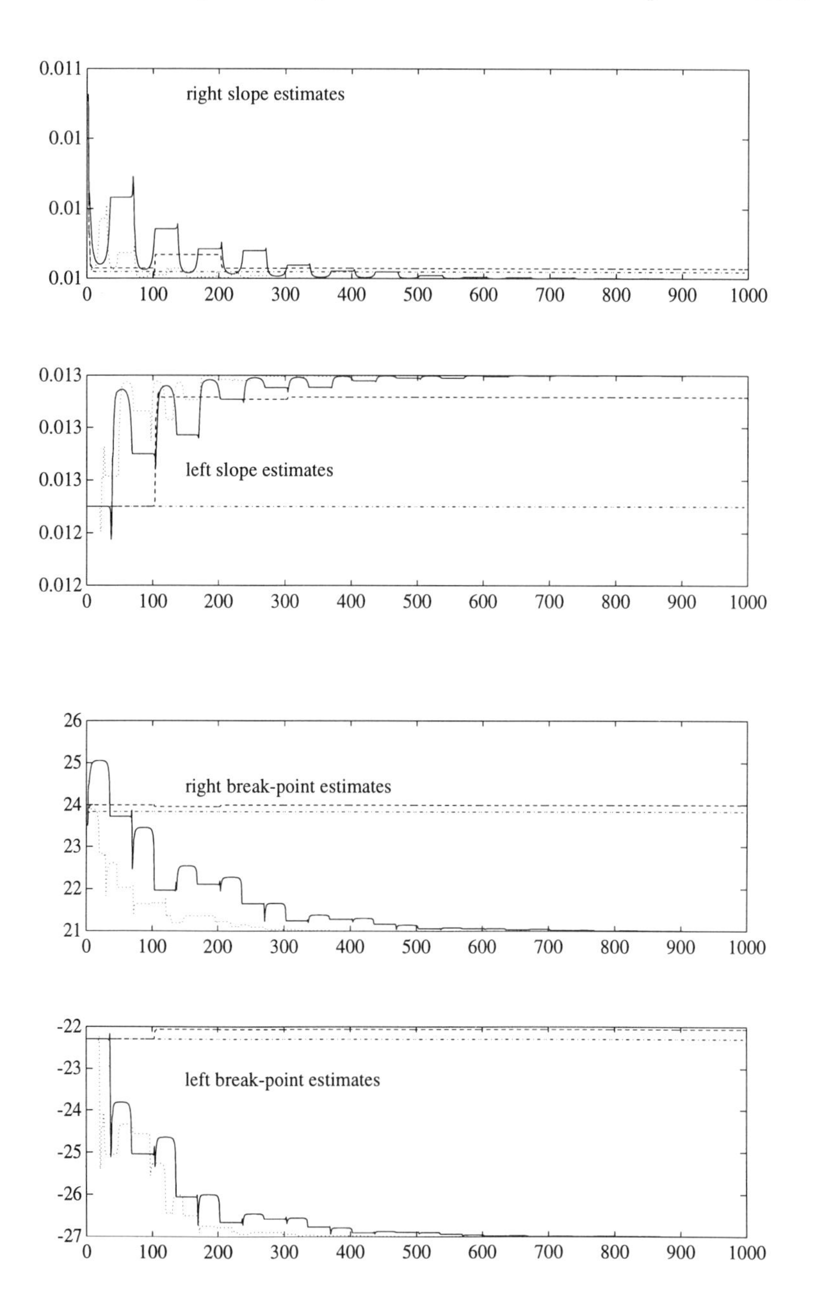

Figure 9.7: Estimates of m_r, m_l, b_r, b_l for the second-order example [solid: $y_m(t) = 10 \sin 0.0942t$; dashed: $y_m(t) = 10\, sgn(\sin 0.0314t)$ (square wave); dotted: $y_m(t) = 10 \sin 0.0942t + 11 \cos 0.157t$; dash-dotted: $y_m(t) = 15$].

parameter errors do not need to be small in order for the tracking error to be small. If, however, a fixed or an adaptive linear controller without a dead-zone inverse is applied to the same plant, then it was observed that the closed-loop system may have unbounded responses.

9.5 Examples: Adaptive Backlash Inverse

For a first-order plant and a second-order plant with an unknown output backlash, simulations show that the adaptive backlash inverse control schemes developed in Sections 9.2.2 and 9.3 lead to significant improvements of system tracking performance.

9.5.1 First-Order Plant

Four different controllers were simulated for the plant:

$$y(t) = B(z(t)),\ z(t) = G(D)[u](t),\ G(D) = \frac{1}{D + a_1},\ a_1 = 1.83$$

with the unknown backlash parameters

$$m = 0.014,\ c_r = 22.5,\ c_l = -24.7.$$

From (8.43) with $\theta_1^* = 0$ we have $\theta_2^* = a_1 = 1.83$. As presented next, the adaptive backlash inverse controller (c) is as in Section 9.2.2, and (d) is as in Section 9.3.1. The controllers (a) and (b) disregard the presence of backlash and serve for comparison.

(a) The fixed linear controller without backlash inverse

$$u(t) = \theta_2^* \theta_m^* y(t) + \theta_m^* y_m(t+1),\ \theta_m^* = \frac{500}{7}$$

would achieve output tracking in the absence of backlash.

(b) The adaptive linear controller without backlash inverse

$$u(t) = \theta_y(t) y(t) + \theta_m(t) y_m(t+1)$$

with the initial estimates $\theta_y(0) = 175$ of $\theta_y^* = \frac{915}{7}$ and $\theta_m(0) = 70$ of $\theta_m^* = \frac{500}{7}$ would achieve asymptotic output tracking if the backlash were absent.

(c) The fixed linear controller with an adaptive backlash inverse is

$$u(t) = \theta_b^T(t) \omega_{bm}(t+1)$$

$$\omega_{bm}(t+1) = (\theta_2^* \widehat{\chi_r}(t) + \widehat{\chi_{rm}}(t+1), \theta_2^* y(t) + y_m(t+1),$$
$$\theta_2^* \widehat{\chi_l}(t) + \widehat{\chi_{lm}}(t+1))^T$$

with $\theta_b(0) = (10, 70, -10)^T$ and $\theta_b^* = (22.5, \frac{500}{7}, -24.7)^T$.

(d) The adaptive linear controller with an adaptive backlash inverse is

$$u(t) = \theta_5^T(t)\omega_5(t) + \theta_b^T(t)\omega_m(t+1)$$

$$\omega_5(t) = (\widehat{\chi_r}(t), y(t), \widehat{\chi_l}(t))^T$$

$$\omega_m(t) = (\widehat{\chi_{rm}}(t+1), y_m(t+1), \widehat{\chi_{lm}}(t+1))^T$$

with the initial estimates $\theta_b(0)$, the same as in the above controller (c), and $\theta_5(0) = (25, 175, -25)^T$, of θ_b^* and $\theta_5^* = (41.175, \frac{915}{7}, -45.201)^T$.

Typical responses in Figure 9.8 for $y_m(t) = 15 \sin 0.0754t$ show that the control schemes (a) and (b) which ignore backlash result in large tracking errors. In particular, the adaptive scheme (b) ignoring backlash exhibits bursting shown in Figure 9.8(b). In contrast, our adaptive inverse schemes (c) and (d), which take into account the effects of the unknown backlash, result in very small tracking errors.

9.5.2 Second-Order Plant

We consider the plant (9.1) with

$$G(D) = \frac{1}{D^2 + a_2 D + a_1}, \ a_2 = -1.3, \ a_1 = -0.3$$

and the unknown output backlash $B(\cdot)$ which is the same as in the above first-order example, and the analogous four controllers are designed and simulated:

Controller (a):

$$u(t) = \theta_1^* u(t-1) + \theta_m^* \theta_2^{*T}(y(t-1), y(t))^T + \theta_m^* y_m(t+2)$$

$$\theta_1^* = -1.3, \ \theta_2^* = (-0.39, -1.99)^T, \ \theta_m^* = \frac{500}{7}.$$

Controller (b):

$$u(t) = \theta_1(t)u(t-1) + \theta_y^T(t)(y(t-1), y(t))^T + \theta_m(t)y_m(t+2)$$

$$(\theta_1(0), \theta_y^T(0), \theta_m(0)) = (\theta_1^*, \theta_2^{*T}\theta_m^*, \theta_m^*).$$

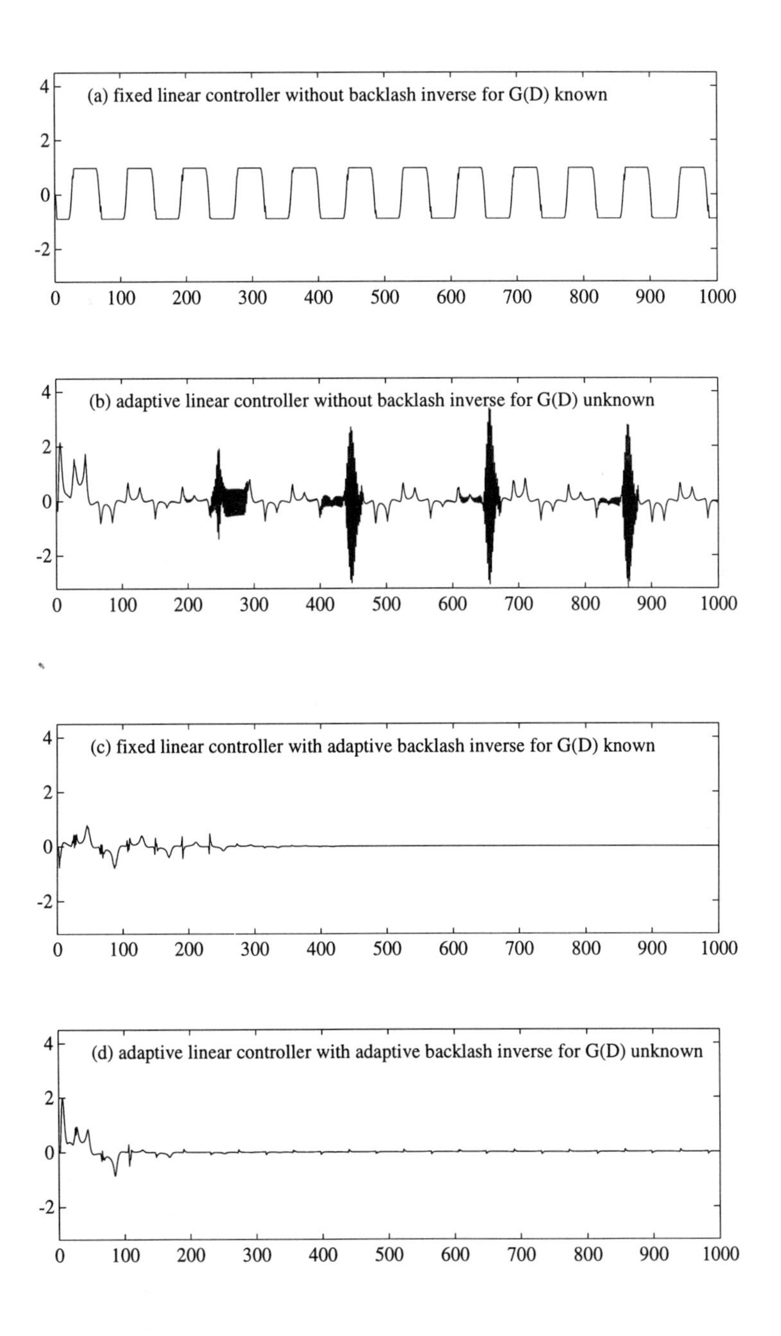

Figure 9.8: Tracking errors with controllers (a) - (d) for the first-order example.

Controller (c):

$$u(t) = \theta_1^* u(t-1) + \theta_b^T(t)\omega_{bm}(t+2)$$

$$\begin{aligned}\omega_{bm}(t+2) = & (\theta_2^{*T}(\widehat{\chi_r}(t-1), \widehat{\chi_r}(t))^T + \widehat{\chi_{rm}}(t+2)) \\ & +\theta_2^{*T}(y(t-1), y(t))^T + y_m(t+2)) \\ & +\theta_2^{*T}(\widehat{\chi_l}(t-1), \widehat{\chi_l}(t))^T + \widehat{\chi_{lm}}(t+2))\end{aligned}$$

$$\theta_b(0) = (21, 80.5, -21).$$

Controller (d):

$$u(t) = \theta_1(t)u(t-1) + \theta_5^T(t)\omega_5(t) + \theta_b^T(t)\omega_m(t+2)$$

$$\omega_5(t) = (\widehat{\chi_r}(t-1), \widehat{\chi_r}(t), y(t-1), y(t), \widehat{\chi_l}(t-1), \widehat{\chi_l}(t))^T$$

$$\omega_m(t+2) = (\chi_{rm}(t+2), y_m(t+2), \chi_{lm}(t+2))^T$$

$$\theta_1(0) = -1.2,\ \theta_5(0) = (24.9\theta_2^T(0), 80.5\theta_2^T(0), -25.3\theta_2^T(0))^T$$

$$\theta_2(0) = (-0.3, -1.8)^T,\ \theta_b(0) = (24.9, 80.5, -25.3).$$

Typical responses for $y_m(t) = 15\sin 0.0754t$ in Figure 9.9 show that again the control schemes (a) and (b) which ignore backlash lead to large tracking errors. The control schemes (c) and (d) which have an adaptive backlash inverse result in very small tracking errors.

9.6 Summary

- For a plant with an output nonlinearity

$$y(t) = N(z(t)),\ z(t) = G(D)[u](t)$$

an adaptive inverse controller employs the adaptive inverse

$$\widehat{z}(t) = \widehat{NI}(y(t)) = \theta_N^T \omega_N(t)$$

explicitly or *implicitly*, to cancel the effects of $N(\cdot)$, where θ_N is an adaptive estimate of θ_N^* which parametrizes the unknown nonlinearity $N(\cdot)$.

- The control scheme with an *explicit* $\widehat{z}(t)$ for $G(D)$ known is

$$u(t) = \theta_1^{*T}\omega_1(t) + \theta_2^{*T} b_\lambda(D)[\widehat{z}](t) + \widehat{z_m}(t+n^*),\ \omega_1(t) = a_\lambda(D)[u](t)$$

$$\widehat{z_m}(t+n^*) = \widehat{NI}(y_m(t+n^*)) = \theta_N^T(t)\omega_m(t+n^*).$$

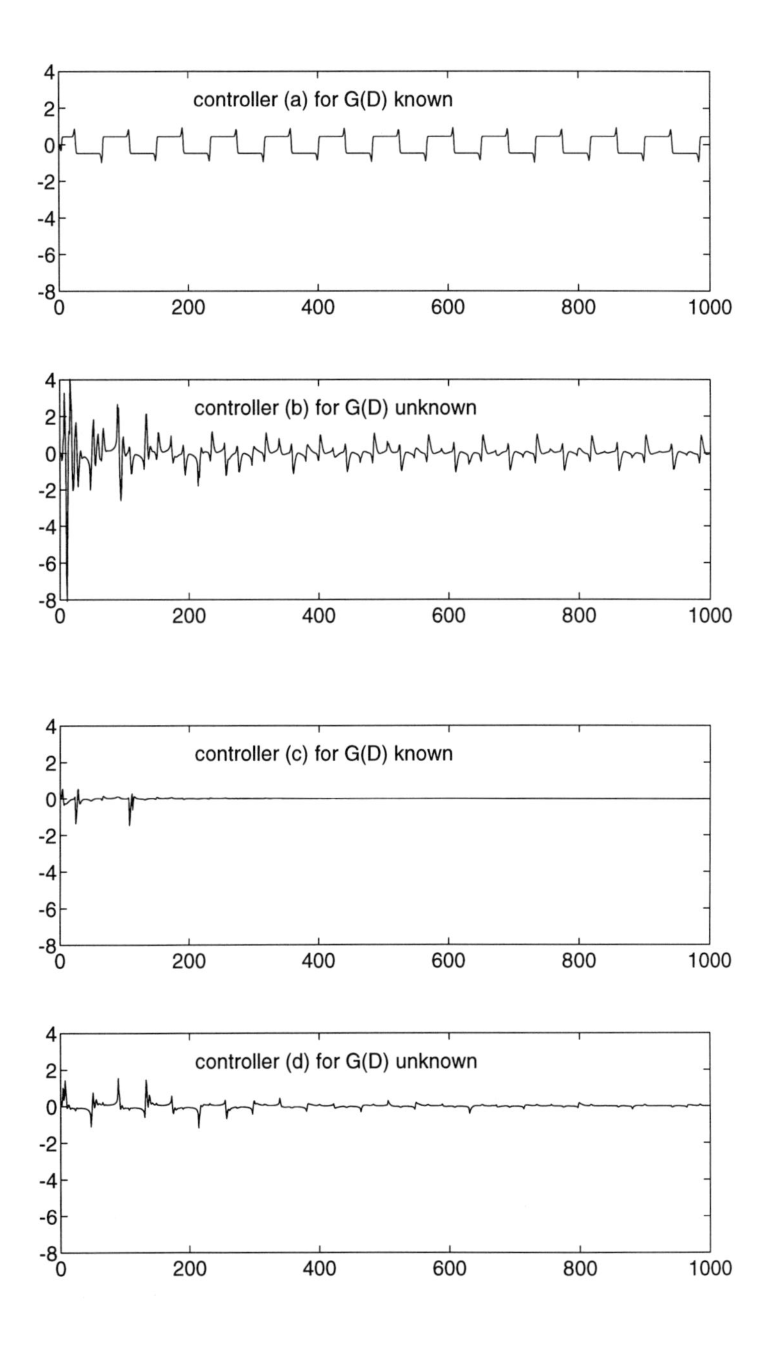

Figure 9.9: Tracking errors for the second-order example.

- The control scheme with an *implicit* $\widehat{z}(t)$ for $G(D)$ known is

$$u(t) = \theta_1^{*T}\omega_1(t) + \theta_N^T(t)\theta_2^{*T}b_\lambda(D)I_{n_N}[\omega_N](t) + \widehat{z_m}(t+n^*)$$
$$\widehat{z_m}(t+n^*) = \widehat{NI}(y_m(t+n^*)) = \theta_N^T(t)\omega_m(t+n^*).$$

- The adaptive update law for $\theta_N(t)$ is

$$\theta_N(t) = \theta_N(t-1) - \frac{\Gamma_N\omega_{NN}(t-n^*)\epsilon_N(t)}{1+\omega_{NN}^T(t-n^*)\omega_{NN}(t-n^*)+\xi_N^2(t)} + f_N(t),$$

where

$$\omega_{NN}(t) = \theta_2^{*T}b_\lambda(D)I_{n_N}[\omega_N](t) + \omega_N(t+n^*)$$
$$\epsilon_N(t) = \theta_1^{*T}\omega_1(t-n^*) + \theta_N^T(t-1)\omega_{NN}(t-n^*) - u(t-n^*)$$

and $\xi_N(t)$ is in (9.16) for an *explicit* inverse design and in (9.19) for an *implicit* inverse design. Projection (9.34) or switching σ-modification with parameter projection (9.35) is used for $f_N(t)$.

- With an *implicit* adaptive inverse $\widehat{z}(t) = \widehat{NI}(y(t))$, the output adaptive inverse scheme for $G(D)$ unknown is

$$u(t) = \theta_1^T(t)\omega_1(t) + \theta_5^T(t)\omega_5(t) + \widehat{z_m}(t+n^*)$$
$$\widehat{z_m}(t+n^*) = \widehat{NI}(y_m(t+n^*) = \theta_N^T(t)\omega_m(t+n^*),$$

where $\theta_5(t)$ is an estimate of $\theta_5^* = \theta_2^* \otimes \theta_N^*$, and

$$\omega_5(t) = B_\lambda(D)[\omega_N](t).$$

- The adaptive update law is

$$\theta(t) = \theta(t-1) - \frac{\Gamma\omega(t-n^*)\epsilon(t)}{1+\omega^T(t-n^*)\omega(t-n^*)+\xi^2(t)} + f(t),$$

where

$$\theta(t) = (\theta_N^T(t), \theta_1^T(t), \theta_5^T(t))^T$$
$$\omega(t) = (\omega_N^T(t+n^*), \omega_1^T(t), \omega_5^T(t))^T$$
$$\epsilon(t) = \theta^T(t-1)\omega(t-n^*) - u(t-n^*)$$
$$\xi(t) = (\theta(t-1) - \theta(t-n^*))^T\omega(t-n^*).$$

Projection (9.52) or switching σ-modification with parameter projection (9.53) is used for $f(t)$.

- Extensive simulations indicate that controllers which ignore the effects of the nonlinearity $N(\cdot)$ usually result in large tracking errors, and controllers with an adaptive inverse lead to asymptotic tracking.

Chapter 10

Adaptive Control of Partially Known Systems

When some partial knowledge about the plant dynamics is available, a possibility may exist to simplify the adaptive controller and improve its transient response. In this chapter we explore the possibilities when some of the plant poles or zeros are known.

We introduce modified linear parametrizations which allow the modified adaptive controllers to reduce the closed-loop system order by $2n_r$ where n_r is the order of the known dynamics. In the presence of dead-zone, backlash, or hysteresis, the order reduction is even more significant. The order of adaptive inverse controllers is reduced by $2n_r n_N$, where n_N is the number of parameters in the input or output nonlinearity.

In Section 10.1, we consider model reference adaptive control of linear plants with partially known zero or pole dynamics. We present modified controller parametrizations guaranteeing output matching and design adaptive laws for updating the controller parameters. In Section 10.2, we develop adaptive inverse controllers for plants with unknown dead-zone, backlash, or hysteresis at the input or output of a linear part whose zero or pole dynamics are partially known.

10.1 Adaptive Linear Control

Throughout this chapter, $R_0(D)$ will represent the known stable pole or zero dynamics. Our objective is to develop reduced-order adaptive controller structures using the knowledge of $R_0(D)$ so that all the closed-loop system signals are bounded and the plant output $y(t)$ tracks the output $y_m(t)$ of the reference model

$$P_m(D)[y_m](t) = r(t), \tag{10.1}$$

where $P_m(D)$ is a monic stable polynomial and $r(t)$ is bounded.

We first solve this problem for the following two linear time-invariant plants

$$R_0(D)P(D)[y](t) = k_p Z(D)[u](t) \tag{10.2}$$

$$P(D)[y](t) = k_p R_0(D)Z(D)[u](t), \tag{10.3}$$

where

$$R_0(D) = D^{n_r} + r_{n_r-1}D^{n_r-1} + \cdots + r_1 D + r_0$$

$$P(D) = D^n + p_{n-1}D^{n-1} + \cdots + p_1 D + p_0$$

$$Z(D) = D^m + z_{m-1}D^{m-1} + \cdots + z_1 D + z_0$$

are polynomials in D which in this section denotes, as the case may be, the Laplace transform variable or the time differentiation operator $D[x](t) = \dot{x}(t)$ in continuous time, and the z-transform variable or the time advance operator $D[x](t) = x(t+1)$ in discrete time, for a unified presentation of adaptive control designs.

To employ the model reference approach, we require that the linear part $G(D)$ and the model dynamics $P_m(D)$ satisfy the assumptions (A1) - (A5) stated in Section 4.1. In particular, the relative degree of $G(D)$ in (10.2) is $n^* = n + n_r - m > 0$ and that of $G(D)$ in (10.3) is $n^* = n - n_r - m > 0$.

10.1.1 Design with Partially Known Poles

When $R_0(D)$ in the plant (10.2) represents n_r known stable poles, then an adaptive controller structure which exploits the knowledge of $R_0(D)$ is

$$u(t) = \theta_1^T \omega_1(t) + \theta_2^T \omega_2(t) + \theta_{20} y(t) + \theta_3 r(t). \tag{10.4}$$

A characteristic of this controller is that $R_0(D)$ is included in the first regressor:

$$\omega_1(t) = \frac{a_1(D)}{\Lambda_0(D)R_0(D)}[u](t),\ \omega_2(t) = \frac{a_2(D)}{\Lambda_0(D)}[y](t), \tag{10.5}$$

where

$$a_1(D) = (1, D, \ldots, D^{n+n_r-2})^T,\ a_2(D) = (1, D, \ldots, D^{n-2})^T$$

$$\theta_1 \in R^{n+n_r-1},\ \theta_2 \in R^{n-1},\ \theta_{20} \in R,\ \theta_3 \in R$$

and $\Lambda_0(D)$ is a monic stable polynomial of degree $n-1$. Note that when $n=1$, then $\theta_2 = 0$; and when $n=0$, then $\theta_2 = 0$, $\theta_{20} = 0$.

The controller (10.4) employs $2n+n_r$ parameters. Without the knowledge of $R_0(D)$, the original model reference controller would have the same structure as (10.4) but its regressors would be

$$\omega_1(t) = \frac{a(D)}{\Lambda(D)}[u](t),\ \omega_2(t) = \frac{a(D)}{\Lambda(D)}[y](t)$$

with $a(D) = (1, D, \ldots, D^{n+n_r-2})^T$, $\Lambda(D)$ of degree $n+n_r-1$, and $\theta_1 \in R^{n+n_r-1}$, $\theta_2 \in R^{n+n_r-1}$, $\theta_{20} \in R$, $\theta_3 \in R$. This would amount to $2n+2n_r$ parameters. Thus, the controller structure (10.4) reduces the order of the adaptive control system by $2n_r$. Furthermore, it has the following properties.

Lemma 10.1 *There exist parameters θ_1^*, θ_2^*, θ_{20}^*, θ_3^* such that for $\theta_1 = \theta_1^*$, $\theta_2 = \theta_2^*$, $\theta_{20} = \theta_{20}^*$, and $\theta_3 = \theta_3^*$, the controller (10.4) ensures closed-loop signal boundedness and exponential tracking of the model output $y_m(t)$ by the plant output $y(t)$.*

Proof: It is sufficient to show that $\theta_1^* \in R^{n+n_r-1}$, $\theta_2^* \in R^{n-1}$, $\theta_{20}^* \in R$, $\theta_3^* \in R$ exist such that the matching equation holds:

$$\theta_1^{*T} a_1(D)P(D) + (\theta_2^{*T} a_2(D) + \theta_{20}^* \Lambda_0(D))k_p Z(D)$$

$$= \Lambda_0(D)(R_0(D)P(D) - k_p\theta_3^* Z(D)P_m(D)). \tag{10.6}$$

Define $Q^*(D) = \Lambda_0(D)(R_0(D)P(D) - k_p\theta_3^* Z(D)P_m(D))$. Choose $\theta_3^* = k_p^{-1}$ so that $Q^*(D)$ has degree $2n+n_r-2$. Since $P(D)$, $Z(D)$ have degrees n, $m \le n+n_r-1$, the Bezout identity (see, for example, [80]) can be used to show that, when $P(D)$ and $Z(D)$ are co-prime, there exist unique polynomials $A^*(D)$, $B^*(D)$ of degrees $n+n_r-2$, $n-1$ such that $A^*(D)P(D) + B^*(D)k_p Z(D) = Q^*(D)$. If $P(D)$ and $Z(D)$ are not co-prime, then solutions $A^*(D)$, $B^*(D)$ still exist but are not unique. Letting $A^*(D) = \theta_1^{*T} a_1(D)$, $B^*(D) = \theta_2^{*T} a_2(D) + \theta_{20}^* \Lambda_0(D)$, we obtain the desired parameters θ_1^*, θ_2^*, θ_{20}^*, θ_3^* for (10.6).

To proceed, we multiply both sides of (10.6) by $R_0(D)$, operate the resulted identity on $y(t)$, and, using (10.4), obtain

$$\begin{aligned}
&\Lambda_0(D)R_0(D)k_p Z(D)[u](t) \\
&\quad = \theta_1^{*T} a_1(D)k_p Z(D)[u](t) + \theta_3^* k_p \Lambda_0(D)R_0(D)Z(D)P_m(D)[y](t) \\
&\quad\quad + (\theta_2^{*T} a_2(D) + \theta_{20}^* \Lambda_0(D))R_0(D)k_p Z(D)[y](t).
\end{aligned} \tag{10.7}$$

Because $\Lambda_0(D)$, $R_0(D)$, and $Z(D)$ are all stable polynomials, (10.7) can be expressed as

$$u(t) = \theta_1^{*T} \frac{a_1(D)}{\Lambda_0(D)R_0(D)}[u](t) + \theta_2^{*T} \frac{a_2(D)}{\Lambda_0(D)}[y](t) + \theta_{20}^* y(t) + \theta_3^* P_m(D)[y](t) + \eta_1(t) \tag{10.8}$$

for some initial condition-dependent exponentially decaying term $\eta_1(t)$. Substituting (10.8) in (10.4) with $\theta_1 = \theta_1^*$, $\theta_2 = \theta_2^*$, $\theta_{20} = \theta_{20}^*$, $\theta_3 = \theta_3^*$ and using (10.1), we obtain the plant-model matching equation

$$\theta_3^* P_m(D)[y - y_m](t) + \eta_1(t) = 0. \tag{10.9}$$

Using (10.2) and (10.9), we get

$$\begin{aligned} P_m(D)Z(D)[u](t) &= k_p^{-1} R_0(D)P(D)P_m(D)[y](t) \\ &= R_0(D)P(D)(\theta_3^* r(t) - \eta_1(t)). \end{aligned} \tag{10.10}$$

Since $P_m(D)$ is a stable polynomial and $y_m(t)$ is bounded, (10.9) implies that $y(t)$ is bounded and $\lim_{t\to\infty}(y(t) - y_m(t)) = 0$ exponentially. Since $r(t)$ is bounded and $P_m(D)Z(D)$ is a stable polynomial of degree $n + n_r$, which is equal to the degree of $R_0(D)P(D)$, (10.10) implies that $u(t)$ is bounded. ∇

10.1.2 Design with Partially Known Zeros

Consider now the case when $R_0(D)$ represents the n_r known stable zeros and the order of the plant (10.3) is n. With the controller structure

$$u(t) = \theta_1^T \omega_1(t) + \theta_2^T \omega_2(t) + \theta_{20} y(t) + \theta_3 r(t) \tag{10.11}$$

the knowledge of the n_r zeros is incorporated in the regressor $\omega_2(t)$:

$$\omega_1(t) = \frac{a_1(D)}{\Lambda_0(D)}[u](t), \ \omega_2(t) = \frac{a_2(D)}{\Lambda_0(D)R_0(D)}[y](t), \tag{10.12}$$

where

$$a_1(D) = (1, D, \ldots, D^{n-n_r-2})^T, \ a_2(D) = (1, D, \ldots, D^{n-2})^T$$

$$\theta_1 \in R^{n-n_r-1}, \ \theta_2 \in R^{n-1}, \ \theta_{20} \in R, \ \theta_3 \in R$$

and $\Lambda_0(D)$ is a monic stable polynomial of degree $n - n_r - 1$. This controller has $2n - n_r$ parameters and reduces the order of the adaptive control system by $2n_r$. Note that when $n_r = n - 1$, then $\theta_1 = 0$.

We now show that Lemma 10.1 also holds for the controller (10.11).

Choose $\theta_3^* = k_p^{-1}$ so that $\Lambda_0(D)(P(D) - k_p\theta_3^* R_0(D)Z(D)P_m(D))$ has degree $2n - n_r - 2$. Since $P(D)$, $Z(D)$ have degrees n, m, it can be shown that there exist constant parameters $\theta_1^* \in R^{n-n_r-1}$, $\theta_2^* \in R^{n-1}$, $\theta_{20}^* \in R$, $\theta_3^* \in R$ which satisfy the matching equation

$$\begin{aligned} &\theta_1^{*T} a_1(D)P(D) + (\theta_2^{*T} a_2(D) + \theta_{20}^* R_0(D)\Lambda_0(D))k_p Z(D) \\ &\qquad = \Lambda_0(D)(P(D) - k_p\theta_3^* R_0(D)Z(D)P_m(D)). \end{aligned} \tag{10.13}$$

Operating (10.13) on $y(t)$ and using (10.3), we obtain

$$\begin{aligned} &\Lambda_0(D)R_0(D)k_p Z(D)[u](t) \\ &\quad = \theta_1^{*T} a_1(D)k_p R_0(D)Z(D)[u](t) + \Lambda_0(D)\theta_3^* k_p R_0(D)Z(D)P_m(D)[y](t) \\ &\qquad + (\theta_2^{*T} a_2(D) + \theta_{20}^* \Lambda_0(D)R_0(D))k_p Z(D)[y](t). \end{aligned} \tag{10.14}$$

Because $\Lambda_0(D)$, $R_0(D)$, and $Z(D)$ are all stable polynomials, (10.14) can be used to obtain

$$\begin{aligned} u(t) = \theta_1^{*T} \frac{a_1(D)}{\Lambda_0(D)}[u](t) + \theta_2^{*T} \frac{a_2(D)}{\Lambda_0(D)R_0(D)}[y](t) \\ + \theta_{20}^* y(t) + \theta_3^* P_m(D)[y](t) + \eta_1(t) \end{aligned} \tag{10.15}$$

for some initial condition-dependent exponentially decaying term $\eta_1(t)$. Substituting (10.15) in (10.11) with $\theta_1 = \theta_1^*$, $\theta_2 = \theta_2^*$, $\theta_{20} = \theta_{20}^*$, $\theta_3 = \theta_3^*$ and using (10.1), we again arrive at the plant-model matching equation (10.9), from which, in view of (10.3), we obtain

$$\begin{aligned} P_m(D)R_0(D)Z(D)[u](t) &= k_p^{-1}P(D)P_m(D)[y](t) \\ &= P(D)(\theta_3^* r(t) - \eta_1(t)). \end{aligned} \tag{10.16}$$

Hence, from (10.9), (10.13), and (10.16), we conclude that with θ_1^*, θ_2^*, θ_{20}^*, θ_3^* satisfying (10.13), the controller (10.11) with $\theta_1 = \theta_1^*$, $\theta_2 = \theta_2^*$, $\theta_{20} = \theta_{20}^*$, and $\theta_3 = \theta_3^*$ ensures signal boundedness and exponential tracking of the model output $y_m(t)$ by the plant output $y(t)$.

10.1.3 Example: Reduced-Order Controllers

Consider the continuous-time plant transfer function

$$G(s) = \frac{y(s)}{u(s)} = \frac{2.5(s+1)(s+3)}{(s+2)^2(s-4)(s-5)}.$$

We first assume that the poles $R_0(s) = (s+2)^2$ are known. The controller (10.4) is

$$u(t) = \frac{\theta_{11} + \theta_{12}s + \theta_{13}s^2}{(s+2)^2\Lambda_0(s)}[u](t) + \frac{\theta_{21}}{\Lambda_0(s)}[y](t) + \theta_{20}y(t) + \theta_3 r(t).$$

For the reference model $P_m(s) = (s+4)(s+5)$ and $\Lambda_0(s) = s+6$, the output matching parameters $\theta_3^* = 0.4$, and θ_{11}^*, θ_{12}^*, θ_{13}^*, θ_{21}^*, θ_{20}^* satisfy

$$\begin{aligned}(\theta_{11}^* + \theta_{12}^*s + \theta_{13}^*s^2)(s-4)(s-5) + (\theta_{21}^* + \theta_{20}^*(s+6))\\ = (s+6)((s+2)^2(s-4)(s-5) - (s+1)(s+2)(s+4)(s+5)).\end{aligned}$$

Solving this equation, we obtain

$$\theta_{11}^* = -48,\ \theta_{12}^* = -71,\ \theta_{13}^* = -18,\ \theta_{21}^* = 1980,\ \theta_{20}^* = -270.$$

Let us now assume that the zeros $R_0(s) = (s+1)(s+3)$ of $G(s)$ are known. The controller structure (10.11) becomes

$$u(t) = \frac{\theta_{11}}{\Lambda_0(s)}[u](t) + \frac{\theta_{21} + \theta_{22}s + \theta_{23}s^2}{(s+1)(s+3)\Lambda_0(s)}[y](t) + \theta_{20}y(t) + \theta_3 r(t).$$

For the same $P_m(s)$ and $\Lambda_0(s)$ above, the plant-model matching parameters $\theta_3^* = 0.4$, and

$$\theta_{11}^* = -18,\ \theta_{21}^* = 6402,\ \theta_{22}^* = 7697,\ \theta_{23}^* = 1985,\ \theta_{20}^* = -269$$

are obtained from the matching equation

$$\begin{aligned}\theta_{11}^*(s+2)^2(s-4)(s-5) + (\theta_{21}^* + \theta_{22}^*s + \theta_{23}^*s^2 + \theta_{20}^*(s+6)^3)\\ = (s+6)((s+2)^2(s-4)(s-5) - (s+1)(s+2)(s+4)(s+5)).\end{aligned}$$

In both cases, when either two poles or two zeros are known, the number of the controller parameters was reduced by 2 and the filter order by 2, so that the total order reduction was 4.

10.1.4 Adaptive Laws

When the plant parameters k_p, z_i, p_j are unknown, we design an adaptive law to update the parameter estimates $\theta_1(t)$, $\theta_2(t)$, $\theta_{20}(t)$, $\theta_3(t)$ of θ_1^*, θ_2^*, θ_{20}^*, θ_3^*. For this, we need a tracking error equation which we derive by introducing

$$\theta^* = (\theta_1^{*T}, \theta_2^{*T}, \theta_{20}^*, \theta_3^*)^T,\ \theta(t) = (\theta_1^T(t), \theta_2^T(t), \theta_{20}(t), \theta_3(t))^T$$

$$\omega(t) = (\omega_1^T(t), \omega_2^T(t), y(t), r(t))^T$$
$$\tilde{\theta}(t) = \theta(t) - \theta^*, \; e(t) = y(t) - y_m(t).$$

Substituting (10.8) in (10.4) or (10.15) in (10.11) and ignoring the exponentially decaying effect of the initial conditions, we obtain the tracking error equation in the familiar form:

$$e(t) = \frac{k_p}{P_m(D)}[\tilde{\theta}^T \omega](t). \tag{10.17}$$

With the help of the auxiliary signals

$$\xi(t) = \theta^T(t)\zeta(t) - \frac{1}{P_m(D)}[\theta^T \omega](t) \tag{10.18}$$

$$\zeta(t) = \frac{1}{P_m(D)}[\omega](t) \tag{10.19}$$

$$m(t) = \sqrt{1 + \zeta^T(t)\zeta(t) + \xi^2(t)}$$

we define the estimation error

$$\epsilon(t) = e(t) + \rho(t)\xi(t), \tag{10.20}$$

where $\rho(t)$ is the estimate of $\rho^* = k_p$. We then employ adaptive update law for $\theta(t)$ and $\rho(t)$:

$$\dot{\theta}(t) = -\frac{sign(k_p)\Gamma_\theta \epsilon(t)\zeta(t)}{m^2(t)} \tag{10.21}$$

$$\dot{\rho}(t) = -\frac{\gamma_\rho \epsilon(t)\xi(t)}{m^2(t)}, \tag{10.22}$$

where the design parameters are $\Gamma_\theta = \Gamma_\theta^T > 0$, $\gamma_\rho > 0$.

The discrete-time counterpart of (10.21) - (10.22) is

$$\theta(t+1) = \theta(t) - \frac{sign(k_p)\Gamma_\theta \epsilon(t)\zeta(t)}{m^2(t)} \tag{10.23}$$

$$\rho(t+1) = \rho(t) - \frac{\gamma_\rho \epsilon(t)\xi(t)}{m^2(t)}, \tag{10.24}$$

where $0 < \Gamma_\theta = \Gamma_\theta^T < \frac{2}{k_p^0}$, $0 < \gamma_\rho < 2$, and $k_p^0 \geq |k_p|$; that is, the knowledge of an upper bound k_p^0 on $|k_p|$ is needed in the discrete-time case.

Substituting (10.17) - (10.19) in (10.20) gives

$$\epsilon(t) = \rho^* \tilde{\theta}^T(t)\zeta(t) + \tilde{\rho}(t)\xi(t).$$

Using this estimation error equation, we obtain the time derivative of the positive definite function

$$V(\tilde{\theta}, \tilde{\rho}) = |\rho^*|\tilde{\theta}^T \Gamma_\theta^{-1} \tilde{\theta} + \gamma_\rho^{-1} \tilde{\rho}^2$$

along the trajectories of the adaptive update law (10.21) - (10.22) as

$$\dot{V} = \frac{-2\epsilon^2(t)}{m^2(t)}.$$

Similarly, the time increment of $V(\tilde{\theta}, \tilde{\rho})$ along the trajectories of the discrete-time adaptive law (10.23) - (10.24) is

$$\begin{aligned} & V(\tilde{\theta}(t+1), \tilde{\rho}(t+1)) - V(\tilde{\theta}(t), \tilde{\rho}(t)) \\ & = -(2 - \frac{|k_p|\zeta^T(t)\Gamma_\theta\zeta(t) + \gamma_\rho\xi^2(t)}{m^2(t)})\frac{\epsilon^2(t)}{m^2(t)}. \end{aligned}$$

With these expressions, results of model reference adaptive control (see Appendix A) can be used to prove the following lemma.

Lemma 10.2 *The adaptive update law (10.21) - (10.22) or (10.23) - (10.24) guarantees that*

- $\theta(t)$, $\rho(t) \in L^\infty$, *and* $\frac{\epsilon(t)}{m(t)}$, $\dot{\theta}(t) \in L^2 \cap L^\infty$, *in continuous time, or*
- $\theta(t)$, $\rho(t) \in l_\infty$, $\frac{\epsilon(t)}{m(t)} \in l^2 \cap l^\infty$, $\theta(t+1) - \theta(t) \in l^2$, *in discrete time.*

With Lemma 10.2, it can be shown, similar to that in Appendix A, that all the signals in the closed-loop system consisting of the plant (10.2) or (10.3), reference model (10.1), controller (10.4) or (10.11), and adaptive law (10.21) - (10.22) or (10.23) - (10.24) are bounded. Moreover, the tracking error $e(t) = y(t) - y_m(t)$ belongs to L_2 (or l_2) and converges to zero as $t \to \infty$.

10.2 Adaptive Inverse Control

We now apply the results of Section 10.1 to design reduced-order adaptive inverse controllers for plants with unknown nonsmooth nonlinearities at the input or at the output of a linear part whose dynamics are partially known.

Consider the plant with an unknown nonlinearity $N(\cdot)$ at the input of a linear part $G(D)$:

$$y(t) = G(D)[u](t), \; u(t) = N(v(t)), \tag{10.25}$$

where $G(D) = k_p \frac{Z(D)R_0(D)}{P(D)}$ with k_p, $R_0(D)$, $P(D)$, and $Z(D)$ defined in (10.3), or the plant with an unknown nonlinearity $N(\cdot)$ at the output of a linear part $G(D)$:

$$y(t) = N(z(t)), \; z(t) = G(D)[u](t), \tag{10.26}$$

where $G(D) = k_p \frac{Z(D)}{R_0(D)P(D)}$ with k_p, $R_0(D)$, $P(D)$, and $Z(D)$ defined in (10.2). The nonlinearity $N(\cdot)$ represents a dead-zone, backlash, or hysteresis in either an actuator or a sensor.

As in in Chapters 4 - 9, we assume that signal $u(t)$ in the input nonlinearity case and the signal $z(t)$ in the output nonlinearity case are not available for measurement. Our goal here is to design reduced-order versions of the adaptive inverse controllers developed in Chapters 6, 7, and 9 when the linear part $G(D)$ of the plant is partially known (that is, $R_0(D)$ is known) and the nonlinear part $N(\cdot)$ is unknown. The assumptions for the linear part $G(D)$ and the reference model dynamics $P_m(D)$ are as in (A1) - (A5) of Section 4.1, and the assumptions for the nonlinear part $N(\cdot)$ are as in (A7a) - (A7c) of Section 6.1 or as in (A8a) - (A8c) of Section 9.1.

10.2.1 Design for Input Nonlinearities

Consider the plant (10.25) where $R_0(D)$ represents the n_r known stable zeros of $G(D)$ and $N(\cdot)$ is an unknown input dead-zone, backlash, or hysteresis. To design an adaptive inverse controller for such a nonlinear plant, we use the approach developed in Chapters 6 and 7, combined with the modified adaptive controller structure shown in Figure 10.1. Recall from Chapter 3 that with the inverse $\widehat{NI}(\cdot)$ of the input nonlinearity $N(\cdot)$,

$$v(t) = \widehat{NI}(u_d(t)),$$

the control error $u(t) - u_d(t)$ is

$$u(t) - u_d(t) = (\theta_N(t) - \theta_N^*)^T \omega_N(t) + d_N(t) \tag{10.27}$$

and that the adaptive inverse $\widehat{NI}(\cdot)$ is parametrized as

$$u_d(t) = -\theta_N^T(t)\omega_N(t) + a_h(t)$$

with signals $\omega_N(t)$, $a_h(t)$ generated by the logic block L, namely, $\bar{\omega}_N(t) = \omega_N(t)$ for *dead-zone* and *backlash* and $\bar{\omega}_N(t) = (\omega_N^T(t), a_h(t))^T$ for *hysteresis*. As before, the unparametrized error $d_N(t)$ is bounded and $d_N(t) = 0$ when

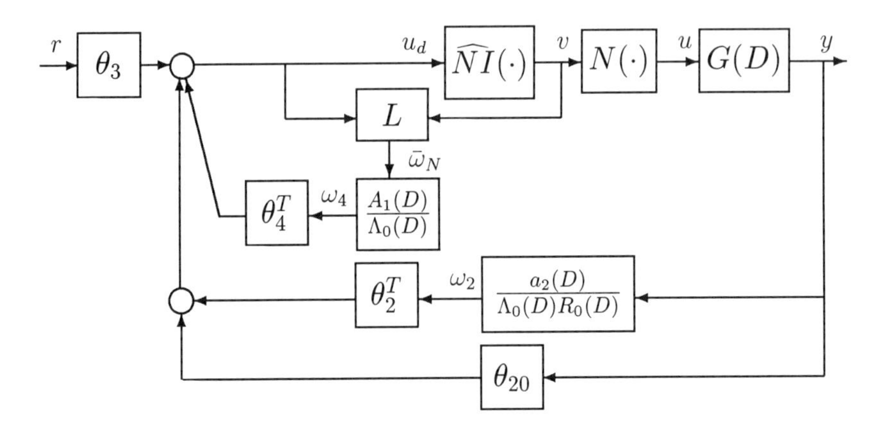

Figure 10.1: Reduced-order controller for input nonlinearities.

$\theta_N(t) = \theta_N^*$ and the inverse $\widehat{NI}(\cdot)$ is correctly initialized. The parameter vector $\theta_N(t)$ is the estimate of θ_N^* defined by

$$\theta_N^* = \begin{cases} (m_r, m_r b_r, m_l, m_l b_l)^T & \text{for } DZ(\cdot) \\ (mc_r, m, mc_l)^T & \text{for } B(\cdot) \\ (m_t, c_t, m_b, c_b, m_r, m_r c_r, m_l, m_l c_l)^T & \text{for } H(\cdot). \end{cases}$$

Since the input nonlinearity $N(\cdot)$ is unknown, we cannot implement the inverse $\widehat{NI}(\cdot)$ with the parameter θ_N^* and we need an adaptive law to update its estimate $\theta_N(t)$. The detailed constructions of the adaptive inverses for dead-zone, backlash, and hysteresis nonlinearities are in Chapters 6, 7, and 9.

The input signal $u_d(t)$ to the adaptive inverse $\widehat{NI}(\cdot)$ is

$$u_d(t) = \theta_2^T \omega_2(t) + \theta_{20} y(t) + \theta_3 r(t) + \theta_4^T \omega_4(t), \tag{10.28}$$

where $\theta_2(t)$, $\omega_2(t)$, $\theta_{20}(t)$, $\theta_3(t)$ are defined in (10.11) - (10.12), the new parameter $\theta_4(t)$ is the estimate of $\theta_4^* = -\theta_1^* \otimes \theta_N^*$, (for hysteresis $\theta_4^* = -\theta_1^* \otimes (\theta_N^{*T}, -1)^T$), and the corresponding regressor is

$$\omega_4(t) = \frac{A_1(D)}{\Lambda_0(D)}[\bar{\omega}_N](t)$$

$$A_1(D) = (I_{n_N}, DI_{n_N}, \ldots, D^{n-n_r-2} I_{n_N})^T$$

with n_N being the dimension of θ_N^* (for hysteresis $(\theta_N^{*T}, -1)^T$) and $\Lambda_0(D)$ being a monic stable polynomial of degree $n - n_r - 1$. Note that when $n_r = n - 1$, then $\theta_4 = 0$.

With the linear parametrization achieved, thanks to θ_4 and $\omega_4(t)$, (10.27) and (10.29) result in the tracking error

$$e(t) = y(t) - y_m(t) = k_p P_m^{-1}(D)[(\theta - \theta^*)^T \omega](t) + d(t), \tag{10.29}$$

where

$$\omega(t) = (\omega_N^T(t), \omega_2^T(t), y(t), r(t), \omega_4^T(t))^T$$
$$\theta(t) = (\theta_N^T(t), \theta_2^T(t), \theta_{20}(t), \theta_3(t), \theta_4^T(t))^T$$
$$\theta^* = (\theta_N^{*T}, \theta_2^{*T}, \theta_{20}^*, \theta_3^*, \theta_4^{*T})^T.$$

and the unknown disturbance term

$$d(t) = (1 - \theta_1^{*T} \frac{a_1(D)}{\Lambda_0(D)})[d_N](t)$$

is bounded because $d_N(t)$ is bounded.

The error equation (10.29) suggests the update law for $\theta(t)$:

$$\dot{\theta}(t) = -\frac{sign(k_p)\Gamma_\theta \zeta(t)\epsilon(t)}{1 + \zeta^T(t)\zeta(t) + \xi^2(t)} + f_\theta(t) \tag{10.30}$$

$$\dot{\rho}(t) = -\frac{\gamma_\rho \xi(t)\epsilon(t)}{1 + \zeta^T(t)\zeta(t) + \xi^2(t)} + f_\rho(t), \tag{10.31}$$

where $\Gamma_\theta = \Gamma_\theta^T > 0$, $\gamma_\rho > 0$, $\rho(t)$ is the estimate of k_p,

$$\epsilon(t) = e(t) + \rho(t)\xi(t)$$
$$\xi(t) = \theta^T(t)\zeta(t) - P_m^{-1}(D)[\theta^T \omega](t)$$
$$\zeta(t) = P_m^{-1}(D)[\omega](t)$$

and $f_\theta(t)$, $f_\rho(t)$ are design signals for parameter projection and robustness with respect to $d(t)$.

The discrete-time version of (10.30) - (10.31) is

$$\theta(t+1) = \theta(t) - \frac{sign(k_p)\gamma_\theta \zeta(t)\epsilon(t)}{1 + \zeta^T(t)\zeta(t) + \xi^2(t)} + f_\theta(t) \tag{10.32}$$

$$\rho(t+1) = \rho(t) - \frac{\gamma_\rho \xi(t)\epsilon(t)}{1 + \zeta^T(t)\zeta(t) + \xi^2(t)} + f_\rho(t), \tag{10.33}$$

where γ_θ and γ_ρ are positive design parameters determined together with the modification terms $f_\theta(t)$ and $f_\rho(t)$.

The order reduction achieved with the controller (10.28) can be seen from the fact that the block $\theta_4^T \frac{A_1(D)}{\Lambda_0(D)}$ has $(n - n_r - 1)n_N$ parameters and its order is $(n - n_r - 1)n_N$, where n_N is the dimension of θ_N^* (for hysteresis $(\theta_N^{*T}, -1)^T$). The analog of $\theta_4^T \frac{A_1(D)}{\Lambda_0(D)}$ without the knowledge of $R_0(D)$ would have $(n-1)n_N$ parameters and order $(n-1)n_N$. Hence, with the new controller (10.28), the order reduction is $n_r n_N + n_r n_N = 2n_r n_N$.

10.2.2 Design for Output Nonlinearities

Now we consider the plant (10.26) in the discrete time where $R_0(D)$ represents the n_r known stable poles of $G(D)$ and $N(\cdot)$ is an unknown output nonlinearity. For a discrete-time design, we choose $P_m(D) = D^{n^*}$ so that the reference model (10.1) becomes $y_m(t) = r(t - n^*)$, where $r(t)$ is bounded. As in Chapter 9, we assume that $k_p = 1$.

In this case, the new adaptive inverse controller structure, based on that proposed in Chapter 9, is shown in Figure 10.2, in which the block $\widehat{NI}(\cdot)$ is an explicit adaptive inverse of $N(\cdot)$ to generate $z_m(t + n^*)$.

The inverses introduced in Chapter 8 are parametrized as

$$\hat{z}(t) = \widehat{NI}(y(t)) = \theta_N^T(t)\omega_N(t) \tag{10.34}$$

$$\widehat{z_m}(t + n^*) = \widehat{NI}(y_m(t + n^*)) = \theta_N^T(t)\omega_m(t + n^*). \tag{10.35}$$

Only $\omega_N(t)$ in (10.34) is explicitly used in the controller. The estimate $\theta_N(t)$ which appears in $\theta_5(t)$ and (10.35) will be updated by an adaptive law. The true value of θ_N is

$$\theta_N^* = \begin{cases} (\frac{1}{m_r}, b_r, \frac{1}{m_l}, b_l)^T & \text{for } DZ(\cdot) \\ (c_r, \frac{1}{m}, c_l)^T & \text{for } B(\cdot) \\ (\frac{1}{m_t}, \frac{c_t}{m_t}, \frac{1}{m_b}, \frac{c_b}{m_b}, \frac{1}{m_r}, c_r, \frac{1}{m_l}, c_l)^T & \text{for } H(\cdot). \end{cases} \tag{10.36}$$

As indicated in Section 8.3, with the inverse (10.34), we have

$$\hat{z}(t) - z(t) = (\theta_N(t) - \theta_N^*)^T\omega_N(t) + d_N(t). \tag{10.37}$$

The input signal $u(t)$ to the linear part $G(D)$ is

$$u(t) = \theta_1^T\omega_1(t) + \theta_5^T\omega_5(t) + \theta_{50}^T\omega_N(t) + \theta_N^T\omega_m(t + n^*), \tag{10.38}$$

where $\theta_1(t)$ and $\omega_1(t) = \frac{a_1(D)}{\Lambda_0(D)R_0(D)}[u](t)$ are defined in Section 10.1.1, while $\omega_m(t)$ is an available signal dependent on the reference signal $y_m(t)$. Similar to the design in Section 9.3, $\theta_5(t)$ is the estimate of the new parameter $\theta_5^* = \theta_2^* \otimes \theta_N^*$ with the corresponding regressor

$$\omega_5(t) = \frac{A_2(D)}{\Lambda_0(D)}[\omega_N](t), \tag{10.39}$$

where

$$A_2(D) = (I_{n_N}, DI_{n_N}, \ldots, D^{n-2}I_{n_N})^T$$

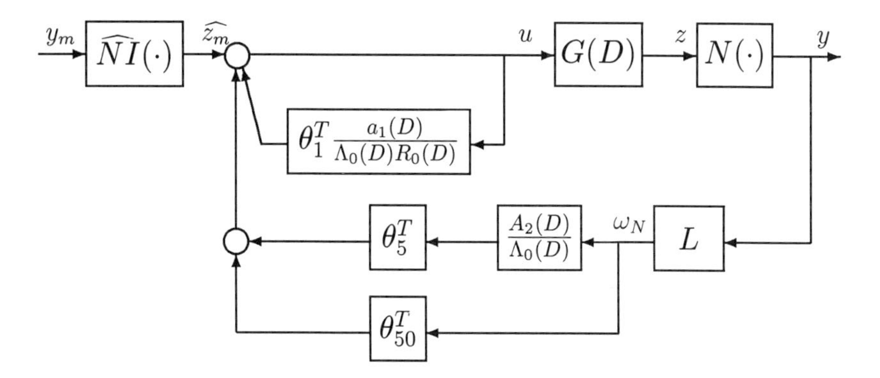

Figure 10.2: Reduced-order controller for output nonlinearities.

and n_N is the dimension of θ_N^* according to (10.36). Furthermore, $\theta_{50}(t)$ is the estimate of $\theta_{50}^* = \theta_{20}^*\theta_N^*$ and its regressor $\omega_N(t)$ is generated by the logic block L. In the discrete-time case, we can choose $\Lambda_0(D) = D^{n-1}$. Note that when $n = 1$, then $\theta_5 = 0$; and when $n = 0$, then $\theta_5 = 0$, $\theta_{50} = 0$.

The inverse (10.34) appears implicitly in the feedback blocks L, $\theta_5^T \frac{A_2(D)}{\lambda_0(D)}$, and θ_{50}; that is, with $\theta_5 = \theta_5^*$, $\theta_{50} = \theta_{50}^*$, it follows that

$$\begin{aligned}(\theta_5^T \frac{A_2(D)}{\lambda_0(D)} + \theta_{50})[\omega_N](t) &= (\theta_2^{*T} \frac{a_2(D)}{\lambda_0(D)} + \theta_{20}^*)[\theta_N^{*T}\omega_N](t) \\ &= (\theta_2^{*T} \frac{a_2(D)}{\lambda_0(D)} + \theta_{20}^*)[NI(y)](t).\end{aligned}$$

To see that with θ_5, θ_{50}, and $\omega_5(t)$ a linear parametrization of the closed-loop adaptive system is achieved, we use (10.38) to obtain

$$u(t) = \theta_N^{*T}\omega_N(t+n^*) + \theta_1^{*T}\omega_1(t) + \theta_5^{*T}\omega_5(t) + \theta_{50}^{*T}\omega_N(t) - d(t), \tag{10.40}$$

where

$$d(t) = d_N(t+n^*) + \theta_2^{*T}\frac{a_2(D)}{\Lambda_0(D)}[d_N](t) + \theta_{20}^* d_N(t)$$

is bounded because $d_N(t)$ is bounded.

In view of (10.40), we define the estimation error

$$\begin{aligned}\epsilon(t) = {} & \theta_N^T(t-1)\omega_N(t) + \theta_1^T(t-1)\omega_1(t-n^*) + \theta_5^T(t-1)\omega_5(t-n^*) \\ & +\theta_{50}^T(t-1)\omega_N(t-n^*) - u(t-n^*),\end{aligned} \tag{10.41}$$

introduce the overall parameter and regressor vectors

$$\theta(t) = (\theta_N^T(t), \theta_1^T(t), \theta_5^T(t), \theta_{50}^T(t))^T$$

$$\theta^* = (\theta_N^{*T}, \theta_1^{*T}, \theta_5^{*T}, \theta_{50}^{*T})^T$$

$$\omega(t) = (\omega_N^T(t+n^*), \omega_1^T(t), \omega_5^T(t), \omega_N^T(t))^T,$$

and use (10.37), (10.39), and (10.41) to obtain the linear error equation

$$\epsilon(t) = (\theta(t-1) - \theta_N^*)^T \omega(t-n^*) + d(t-n^*).$$

This error equation suggests the adaptive update law for $\theta(t)$:

$$\theta(t) - \theta(t-1) = -\frac{\gamma\omega(t-n^*)\epsilon(t)}{1+\omega^T(t-n^*)\omega(t-n^*)} + f(t), \tag{10.42}$$

where $f(t)$ is a modification for robustness with respect to $d(t)$, and γ is a positive design parameter determined together with $f(t)$.

For the adaptive controller (10.38), we see that the block $\theta_5^T \frac{A_2(D)}{\Lambda_0(D)}$ has $(n-1)n_N$ parameters and its dynamic order is $(n-1)n_N$, where n_N is the dimension of the parameter vector θ_N^*. Without using $R_0(D)$ the analogous block would have $(n+n_r-1)n_N$ parameters and order $(n+n_r-1)n_N$. Hence, with the new controller (10.38), the order reduction is again $2n_r n_N$.

10.2.3 Example: Adaptive Output Dead-Zone Inverse

We now present an example to illustrate the design and performance of a reduced-order adaptive output dead-zone inverse controller, as compared with the corresponding full-order controller in Section 9.3.1.

Consider the plant with an output dead-zone:

$$y(t) = DZ(z(t)),\ z(t) = G(D)[u](t) \tag{10.43}$$

$$G(D) = \frac{1}{(D-p_1)(D-p_2)},\ p_1 = 0.5,\ p_2 = 1.2$$

$$DZ(\cdot) : m_r = 0.1,\ m_l = 0.13,\ b_r = 8,\ b_l = -6,$$

where p_1 is known—that is, $R_0(D) = D - p_1$ is known—and p_2, m_r, m_l, b_r, and b_l are unknown. For this $G(D)$, $n = 1$, $n_r = 1$.

Using $R_0(D)$, the reduced-order controller structure (10.38) is

$$u(t) = \theta_1(t)\omega_1(t) + \theta_{50}^T(t)\omega_d(t) + \theta_d^T(t)\omega_m(t+2), \tag{10.44}$$

where

$$\omega_1(t) = \frac{1}{R_0(D)}[u](t) = \frac{1}{D-p_1}[u](t)$$

and $\theta_1(t) \in R$, $\theta_{50}(t) \in R^4$, and $\theta_d(t) \in R^4$ are the estimates of

$$\theta_1^* = -p_1 - p_2 = -1.7$$

$$\theta_{50}^* = \theta_{20}^* \theta_d^* = (-14.4, -11.52, -11.077, 8.64)^T$$

$$\theta_d^* = (\frac{1}{m_r}, b_r, \frac{1}{m_l}, b_l)^T = (10, 8, \frac{100}{13}, -6),$$

respectively, with $\theta_{20}^* = -p_2^2 = -1.44$. Here, θ_1^* and θ_{20}^* satisfy the matching equation (10.6) with $\Lambda_0(D) = 1$, $a_1(D) = 1$, $P(D) = D - p_2$, $k_p = \theta_3^* = 1$, $Z(D) = 1$, and $P_m(D) = D^2$:

$$\theta_1^*(D - p_2) + \theta_{20}^* = -(p_1 + p_2)D + p_1 p_2.$$

Without using $R_0(D)$, the full-order controller structure (9.43) is

$$u(t) = \theta_1(t)u(t-1) + \theta_5^T(t)(\omega_d^T(t-1), \omega_d^T(t))^T + \theta_d^T(t)\omega_m(t+2), \quad (10.45)$$

where $\theta_1(t) \in R$, $\theta_5(t) \in R^8$, and $\theta_d(t) \in R^4$ are the estimates of

$$\theta_1^* = -p_1 - p_2 = -1.7$$

$$\theta_5^* = \theta_2^* \otimes \theta_d^* = (\theta_{21}^* \theta_d^{*T}, \theta_{22}^* \theta_d^{*T})^T$$

and θ_d^*, respectively, with

$$\theta_2^* = (\theta_{21}^*, \theta_{22}^*)^T = (p_1 p_2(p_1 + p_2), p_1 p_2 - (p_1 + p_2)^2) = (1.02, -2.29)^T.$$

Now, θ_1^*, θ_{21}^*, and θ_{22}^* satisfy the matching equation (8.43) with $a_\lambda(D) = D^{-1}$, $b_\lambda(D) = (D^{-1}, 1)^T$, $P(D) = (D - p_1)(D - p_2)$, $Z(D) = 1$, and $n^* = 2$:

$$\theta_1^* D^{-1}(D - p_1)(D - p_2) + (\theta_{21}^* D^{-1} + \theta_{22}^*) = -(p_1 + p_2)D + p_1 p_2.$$

The controller (10.45) is updated by (9.50) with $f(t)$ implemented as the switching σ-modification with parameter projection in (9.52). The controller (10.44) is updated by (10.42) with $f(t)$ being a switching σ-modification with parameter projection similar to that in (9.52). For simulations, $\gamma = 0.75$, $\sigma_0 = 0.1$, $M_\theta = \|\theta^*\|_2 + 10$ in (9.52) for the full-order controller (10.45), and $\gamma = 0.75$, $\sigma_0 = 0.1$, $M_\theta = \|\theta^*\|_2 + 5$ for the reduced-order controller (10.44). For parameter projection in both controllers, the parameter bounds were $m_{r1} = m_{l1} = 1$, as defined in the assumption (A8a) in Section 9.1.

Tracking errors $e(t) = y(t) - y_m(t)$ are shown in Figure 10.3 for $y_m(t) = 15$ and in Figure 10.4 for $y_m(t) = 10 \sin 0.0785t$. The initial parameter estimates for the reduced-order controller (10.44) were

$$\theta_1(0) = -1.5, \ \theta_{20}(0) = -1.64$$

$$\theta_{50}(0) = \theta_{20}(0)\theta_d(0) = (-17.22, -13.94, 11.808, 10.66)$$
$$\theta_d(0) = (10.5, 8.5, 7.2, -6.5)$$

and for the full-order controller (10.45) were

$$\theta_1(0) = -1.5,\ \theta_2(0) = (0.82, -2.09)^T$$

$$\theta_5(0) = \theta_2(0) \otimes \theta_d(0) = (\theta_{21}(0)\theta_d^T(0), \theta_{22}(0)\theta_d^T(0))^T$$

and the same $\theta_d(0)$ above.

Compared with the controller (10.45), the controller (10.44) has less parameters to be updated. The simulation results in Figures 10.3 and 10.4 indicate that with a reduced-order controller, the adaptive control system has better transient responses. We should note that, without a dead-zone inverse, each of the two controllers, the reduced-order controller

$$u(t) = \theta_1^* \omega_1(t) + \theta_{20}^* y(t) + y_m(t+2)$$

and the full-order controller

$$u(t) = \theta_1^* u(t-1) + \theta_{21}^* y(t-1) + \theta_{22}^* y(t) + y_m(t+2),$$

is good for the plant (10.43) without the output dead-zone (that is, when $y(t) = z(t)$). However, both of them lead to unbounded tracking errors in the presence of the output dead-zone, as indicated by simulations not shown here.

10.3 Summary

- For the plant with known stable poles $R_0(D)$:

$$R_0(D)P(D)[y](t) = k_p Z(D)[u](t)$$

the reduced-order controller is

$$u(t) = \theta_1^T \frac{a_1(D)}{\Lambda_0(D)R_0(D)}[u](t) + \theta_2^T \frac{a_2(D)}{\Lambda_0(D)}[y](t) + \theta_{20} y(t) + \theta_3 r(t).$$

- For the plant with known stable zeros $R_0(D)$:

$$P(D)[y](t) = k_p R_0(D) Z(D)[u](t)$$

the reduced-order controller is

$$u(t) = \theta_1^T \frac{a_1(D)}{\Lambda_0(D)}[u](t) + \theta_2^T \frac{a_2(D)}{\Lambda_0(D)R_0(D)}[y](t) + \theta_{20} y(t) + \theta_3 r(t).$$

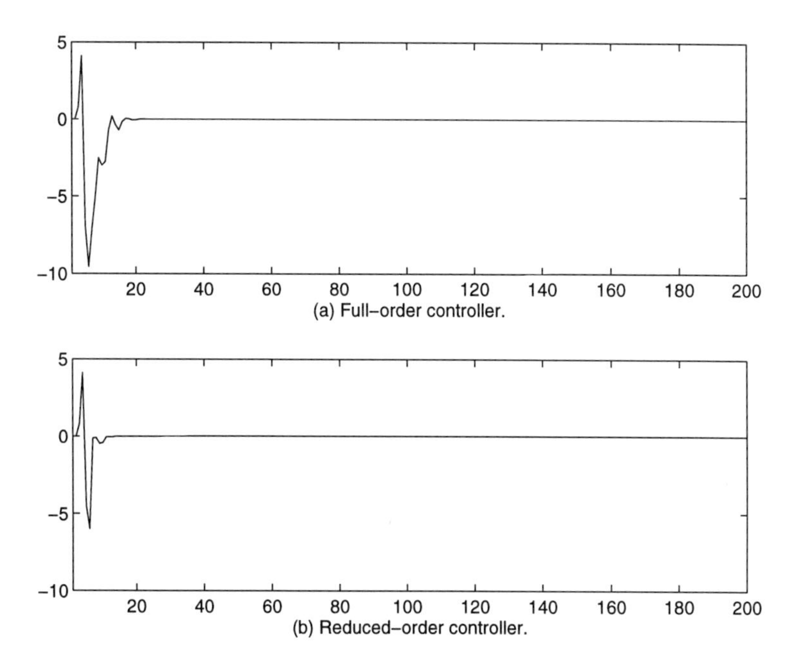

(a) Full-order controller.

(b) Reduced-order controller.

Figure 10.3: Tracking errors for $y_m(t) = 15$.

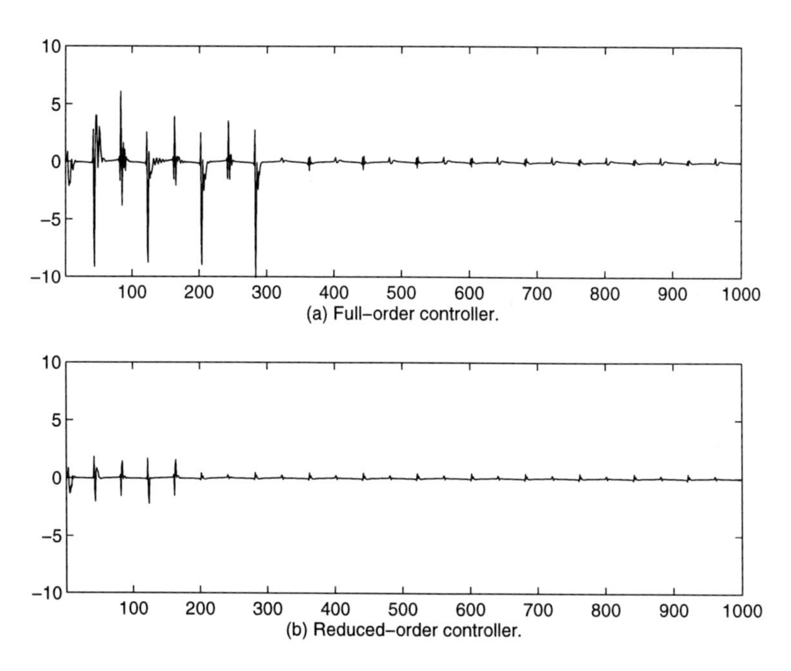

(a) Full-order controller.

(b) Reduced-order controller.

Figure 10.4: Tracking errors for $y_m(t) = 10 \sin 0.0785t$.

- The order reduction is $2n_r$, where n_r is the degree of $R_0(D)$.

- The above controllers implemented with the exact parameters, as well as their adaptive versions implemented with parameter estimates, ensure output tracking of the reference output $y_m(t) = W_m(D)[r](t)$:

$$\lim_{t\to\infty} (y(t) - y_m(t)) = 0.$$

- For plants with partially known zeros $R_0(D)$ and an unknown input nonlinearity $N(\cdot)$:

$$P(D)[y](t) = k_p R_0(D) Z(D)[u](t),\ u(t) = N(v(t))$$

  the adaptive inverse controller is

$$v(t) = \widehat{NI}(u_d(t))$$

$$u_d(t) = \theta_2^T \frac{a_2(D)}{\Lambda_0(D)R_0(D)}[y](t) + \theta_{20} y(t) + \theta_3 r(t) + \theta_4^T \omega_4(t),$$

  where $\omega_4(t)$ is a regressor vector for the estimate $\theta_4(t)$ of $\theta_4^* = -\theta_1^* \otimes \theta_N^*$ ($-\theta_1^* \otimes (\theta_N^{*T}, -1)^T$, for hysteresis). The adaptive law is (10.30) - (10.31) or (10.32) - (10.33).

- For plants with partially known poles $R_0(D)$ and an unknown output nonlinearity $N(\cdot)$:

$$R_0(D)P(D)[z](t) = k_p Z(D)[u](t),\ y(t) = N(z(t))$$

  the discrete-time adaptive inverse controller is

$$u(t) = \theta_1^T \frac{a_1(D)}{\Lambda_0(D)R_0(D)}[u](t) + \theta_5^T \omega_5(t) + \theta_{50}^T \omega_N(t) + \widehat{z_m}(t+n^*)$$

$$\widehat{z_m}(t+n^*) = \widehat{NI}(y_m(t+n^*)) = \theta_N^T(t)\omega_m(t+n^*),$$

  where $\omega_5(t)$ depends on $\omega_N(t)$, and θ_5, θ_{50}, θ_N are estimates of $\theta_5^* = \theta_2^* \otimes \theta_N^*$, $\theta_{50}^* = \theta_{20}^* \theta_N^*$, θ_N^*. The adaptive law is (10.42).

- The order reduction with the above inverse controllers is $2n_r n_N$, where n_N is the dimension of θ_N.

Chapter 11

Adaptive Control with Input and Output Nonlinearities

In this final chapter of the book we consider the plant with sandwich nonlinearities shown in Figure 11.1; that is, with a linear part $G(D)$, an input nonlinearity $N_i(\cdot)$, and an output nonlinearity $N_o(\cdot)$:

$$y(t) = N_o(z(t)),\ z(t) = G(D)[u](t),\ u(t) = N_i(v(t)), \tag{11.1}$$

where $G(D) = k_p \frac{Z(D)}{P(D)}$, $v(t)$ is the accessible control input, and $y(t)$ is the measured output. This formulation encompasses plants with unknown nonlinearities appearing in both actuators and sensors. In many realistic situations, the signals $u(t)$ and $z(t)$ are not accessible for either measurement or control.

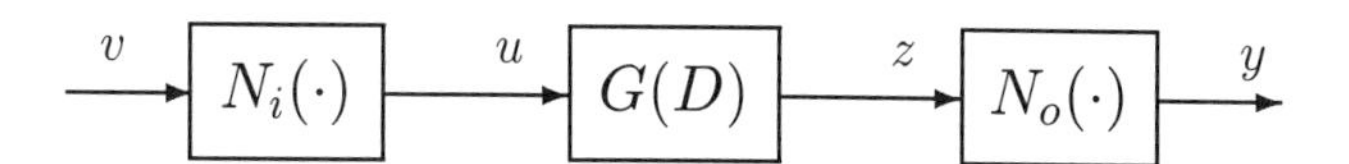

Figure 11.1: Plant with input and output nonlinearities.

Our objective is to further develop the adaptive control schemes of Chapters 7 and 9 to design adaptive control schemes for the plant (11.1).

As before, the linear part $G(D)$ satisfies the assumptions (A1) - (A5) stated in Section 4.1 (without loss of generality, we assume that $k_p = 1$), the input nonlinearity $N_i(\cdot)$ satisfies the assumptions (A7a) - (A7c) stated in Section 6.1, and the output nonlinearity $N_o(\cdot)$ satisfies the assumptions (A8a) - (A8c) stated in Section 8.3.

We will first introduce an output matching inverse controller which, using the knowledge of $G(D)$, $N_i(\cdot)$, and $N_o(\cdot)$, can achieve the tracking of a given reference signal $y_m(t)$ by the plant output $y(t)$. Our adaptive inverse controllers for plants with $N_i(\cdot)$ and $N_o(\cdot)$ unknown are first presented with $G(D)$ known. Then we proceed to the general case when $N_i(\cdot)$, $N_o(\cdot)$, and $G(D)$ are all unknown. These designs will be in discrete time, and the symbol D will denote, as the case may be, the z-transform variable or the time advance operator $D[x](t) = x(t+1)$.

11.1 Output Matching

We begin with a fixed inverse controller structure, shown in Figure 11.2, for the case when $G(D)$, $N_i(\cdot)$, and $N_o(\cdot)$ are all known. The inverse controller to be designed consists of an inverse $NI_i(\cdot)$ for the input nonlinearity $N_i(\cdot)$, a pair of inverses $NI_o(\cdot)$ for the output nonlinearity $N_o(\cdot)$, and a linear controller structure which would achieve output tracking if these nonlinearities were absent.

The inverse for $N_i(\cdot)$ is

$$v(t) = NI_i(u_d(t)), \tag{11.2}$$

which is parametrized by the input nonlinearity parameter vector $\theta^*_{N_i}$ defined in Chapter 3 as θ^*_d, θ^*_b, or θ^*_h, that is,

$$u_d(t) = -\theta^{*T}_{N_i}\omega_{N_i}(t) + a_h(t)$$

with $\omega_{N_i}(t)$ standing for $\omega_d(t)$, $\omega_b(t)$ or $\omega_h(t)$.

The linear controller structure is

$$u_d(t) = \theta_1^{*T} a_\lambda(D)[u_d](t) + \theta_2^{*T} b_\lambda(D)[\bar{z}](t) + z_m(t+n^*), \tag{11.3}$$

where $\theta_1^* \in R^{n-1}$, $\theta_2^* \in R^n$, $a_\lambda(D)$, and $b_\lambda(D)$ are defined by

$$\theta_1^{*T} a_\lambda(D)P(D) + \theta_2^{*T} b_\lambda(D)Z(D) = P(D) - Z(D)D^{n^*} \tag{11.4}$$

$$a_\lambda(D) = (D^{-n+1}, \ldots, D^{-1})^T$$

$$b_\lambda(D) = (D^{-n+1}, \ldots, D^{-1}, 1)^T$$

and the inverses for $N_o(\cdot)$ are

$$z_m(t) = NI_o(y_m(t)) = \theta^{*T}_{N_o}\omega_m(t) \tag{11.5}$$

$$\bar{z}(t) = NI_o(y(t)) = \theta^{*T}_{N_o}\omega_{N_o}(t), \tag{11.6}$$

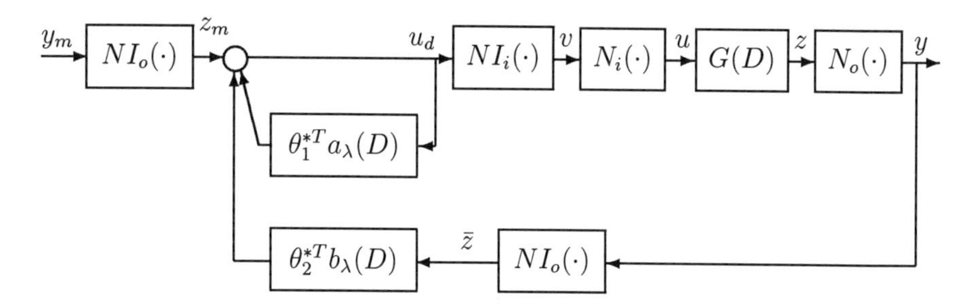

Figure 11.2: Output matching controller structure.

where $\omega_{N_o}(t)$ stands for $\omega_d(t)$, $\omega_b(t)$, or $\omega_h(t)$ in Section 8.3, $\omega_m(t)$ for $\omega_{dm}(t)$, $\omega_{bm}(t)$, or $\omega_{hm}(t)$ in Section 8.4, and $\theta^*_{N_o}$ for θ^*_d, θ^*_b, or θ^*_h in Section 8.3.

The blocks $NI_i(\cdot)$ and $NI_o(\cdot)$ are the inverses of the input nonlinearity $N_i(\cdot)$ and the output nonlinearity $N_o(\cdot)$, respectively, implemented with the true parameters $\theta^*_{N_i}$, $\theta^*_{N_o}$:

$$NI_i(\cdot) = \widehat{NI}_i(\cdot)|_{\theta_{N_i}(t)=\theta^*_{N_i}},\ NI_o(\cdot) = \widehat{NI}_o(\cdot)|_{\theta_{N_o}(t)=\theta^*_{N_o}},$$

where $\widehat{NI}_i(\cdot)$, $\widehat{NI}_o(\cdot)$ have been defined in Chapters 3 and 8, respectively.

When $G(D)$, $N_i(\cdot)$, $N_o(\cdot)$ are all known, we can use the parameters $\theta^*_{N_i}$, $\theta^*_{N_o}$, θ^*_1, θ^*_2 to implement the inverses $NI_i(\cdot)$, $NI_o(\cdot)$ (as the dead-zone inverse $DI(\cdot)$, backlash inverse $BI(\cdot)$, or hysteresis inverse $HI(\cdot)$ implemented with the true parameter θ^*_d, θ^*_b, or θ^*_h) and the linear controller in (11.2) - (11.6).

The results of Chapter 3, Section 4.2.1, and Section 8.4 reveal the properties of this inverse controller. In particular, for output tracking, these inverses are to be initialized as follows:

$$u_d(\tau) = N_i(NI_i(u_d(\tau)) \begin{cases} \text{not needed} & \text{for } DI(\cdot) \\ \tau = t_0 & \text{for } BI(\cdot) \\ \tau = t_0 & \text{for } HI(\cdot) \end{cases} \tag{11.7}$$

$$NI_o(N_o(z(\tau))) = z(\tau) \begin{cases} \tau = t_0, \ldots, t_0 + n^* + n - 2 & \text{for } DI(\cdot) \\ \tau = t_0, \ldots, t_0 + n^* + n - 1 & \text{for } BI(\cdot) \\ \tau = t_0 & \text{for } HI(\cdot) \end{cases} \tag{11.8}$$

$$y(\tau) = y_m(\tau) \begin{cases} \text{not needed} & \text{for } DI(\cdot) \\ \tau = t_0 + n^* + n - 1 & \text{for } BI(\cdot) \\ \tau = t_0 + n^* + n - 1 & \text{for } HI(\cdot). \end{cases} \tag{11.9}$$

The condition (11.7) is for the initialization of the input inverse $NI_i(\cdot)$. The conditions (11.8) and (11.9) are for initializing the inverses (11.5), (11.6) and the linear controller structure (11.3) for the output nonlinearity $N_o(\cdot)$, according to Section 8.4.

Theorem 11.1 *Under the initialization conditions (11.7) - (11.9), the controller (11.2) - (11.6) ensures that all closed-loop signals are bounded and that the output tracking $y(t) = y_m(t)$, $t \geq t_0 + n^* + n - 1$ is achieved.*

Proof: With (11.4) we derive

$$u(t) = \theta_1^{*T} a_\lambda(D)[u](t) + \theta_2^{*T} b_\lambda(D)[z](t) + z(t + n^*) \tag{11.10}$$

as we did for (8.58). In the present notation the control error equation (3.76) for the input nonlinearity problem is

$$u(t) = u_d(t) + (\theta_{N_i}(t) - \theta_{N_i}^*)^T \omega_{N_i}(t) + d_{N_i}(t). \tag{11.11}$$

Substituting (8.42) in (8.43) in the present notation with $\theta_{N_o} = \theta_{N_o}^*$ and $\hat{z}(t) = \bar{z}(t)$, we get

$$\bar{z}(t) = z(t) + d_{N_o}(t).$$

Using this equality and (11.11) with $\theta_{N_i}(t) = \theta_{N_i}^*$, we express (11.10) as

$$\begin{aligned} u_d(t) = {} & \theta_1^{*T} a_\lambda(D)[u_d](t) + \theta_2^{*T} b_\lambda(D)[\bar{z}](t) + \bar{z}(t + n^*) \\ & -(1 - \theta_1^{*T} a_\lambda(D))[d_{N_i}](t) - \theta_2^{*T} b_\lambda(D)[d_{N_o}](t) + d_{N_o}(t + n^*), \end{aligned}$$

which, together with (11.3), implies that

$$\begin{aligned} \bar{z}(t + n^*) = {} & z_m(t + n^*) + (1 - \theta_1^{*T} a_\lambda(D))[d_{N_i}](t) \\ & + \theta_2^{*T} b_\lambda(D)[d_{N_o}](t) + d_{N_o}(t + n^*). \end{aligned}$$

Since $z_m(t)$, $d_{N_i}(t)$ and $d_{N_o}(t)$ are bounded, $\bar{z}(t)$ is bounded, and so are $z(t)$ and $u(t)$ from the minimum phase assumption (A1).

To show the output tracking $y_m(t) = y(t)$, $t \geq t_0 + n^* + n - 1$, we note that under the condition (11.7), the effect of the input nonlinearity $N_i(\cdot)$ is cancelled: $u(t) = u_d(t)$, $t \geq t_0$. Then the situation is the same as that in Section 8.4 for the output nonlinearity alone. The desired tracking follows from the proof of Theorem 8.1. ∇

Our main objective is to adaptively control the plant (11.1) with unknown actuator nonlinearity $N_i(\cdot)$ and sensor nonlinearity $N_o(\cdot)$. This will be achieved in two steps—first for the linear part $G(D)$ known, and then for $G(D)$ unknown—in the next sections.

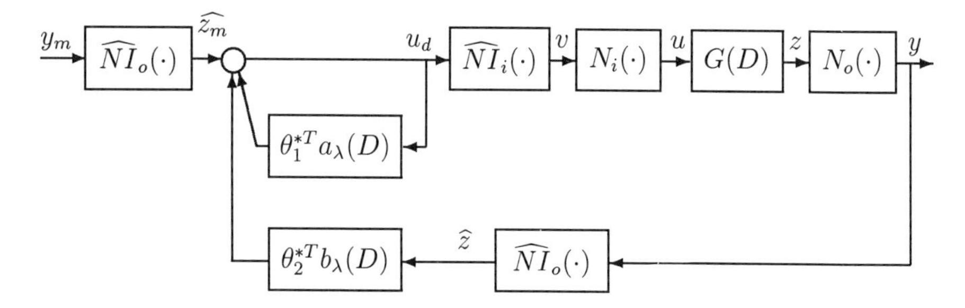

Figure 11.3: Explicit inverse controller for $G(D)$ known.

11.2 Designs for $G(D)$ Known

When the linear part $G(D)$ of the plant (11.1) is known and the nonlinearities $N_i(\cdot)$, $N_o(\cdot)$ are unknown, we use the adaptive estimates $\theta_{N_i}(t)$, $\theta_{N_o}(t)$ of the unknown parameters $\theta^*_{N_i}$, $\theta^*_{N_o}$ to implement the adaptive versions $\widehat{NI}_i(\cdot)$, $\widehat{NI}_o(\cdot)$ of $NI_i(\cdot)$, $NI_o(\cdot)$. As in Section 9.2, there are two designs when the linear part $G(D)$ is known.

11.2.1 Explicit Inverse Scheme

The controller structure is shown in Figure 11.3. It employs adaptive versions of (11.2), (11.5) and an explicit adaptive version of (11.6):

$$v(t) = \widehat{NI}_i(u_d(t)) \tag{11.12}$$

$$u_d(t) = \theta_1^{*T} a_\lambda(D)[u_d](t) + \theta_2^{*T} b_\lambda(D)[\widehat{z}](t) + \widehat{z_m}(t+n^*) \tag{11.13}$$

$$\widehat{z_m}(t+n^*) = \widehat{NI}_o(y_m(t+n^*)) = \theta^T_{N_o}(t)\omega_m(t+n^*) \tag{11.14}$$

$$\widehat{z}(t) = \widehat{NI}_o(y(t)) = \theta^T_{N_o}(t)\omega_{N_o}(t), \tag{11.15}$$

where θ_{N_o} is an estimate of the unknown parameter vector $\theta^*_{N_o}$. Note that the inverse (10.12) is now parametrized as

$$u_d(t) = -\theta^T_{N_i}\omega_{N_i}(t) + a_h(t), \tag{11.16}$$

where θ_{N_i} is the estimate of the unknown parameter vector $\theta^*_{N_i}$.

11.2.2 Implicit Inverse Scheme

The controller structure is shown in Figure 11.4. It employs adaptive versions of (11.2), (11.5) and an implicit adaptive version of (11.6):

$$v(t) = \widehat{NI}_i(u_d(t)) \tag{11.17}$$

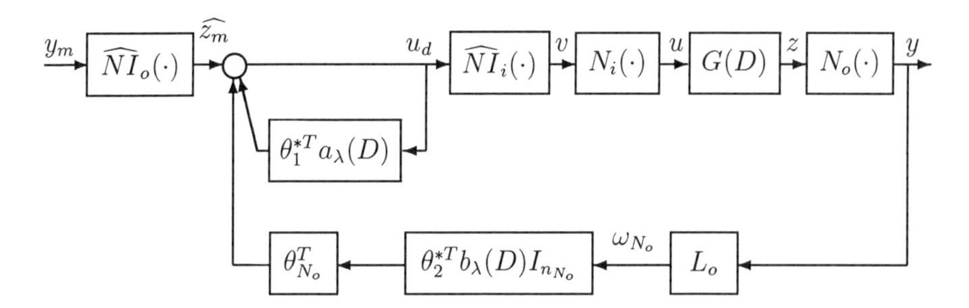

Figure 11.4: Implicit inverse controller for $G(D)$ known.

$$\widehat{z_m}(t+n^*) = \widehat{NI}_o(y_m(t+n^*)) = \theta_{N_o}^T(t)\omega_m(t+n^*) \tag{11.18}$$

$$\begin{aligned} u_d(t) &= \theta_1^{*T}a_\lambda(D)[u_d](t) + \theta_{N_o}^T(t)\theta_2^{*T}b_\lambda(D)I_{n_{N_o}}[\omega_{N_o}](t) + \widehat{z_m}(t+n^*) \\ &= \theta_1^{*T}a_\lambda(D)[u_d](t) + \theta_{N_o}^T(t)\omega_{Nm}(t), \end{aligned} \tag{11.19}$$

where $I_{n_{N_o}}$ is the $n_{N_o} \times n_{N_o}$ identity matrix for $\theta_{N_o} \in R^{n_{N_o}}$, $\omega_{N_o}(t)$ is the regressor of the inverse $\widehat{NI}_o(y(t)) = \theta_{N_o}^T(t)\omega_{N_o}(t)$, and

$$\omega_{Nm}(t) = \theta_2^{*T}b_\lambda(D)I_{n_{N_o}}[\omega_{N_o}](t) + \omega_m(t+n^*).$$

Again, the term $\theta_{N_o}^T(t)\theta_2^{*T}b_\lambda(D)I_{n_{N_o}}[\omega_{N_o}](t)$ contains an implicit inverse because if $\theta_{N_o} = \theta_{N_o}^*$, then

$$\theta_{N_o}^T\theta_2^{*T}b_\lambda(D)I_{n_{N_o}}[\omega_{N_o}](t) = \theta_2^{*T}b_\lambda(D)[NI_o(y)](t) = \theta_2^{*T}b_\lambda(D)[\bar{z}](t)$$

for $\bar{z}(t) = NI_o(y)(t) = \theta_{N_o}^{*T}\omega_{N_o}(t)$ in (11.6) and $\theta_2^{*T}b_\lambda(D)[\bar{z}](t)$ in (11.3).

11.2.3 Adaptive Laws

To develop an adaptive update law for $\theta_{N_i}(t)$ and $\theta_{N_o}(t)$ in the control schemes (11.12) - (11.15) and (11.17) - (11.19), we use (11.12) and (11.13) to obtain

$$\begin{aligned} u_d(t) &= \theta_1^{*T}a_\lambda(D)[u_d](t) - (1-\theta_1^{*T}a_\lambda(D))[(\theta_{N_i}-\theta_{N_i}^*)^T\omega_{N_i}](t) \\ &\quad -(1-\theta_1^{*T}a_\lambda(D))[d_{N_i}](t) + \theta_2^{*T}b_\lambda(D)[z](t) + z(t+n^*). \end{aligned} \tag{11.20}$$

With the definition (9.9) of the regressor

$$\omega_{NN}(t) = \theta_2^{*T}b_\lambda(D)I_{n_{N_o}}[\omega_{N_o}](t) + \omega_{N_o}(t+n^*)$$

and the relationship (9.8):

$$z(t) = \theta_{N_o}^{*T}\omega_{N_o}(t) - d_{N_o}(t) \tag{11.21}$$

we express

$$\theta_2^{*T} b_\lambda(D)[z](t) + z(t+n^*) \\ = \theta_{N_o}^{*T} \omega_{NN}(t) - (d_{N_o}(t+n^*) + \theta_2^{*T} b_\lambda(D)[d_{N_o}](t)). \quad (11.22)$$

Substituting (11.22) in (11.20) and defining

$$d(t+n^*) = (1 - \theta_1^{*T} a_\lambda(D))[d_{N_i}](t) \\ + d_{N_o}(t+n^*) + \theta_2^{*T} b_\lambda(D)[d_{N_o}](t) \quad (11.23)$$

we obtain the linear parametrization:

$$u_d(t) = \theta_1^{*T} a_\lambda(D)[u_d](t) - (1 - \theta_1^{*T} a_\lambda(D))[(\theta_{N_i} - \theta_{N_i}^*)^T \omega_{N_i}](t) \\ + \theta_{N_o}^{*T} \omega_{NN}(t) - d(t+n^*). \quad (11.24)$$

As in all previous designs, the error term $d(t+n^*)$ is bounded for any $t \geq 0$ (when bounded estimates of c_t, c_b, c_l and c_r are used for the hysteresis case). In view of (11.24) and with the estimates $\theta_{N_i}(t)$, $\theta_{N_o}(t)$ of $\theta_{N_i}^*$, $\theta_{N_o}^*$, we define the estimation error as

$$\epsilon_N(t) = \theta_1^{*T} a_\lambda(D)[u_d](t-n^*) + \theta_{N_o}^T(t-1)\omega_{NN}(t-n^*) \\ - u_d(t-n^*) + \xi_{N_i}(t), \quad (11.25)$$

where

$$\xi_{N_i}(t) = \theta_{N_i}^T(t-1)\zeta_{N_i}(t) - W(D)[\theta_{N_i}^T \omega_{N_i}](t) \quad (11.26)$$

$$\zeta_{N_i}(t) = W(D)[\omega_{N_i}](t) \quad (11.27)$$

$$W(D) = (1 - \theta_1^{*T} a_\lambda(D))D^{-n^*}.$$

It then follows from (11.24) - (11.27) that

$$\epsilon_N(t) = (\theta_{N_o}(t-1) - \theta_{N_o}^*)^T \omega_{NN}(t-n^*) \\ + (\theta_{N_i}(t-1) - \theta_{N_i}^*)^T \zeta_{N_i}(t) + d(t), \quad (11.28)$$

which is again a linear equation with a bounded error $d(t)$.

Based on this linear error equation, we choose the adaptive update laws for $\theta_{N_i}(t)$ and $\theta_{N_o}(t)$ to be

$$\theta_{N_i}(t) = \theta_{N_i}(t-1) - \frac{\Gamma_{N_i}\zeta_{N_i}(t)\epsilon_N(t)}{m_N^2(t)} + f_{N_i}(t) \quad (11.29)$$

$$\theta_{N_o}(t) = \theta_{N_o}(t-1) - \frac{\Gamma_{N_o}\omega_{NN}(t-n^*)\epsilon_N(t)}{m_N^2(t)} + f_{N_o}(t), \quad (11.30)$$

where

$$m_N(t) = \sqrt{1 + \zeta_{N_i}^T(t)\zeta_{N_i}(t) + \xi_{N_i}^2(t) + \omega_{NN}^T(t-n^*)\omega_{NN}(t-n^*) + \xi_{N_o}^2(t)}$$

with different $\xi_{N_o}(t)$ for the two different designs:

$$\xi_{N_o}(t) = \theta_N^T(t-1)\omega_{NN}(t-n^*) \\ -\theta_2^{*T} b_\lambda(D)[\theta_N^T\omega_N](t-n^*) - \theta_N^T(t-n^*)\omega_N(t)$$

for the *explicit inverse* design, and

$$\xi_{N_o}(t) = (\theta_{N_o}(t-1) - \theta_{N_o}(t-n^*))^T\omega_{NN}(t-n^*)$$

for the *implicit inverse* design. As before, $f_{N_i}(t)$, $f_{N_o}(t)$ are modifications for robustness with respect to $d(t)$ and for parameter projection of $\theta_{N_i}(t)$ and $\theta_{N_o}(t)$ to meet the conditions of the assumptions (A7a) - (A7c) and assumptions (A8a) - (A8c), and Γ_{N_i}, Γ_{N_o} are positive diagonal step-size matrices.

As in Chapters 7 and 9, $f_{N_i}(t)$, $f_{N_o}(t)$, Γ_{N_i} and Γ_{N_o} are chosen either for parameter projection or for switching σ-modification with parameter projection. With these designs, the adaptive law (11.19) - (11.20) ensures that $\theta_{N_i}(t)$, $\theta_{N_o}(t)$, and $\frac{\epsilon_N(t)}{m_N(t)}$ are bounded and that for $x_1(t) = \frac{\epsilon_N^2(t)}{m_N^2(t)}$, $x_2(t) = \|\theta_{N_i}(t) - \theta_{N_i}(t-1)\|_2^2$, and $x_3(t) = \|\theta_{N_o}(t) - \theta_{N_o}(t-1)\|_2^2$, there exist constants $a_i > 0$, $b_i > 0$, $i = 1, 2, 3$, such that

$$\sum_{t=t_1}^{t_2} x_i(t) \le a_i + b_i \sum_{t=t_1}^{t_2} \frac{d^2(t)}{m_N^2(t)}$$

for all $t_2 > t_1 \ge 0$. These properties can be used to prove the closed-loop signal boundedness.

The explicit and implicit designs, although with different structures, are both based on the same estimation error (11.26). The implicit design is suitable when $G(D)$ is unknown.

11.3 Design for $G(D)$ Unknown

When $G(D)$, $N_i(\cdot)$, and $N_o(\cdot)$ are all unknown a new controller structure is similar to those in Sections 6.3 and 9.3.

For the input nonlinearity $N_i(\cdot)$, an inverse with the true parameter $\theta_{N_i}^*$ (see (3.80)) has the expression

$$u_d(t) = -\theta_{N_i}^{*T}\omega_{N_i}(t) + a_h(t).$$

Following (6.72) - (6.73), we reparametrize the term $\theta_1^{*T} a_\lambda(D)[u_d](t)$, which appears in the linear controller (11.19), as

$$\theta_1^{*T} a_\lambda(D)[u_d](t) = \theta_4^{*T} \omega_4(t), \tag{11.31}$$

where for the *dead-zone* or *backlash* case we have

$$\theta_4^* = -\theta_1^* \otimes \theta_{N_i}^*, \ \omega_4(t) = A_\lambda(D)[\omega_{N_i}](t)$$

and for the *hysteresis* case we have

$$\theta_4^* = -\theta_1^* \otimes (\theta_h^{*T}, -1)^T, \ \omega_4(t) = A_\lambda(D)[(\omega_{N_i}^T, a_h)^T](t)$$

with $A_\lambda(D) = (D^{-n+1} I_{n_{N_i}}, \ldots, D^{-1} I_{n_{N_i}})^T$ and $n_{N_i} = 4, 3, 9$ for dead-zone, backlash, and hysteresis, respectively.

Similarly, for an output inverse with $\theta_{N_o}^*$ (see (8.73)):

$$\widehat{z}(t) = \bar{z}(t) = \theta_{N_o}^{*T} \omega_{N_o}(t)$$

we reparametrize the term $\theta_2^{*T} b_\lambda(D)[\widehat{z}](t)$ in (11.19) as

$$\theta_2^{*T} b_\lambda(D)[\bar{z}](t) = \theta_5^{*T} \omega_5(t) \tag{11.32}$$

where, as in (9.45),

$$\theta_5^* = \theta_2^* \otimes \theta_{N_o}^*, \ \omega_5(t) = B_\lambda(D)[\omega_{N_o}](t)$$

with $B_\lambda(D) = (D^{-n+1} I_{n_{N_o}}, \ldots, D^{-1} I_{n_{N_o}}, I_{n_{N_o}})^T$ and $n_{N_o} = 4, 3, 8$ for dead-zone, backlash, and hysteresis, respectively.

With the new parametrization (11.30) - (11.31), we design the following adaptive controller:

$$v(t) = \widehat{NI}_i(u_d(t)) = \widehat{NI}_i(\theta_{N_i}(t); u_d(t)) \tag{11.33}$$

$$u_d(t) = \theta_4^T(t)\omega_4(t) + \theta_5^T(t)\omega_5(t) + \widehat{z_m}(t + n^*) \tag{11.34}$$

$$\widehat{z_m}(t + n^*) = \widehat{NI}_o(y_m(t + n^*)) = \theta_{N_o}^T(t)\omega_m(t + n^*), \tag{11.35}$$

where $\theta_4(t)$, $\theta_5(t)$ are the adaptive estimates of θ_4^*, θ_5^*, respectively, and $\omega_m(t)$ is defined in (8.52). This controller structure is shown in Figure 11.5, where the logic block L_i generates $\bar{\omega}_{N_i}(t) = \omega_{N_i}(t)$ for the input dead-zone or backlash and $\bar{\omega}_{N_i}(t) = (\omega_h^T(t), a_h(t))^T$ for the input hysteresis, where the logic block L_o generates $\omega_{N_o}(t)$ for the output dead-zone, backlash, or hysteresis.

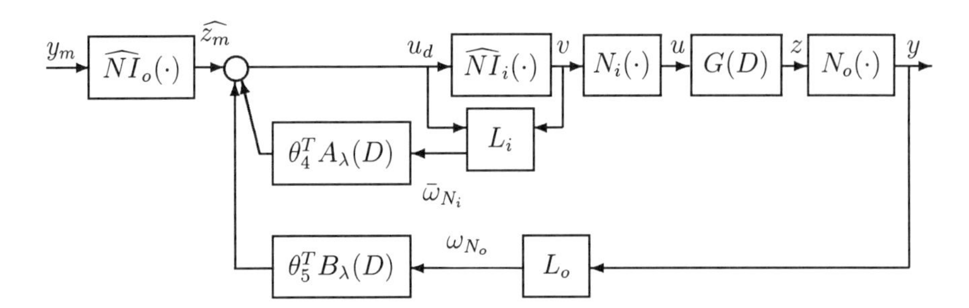

Figure 11.5: Adaptive controller structure for $G(D)$ unknown.

To develop an adaptive law for updating $\theta_{N_i}(t)$, $\theta_{N_o}(t)$, $\theta_4(t)$, and $\theta_5(t)$, we substitute (11.12) in (11.11) to express $u(t)$ as

$$u(t) = -\theta_{N_i}^{*T}\omega_{N_i}(t) + d_{N_i}(t) + a_h(t). \tag{11.36}$$

We then use (11.11) for the left-hand side of (11.10) and (11.36) for the right-hand side of (11.10) and substitute (11.21) in (11.10) to obtain the linear parametrization of $u_d(t)$:

$$\begin{aligned} u_d(t) = \theta_4^{*T}\omega_4(t) - (\theta_{N_i}(t) - \theta_{N_i}^*)^T\omega_{N_i}(t) \\ +\theta_5^{*T}\omega_5(t) + \theta_{N_o}^{*T}\omega_{N_o}(t+n^*) - d(t+n^*) \end{aligned} \tag{11.37}$$

in which θ_4^*, θ_5^*, $\theta_{N_i}^*$, and $\theta_{N_o}^*$ are unknown, $\omega_4(t)$, $\omega_5(t)$, $\omega_{N_i}(t)$, and $\omega_{N_o}(t)$ are measured signals, and $d(t+n^*)$ is bounded.

Let $\theta_4(t)$, $\theta_5(t)$, $\theta_{N_i}(t)$, and $\theta_{N_o}(t)$ be the estimates of θ_4^*, θ_5^*, $\theta_{N_i}^*$, and $\theta_{N_o}^*$. In view of (11.37), we define the estimation error

$$\begin{aligned} \epsilon(t) = \theta_4^T(t-1)\omega_4(t-n^*) + \theta_5^T(t-1)\omega_5(t-n^*) \\ +\theta_{N_o}^T(t-1)\omega_{N_o}(t) - u_d(t-n^*) + \xi_{N_i}(t), \end{aligned} \tag{11.38}$$

where

$$\xi_{N_i}(t) = (\theta_{N_i}(t-1) - \theta_{N_i}(t-n^*))^T\omega_{N_i}(t-n^*). \tag{11.39}$$

Using (11.37) - (11.39), we obtain the error equation

$$\begin{aligned} \epsilon(t) = (\theta_4(t-1) - \theta_4^*)^T\omega_4(t-n^*) + (\theta_5(t-1) - \theta_5^*)^T\omega_5(t-n^*) \\ +(\theta_{N_i}(t-1) - \theta_{N_i}^*)^T\omega_{N_i}(t-n^*) \\ +(\theta_{N_o}(t-1) - \theta_{N_o}^*)^T\omega_{N_o}(t) + d(t), \end{aligned} \tag{11.40}$$

which has the familiar linear form with bounded $d(t)$.

Defining the overall parameter $\theta(t)$ and regressor $\zeta(t)$ as

$$\theta(t) = (\theta_{N_i}^T(t), \theta_{N_o}^T(t), \theta_4^T(t), \theta_5^T(t))^T$$

$$\zeta(t) = (\omega_{N_i}^T(t-n^*), \omega_{N_o}^T(t), \omega_4^T(t-n^*), \omega_5^T(t-n^*))^T$$

we choose the adaptive update law:

$$\theta(t) = \theta(t-1) - \frac{\Gamma\zeta(t)\epsilon(t)}{m^2(t)} + f(t), \tag{11.41}$$

where

$$m(t) = \sqrt{1 + \zeta^T(t)\zeta(t)}.$$

As before, $f(t)$ and Γ can be chosen as parameter projection or as switching σ-modification with parameter projection. With such choices, the adaptive law (11.41) ensures the boundedness of $\theta(t)$, $\frac{\epsilon(t)}{m(t)}$, the smallness in the mean of $\frac{\epsilon^2(t)}{m^2(t)}$, $||\theta(t) - \theta(t-1)||_2^2$, and in turn, the closed-loop signal boundedness.

11.4 Designs for $G(D)$ Partially Known

When the linear part $G(D)$ in the plant (11.1) is partially known, a lower-order adaptive inverse controller can be designed to reduce the adaptive system complexity. The following two reduced-order designs correspond to known zeros or poles.

11.4.1 Design with Partially Known Zeros

Following the notation of (10.3), we consider the plant (11.1) with $k_p = 1$ in the form

$$y(t) = N_o(z(t)),\ z(t) = \frac{Z(D)R_0(D)}{P(D)}[u](t),\ u(t) = N_i(v(t)),$$

where $R_0(D)$ is a known stable polynomial of degree n_r, and $P(D)$, $Z(D)$ are unknown polynomials of degrees n, m satisfying the assumptions in Section 10.1. In particular, $Z(D)$ is stable, and n, m are known.

The adaptive inverse controller for plants with unknown actuator and sensor nonlinearities and partially known zero dynamics is

$$v(t) = \widehat{NI}_i(u_d(t)) = \widehat{NI}_i(\theta_{N_i}(t); u_d(t)) \tag{11.42}$$

$$u_d(t) = \theta_4^T(t)\omega_4(t) + \theta_5^T(t)\omega_5(t) + \widehat{z_m}(t+n^*) \tag{11.43}$$

$$\widehat{z_m}(t+n^*) = \widehat{NI}_o(y_m(t+n^*)) = \theta_{N_o}^T(t)\omega_m(t+n^*), \tag{11.44}$$

where $\theta_4(t)$, $\theta_5(t)$ are the adaptive estimates of

$$\theta_4^* = -\theta_1^* \otimes \theta_{N_i}^*,\ \theta_5^* = ((\theta_2^* \otimes \theta_{N_o}^*)^T, (\theta_{20}^* \theta_{N_o}^*)^T)^T$$

with $\theta_1^* \in R^{n-n_r-1}$, $\theta_2^* \in R^{n-1}$, $\theta_{20}^* \in R$ which satisfy

$$\theta_1^{*T} a_1(D)P(D) + (\theta_2^{*T} a_2(D) + \theta_{20}^* R_0(D) D^{n-n_r-1}) Z(D)$$
$$= D^{n-n_r-1}(P(D) - R_0(D)Z(D)D^{n-n_r-m})$$
$$a_1(D) = (1, D, \ldots, D^{n-n_r-2})^T$$
$$a_2(D) = (1, D, \ldots, D^{n-2})^T.$$

The regressors corresponding θ_4 and θ_5 are

$$\omega_4(t) = A_{\lambda_0}(D)[\bar{\omega}_{N_i}](t)$$
$$\omega_5(t) = \bar{B}_\lambda(D)[\omega_{N_o}](t),$$

where

$$A_{\lambda_0}(D) = (D^{-n+n_r+1} I_{n_{N_i}}, \ldots, D^{-1} I_{n_{N_i}})^T$$
$$\bar{B}_\lambda(D) = (\frac{A_2^T(D)}{D^{n-n_r-1} R_0(D)}, I_{n_{N_o}})^T$$
$$A_2(D) = (I_{n_{N_o}}, D I_{n_{N_o}}, \ldots, D^{n-n_r-2} I_{n_{N_o}})^T.$$

This controller has the same structure as that shown in Figure 11.5 with $A_\lambda(D)$, $B_\lambda(D)$ replaced by $A_{\lambda_0}(D)$, $\bar{B}_\lambda(D)$, respectively. The development of an adaptive law for updating $\theta_4(t)$, $\theta_5(t)$, $\theta_{N_i}(t)$, and $\theta_{N_o}(t)$ applies to this reduced-order controller structure.

11.4.2 Design with Partially Known Poles

With $R_0(D)$ representing n_r known stable poles, the plant (11.1) with $k_p = 1$ is rewritten in the form:

$$y(t) = N_o(z(t)),\ z(t) = \frac{Z(D)}{R_0(D)P(D)}[u](t),\ u(t) = N_i(v(t)),$$

where $P(D)$, $Z(D)$ are unknown polynomials of degrees n, m, respectively, $Z(D)$ is stable, and n, m are known.

The adaptive inverse controller for plants with unknown input and output nonlinearities and partially known poles has the same form as (11.42) - (11.44):

$$v(t) = \widehat{NI}_i(u_d(t)) = \widehat{NI}_i(\theta_{N_i}(t); u_d(t)) \tag{11.45}$$

$$u_d(t) = \theta_4^T(t)\omega_4(t) + \theta_5^T(t)\omega_5(t) + \widehat{z_m}(t+n^*) \tag{11.46}$$

$$\widehat{z_m}(t+n^*) = \widehat{NI}_o(y_m(t+n^*)) = \theta_{N_o}^T(t)\omega_m(t+n^*), \tag{11.47}$$

where $\theta_4(t)$, $\theta_5(t)$ are the adaptive estimates of

$$\theta_4^* = -\theta_1^* \otimes \theta_{N_i}^*,\ \theta_5^* = ((\theta_2^* \otimes \theta_{N_o}^*)^T, (\theta_{20}^* \theta_{N_o}^*)^T)^T$$

but for $\theta_1^* \in R^{n+n_r-1}$, $\theta_2^* \in R^{n-1}$, $\theta_{20}^* \in R$ satisfying

$$\theta_1^{*T} a_1(D)P(D) + (\theta_2^{*T} a_2(D) + \theta_{20}^* D^{n-1})Z(D)$$
$$= D^{n-1}(R_0(D)P(D) - Z(D)D^{n+n_r-m})$$
$$a_1(D) = (1, D, \ldots, D^{n+n_r-2})^T$$
$$a_2(D) = (1, D, \ldots, D^{n-2})^T.$$

The corresponding regressors are

$$\omega_4(t) = \bar{A}_{\lambda_0}(D)[\bar{\omega}_{N_i}](t)$$

$$\omega_5(t) = B_{\lambda_0}(D)[\omega_{N_o}](t),$$

where

$$\bar{A}_\lambda(D) = \frac{A_1(D)}{D^{n-1}R_0(D)}$$

$$A_1(D) = (I_{n_{N_i}}, DI_{n_{N_o}}, \ldots, D^{n+n_r-2} I_{n_{N_i}})^T$$

$$B_{\lambda_0}(D) = (D^{-n+1} I_{n_{N_o}}, \ldots, D^{-1} I_{n_{N_o}}, I_{n_{N_o}})^T.$$

Figure 11.5 with $A_\lambda(D)$, $B_\lambda(D)$ replaced by $\bar{A}_\lambda(D)$, $B_{\lambda_0}(D)$ also shows the control scheme (11.45) - (11.46) whose parameters can be updated by an adaptive law similar to (11.41), based on the parametrization (11.37).

11.5 Summary

- For unknown plants with an input nonlinearity $N_i(\cdot)$, an output nonlinearity $N_o(\cdot)$, and a linear part $G(D)$:

$$y(t) = N_o(z(t)),\ z(t) = G(D)[u](t),\ u(t) = N_i(v(t))$$

  the inverse control scheme is

$$v(t) = NI_i(u_d(t))$$

$$\bar{z}(t) = NI_o(y(t)) = \theta_{N_o}^{*T}\omega_{N_o}(t)$$

$$z_m(t) = NI_o(y_m(t)) = \theta_{N_o}^{*T}\omega_m(t)$$
$$u_d(t) = \theta_1^{*T}a_\lambda(D)[u_d](t) + \theta_2^{*T}b_\lambda(D)[\bar{z}](t) + z_m(t+n^*).$$

When properly initialized, this inverse controller ensures closed-loop signal boundedness and output tracking:

$$y(t) = y_m(t), \ t \geq t_0 + n^* + n - 1.$$

- When only $G(D)$ is known, there are two inverse control schemes:

  – The *explicit* adaptive inverse scheme is

$$v(t) = \widehat{NI}_i(u_d(t))$$

$$\hat{z}(t) = \widehat{NI}_o(y(t)) = \theta_{N_o}^T(t)\omega_{N_o}(t)$$
$$\widehat{z_m}(t+n^*) = \widehat{NI}_o(y_m(t+n^*)) = \theta_{N_o}^T(t)\omega_m(t+n^*)$$
$$u_d(t) = \theta_1^{*T}a_\lambda(D)[u_d](t) + \theta_2^{*T}b_\lambda(D)[\hat{z}](t) + \widehat{z_m}(t+n^*).$$

  – The *implicit* adaptive inverse scheme is

$$v(t) = \widehat{NI}_i(u_d(t))$$

$$\widehat{z_m}(t+n^*) = \widehat{NI}_o(y_m(t+n^*)) = \theta_{N_o}^T(t)\omega_m(t+n^*)$$
$$u_d(t) = \theta_1^{*T}a_\lambda(D)[u_d](t) + \theta_{N_o}^T(t)\theta_2^{*T}b_\lambda(D)I_{n_{N_o}}[\omega_{N_o}](t) + \widehat{z_m}(t+n^*).$$

- Adaptive update laws for $\theta_{N_i}(t)$, $\theta_{N_o}(t)$ are (11.29), (11.30).

- For $N_i(\cdot)$, $N_o(\cdot)$, and $G(D)$ all unknown, the adaptive inverse control scheme is
$$v(t) = \widehat{NI}_i(u_d(t))$$
$$\widehat{z_m}(t+n^*) = \widehat{NI}_o(y_m(t+n^*)) = \theta_{N_o}^T(t)\omega_m(t+n^*)$$
$$u_d(t) = \theta_4^T(t)\omega_4(t) + \theta_5^T(t)\omega_5(t) + \widehat{z_m}(t+n^*),$$
where θ_4 is the estimate of $\theta_4^* = -\theta_1^* \otimes \theta_N^*$ $(-\theta_1^* \otimes (\theta_N^{*T}, -1)^T$ for hysteresis) and $\theta_5(t)$ is the estimate of $\theta_5^* = \theta_2^* \otimes \theta_{N_o}^*$, with the corresponding regressors $\omega_4(t)$, $\omega_5(t)$. The adaptive law is (11.41).

- With partial knowledge of zeros or poles, the order of the adaptive inverse control scheme can be reduced by $2n_r n_{N_i}$ when n_r stable zeros of $G(D)$ are known, or by $2n_r n_{N_o}$ when n_r stable poles of $G(D)$ are known.

Appendix A

Model Reference Adaptive Control

For completeness we give a brief review of model reference adaptive control (MRAC) for linear time-invariant plants in a unified framework for both continuous-time and discrete-time designs. For a complete and detailed treatment we recommend the text by Ioannou and Sun [39].

Consider the linear time-invariant plant

$$y(t) = G(D)[u](t), \tag{A.1}$$

where $G(D) = k_p \frac{Z(D)}{P(D)}$, k_p is the "high frequency" gain, and $Z(D)$, $P(D)$ are monic polynomials of degrees n, m. We use D to denote, as the case may be, the Laplace transform variable s or the time-differentiation operator $D[x](t) = \dot{x}(t)$, in the continuous time, and the z-transform variable z or the time-advance operator $D[x](t) = x(t+1)$, in the discrete time.

Given a reference model

$$y_m(t) = W_m(D)[r](t), \tag{A.2}$$

where $W_m(D)$ is stable and $r(t)$ is bounded, the control objective is to find a feedback control u such that all closed-loop signals are bounded and the plant output y tracks the reference output y_m as close as possible.

In model reference adaptive control, the following assumptions for $G(D)$ and $W_m(D)$ are common:

(M1) $Z(D)$ is a stable polynomial.

(M2) The degree n of $P(D)$ is known.[1]

[1] The knowledge of an upper bound $\bar{n}$ on n: $\bar{n} \geq n$, is sufficient for constructing a model reference controller structure.

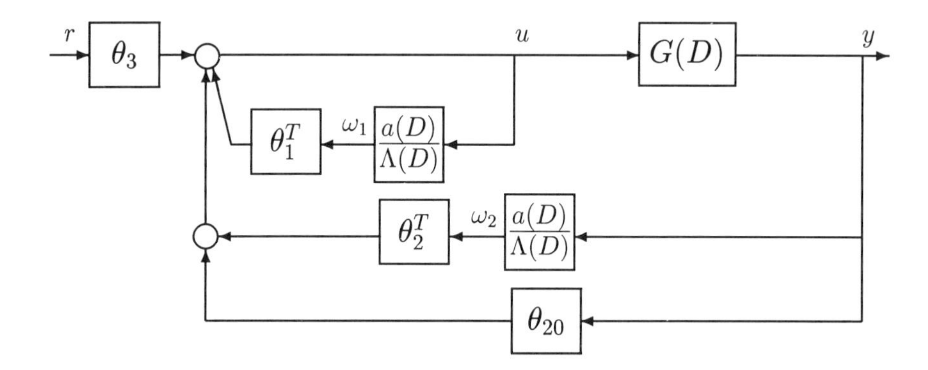

Figure A.6: Model reference controller structure.

(**M3**) The relative degree $n^* = n - m$ of $G(D)$ is known.

(**M4**) The sign of k_p is known.

(**M5**) $W_m(D) = P_m^{-1}(D)$ for a stable polynomial $P_m(D)$ of degree n^* (for a discrete-time design, $P_m(D) = D^{n^*}$).

The model reference controller structure in Figure A.1 is described by

$$u(t) = \theta_1^T \omega_1(t) + \theta_2^T \omega_2(t) + \theta_{20} y(t) + \theta_3 r(t), \tag{A.3}$$

where

$$\omega_1(t) = \frac{a(D)}{\Lambda(D)}[u](t), \ \omega_2(t) = \frac{a(D)}{\Lambda(D)}[y](t) \tag{A.4}$$

$$a(D) = (1, D, \cdots, D^{n-2})^T$$

$$\theta_1, \ \theta_2 \in R^{n-1}, \ \theta_{20}, \ \theta_3 \in R$$

and $\Lambda(D)$ is a monic stable polynomial of degree $n-1$ (for a discrete-time design, $\Lambda(D) = D^{n-1}$).[2]

We first present the solution to the nonadaptive control problem when $G(D)$ is known.

Proposition A.1 *With* $\theta_3^* = k_p^{-1}$, *there exist constant parameters* $\theta_1^*, \ \theta_2^* \in R^{n-1}, \ \theta_{20}^* \in R$ [3]*such that*

$$\theta_1^{*T} a(D) P(D) + (\theta_2^{*T} a(D) + \theta_{20}^* \Lambda(D)) k_p Z(D)$$
$$= \Lambda(D)(P(D) - k_p \theta_3^* Z(D) P_m(D)). \tag{A.5}$$

[2]When only an upper bound $\bar{n}$ on n: $\bar{n} \geq n$, is known, then the controller structure (A.3) is designed with n replaced by $\bar{n}$.

[3]The parameters θ_1^*, θ_2^*, and θ_{20}^* are not unique when $P(D)$ and $Z(D)$ are not co-prime or/and $\bar{n} > n$ is used in designing (A.3) (that is, when $\theta_1^*, \ \theta_2^* \in R^{\bar{n}-1}$).

The controller (A.3) using θ_1^*, θ_2^*, θ_{20}^*, and θ_3^*:

$$u(t) = \theta_1^{*T}\omega_1(t) + \theta_2^{*T}\omega_2(t) + \theta_{20}^* y(t) + \theta_3^* r(t)$$

ensures that all the closed-loop signals are bounded and the output tracking is achieved:

$$y(t) = y_m(t) + \eta_0(t)$$

for some initial condition-dependent exponentially decaying $\epsilon_0(t)$.

When $G(D)$ is unknown, an adaptive version of the controller (A.3) is implemented with $\theta_1 = \theta_1(t)$, $\theta_2 = \theta_2(t)$, $\theta_{20} = \theta_{20}(t)$, and $\theta_3 = \theta_3(t)$, where $\theta_1(t)$, $\theta_2(t)$, $\theta_{20}(t)$, and $\theta_3(t)$ are the estimates of θ_1^*, θ_2^*, θ_{20}^*, and θ_3^*, to be updated with an adaptive law.

To design such an adaptive law, we need to derive an error equation. Operating both sides of (A.5) on $y(t)$ and using (A.1): $P(D)[y](t) = k_p Z(D)[u](t)$, we obtain

$$\begin{aligned}&\theta_1^{*T} a(D) k_p Z(D)[u](t) + (\theta_2^{*T} a(D) + \theta_{20}^* \Lambda(D)) k_p Z(D)[y](t)\\&= \Lambda(D) k_p Z(D)[u](t) - \Lambda(D) Z(D) P_m(D)[y](t).\end{aligned}$$

This equality, with $Z(D)$ and $\Lambda(D)$ stable and with the effect of initial conditions ignored, leads to the parametrized plant model

$$u(t) = \theta_1^{*T}\frac{a(D)}{\Lambda(D)}[u](t) + \theta_2^{*T}\frac{a(D)}{\Lambda(D)}[y](t) + \theta_{20}^* y(t) + \theta_3^* W_m^{-1}(D)[y](t). \quad \text{(A.6)}$$

Introducing $\rho^* = \frac{1}{\theta_3^*} = k_p$ and

$$\omega(t) = (\omega_1^T(t), \omega_2^T(t), y(t), r(t))^T \quad \text{(A.7)}$$

$$e(t) = y(t) - y_m(t),\ \tilde{\theta}(t) = \theta(t) - \theta^*, \quad \text{(A.8)}$$

where

$$\begin{aligned}\theta(t) &= (\theta_1^T(t), \theta_2^T(t), \theta_{20}(t), \theta_3(t))^T\\ \theta^* &= (\theta_1^{*T}, \theta_2^{*T}, \theta_{20}^*, \theta_3^*)^T,\end{aligned}$$

substituting (A.6) in (A.3) with adaptive parameter estimates, ignoring the effect of the initial conditions, and using (A.7) and (A.8), we arrive at the tracking error equation

$$\begin{aligned}e(t) &= \rho^* W_m(D)[\tilde{\theta}^T\omega](t)\\ &= -\rho^*(\theta^{*T} W_m(D)[\omega](t) - W_m(D)[\theta^T\omega](t)). \quad \text{(A.9)}\end{aligned}$$

Since both θ^* and ρ^* are unknown, the second equality of (A.9) suggests that we define the estimation error

$$\epsilon(t) = e(t) + \rho(t)\xi(t), \tag{A.10}$$

where $\rho(t)$ is the estimate of ρ^*, and

$$\xi(t) = \theta^T(t)\zeta(t) - W_m(D)[\theta^T\omega](t) \tag{A.11}$$

$$\zeta(t) = W_m(D)[\omega](t). \tag{A.12}$$

Substituting (A.9), (A.11), and (A.12) into (A.10) and introducing $\tilde{\rho}(t) = \rho(t) - \rho^*$, we express $\epsilon(t)$ as

$$\epsilon(t) = \rho^*\tilde{\theta}^T(t)\zeta(t) + \tilde{\rho}(t)\xi(t), \tag{A.13}$$

which is linear in the parameter errors $\tilde{\theta}(t)$, $\tilde{\rho}(t)$ with a measure characterized by the positive definite function

$$V(\tilde{\theta}, \tilde{\rho}) = \frac{1}{2}(|\rho^*|\tilde{\theta}^T\Gamma^{-1}\tilde{\theta} + \gamma^{-1}\tilde{\rho}^2) \tag{A.14}$$

for $\Gamma = \Gamma^T > 0$ and $\gamma > 0$.

Finally, we introduce the normalization signal

$$m(t) = \sqrt{1 + \zeta^T(t)\zeta(t) + \xi^2(t)}$$

and formulate the adaptive laws to update the controller parameter $\theta(t)$ for continuous-time and discrete-time designs.

Continuous-time design:

$$\dot{\theta}(t) = \frac{-sign(k_p)\Gamma\epsilon(t)\zeta(t)}{m^2(t)} \tag{A.15}$$

$$\dot{\rho}(t) = \frac{-\gamma\epsilon(t)\xi(t)}{m^2(t)}, \tag{A.16}$$

where $\Gamma = \Gamma^T > 0$ and $\gamma > 0$ are constant adaptation gains.

Discrete-time design:

$$\theta(t+1) = \theta(t) - \frac{sign(k_p)\Gamma\epsilon(t)\zeta(t)}{m^2(t)} \tag{A.17}$$

$$\rho(t+1) = \rho(t) - \frac{\gamma\epsilon(t)\xi(t)}{m^2(t)}, \tag{A.18}$$

where

$$0 < \Gamma = \Gamma^T < \frac{2}{k_p^0} I, \; 0 < \gamma < 2 \tag{A.19}$$

are constant adaptation step sizes, for $k_p^0 \geq |k_p|$; that is, the knowledge of an upper bound k_p^0 on $|k_p|$ is needed for the discrete-time design.

The continuous-time adaptive update law (A.15) - (A.16) has the following desired stability properties.

Lemma A.1 *The adaptive update law (A.15) - (A.16) ensures that* $\theta(t) \in L^\infty$, $\rho(t) \in L^\infty$, $\frac{\epsilon(t)}{m(t)} \in L^2 \cap L^\infty$, *and* $\dot{\theta}(t) \in L^2 \cap L^\infty$.

Proof: In view of (A.13), the time derivative of $V(\tilde{\theta}, \tilde{\rho})$ in (A.14), along the trajectories of (A.15) - (A.16), satisfies

$$\dot{V} = \frac{-\epsilon^2(t)}{m^2(t)}.$$

This, together with (A.13), (A.15), proves the lemma. ∇

The analogous properties are achieved in discrete time.

Lemma A.2 *The adaptive update law (A.17) - (A.18) ensures that* $\theta(t) \in l^\infty$, $\rho(t) \in l^\infty$, $\frac{\epsilon(t)}{m(t)} \in l^2 \cap l^\infty$, *and* $\theta(t+1) - \theta(t) \in l^2$.

Proof: In view of (A.13), the time increment of $V(\tilde{\theta}, \tilde{\rho})$ in (A.14) with Γ and γ now satisfying (A.19), along the trajectories of (A.17) - (A.18), satisfies

$$\begin{aligned} & V(\tilde{\theta}(t+1), \tilde{\rho}(t+1)) - V(\tilde{\theta}(t), \tilde{\rho}(t)) \\ & \quad = -(2 - \frac{|k_p|\zeta^T(t)\Gamma\zeta(t) + \gamma\xi^2(t)}{m^2(t)})\frac{\epsilon^2(t)}{m^2(t)}. \end{aligned}$$

This, together with (A.13), (A.17), proves the lemma. ∇

For both the continuous-time and discrete-time MRAC designs, the stability and tracking performance are stated as follows.

Theorem A.2 *All signals in the closed-loop system with the plant (A.1), reference model (A.2), and controller (A.3) - (A.4) updated by the adaptive law (A.15) - (A.16) or (A.17) - (A.18) are bounded, and*

$$\lim_{t \to \infty} (y(t) - y_m(t)) = 0.$$

Proof: Continuous-time case. Starting with a single-input, single-output linear operator $T(s,t)$ with input $x(t)$, we define two classes of operators:

- $T(s,t)$ is stable and proper if

$$|T(s,\cdot)[x](t)| \leq \beta_1 \int_0^t e^{-\alpha(t-\tau)}|x(\tau)|d\tau + \beta_2|x(t)|$$

  for some constants $\beta_1 \geq 0$, $\alpha > 0$, $\beta_2 > 0$, and any $t \geq 0$.

- $T(s,t)$ is stable and strictly proper if it is stable and proper with $\beta_2 = 0$.

Next, we introduce two fictitious signals $z_0(t)$, $z_1(t)$ as

$$z_0(t) = \frac{1}{s+a_0}[u](t),\ z_1(t) = \frac{1}{s+a_0}[y](t) \tag{A.20}$$

for any constant $a_0 > 0$, and two fictitious filters $K(s)$, $K_1(s)$ such that

$$sK_1(s) = 1 - K(s),\ K(s) = \frac{a^{n^*}}{(s+a)^{n^*}} \tag{A.21}$$

for a constant $a > 0$ to be specified. We use these filters to obtain

$$z_0(t) + a_0K_1(s)[z_0](t) - K_1(s)[u](t) = K(s)G^{-1}(s)[z_1](t). \tag{A.22}$$

Now we return to the adaptive version of the controller (A.3) and operate its both sides by $K_1(s)$. Upon the substitution of (A.20), we obtain

$$\begin{aligned} K_1(s)[u](t) &= K_1(s)[\theta_1^T(\cdot)\frac{a(s)}{\Lambda(s)}(s+a_0)[z_0]](t) \\ &\quad + K_1(s)[\theta_2^T(\cdot)\frac{a(s)}{\Lambda(s)}(s+a_0)][z_1](t) \\ &\quad + K_1(s)[\theta_{20}(\cdot)(s+a_0)[z_1]](t) + K_1(s)[\theta_3 r](t). \end{aligned} \tag{A.23}$$

Substituting (A.23) in (A.22), we get

$$\begin{aligned} &(1 + K_1(s)(a_0 - \theta_1^T(\cdot)\frac{a(s)}{\Lambda(s)}(s+a_0)))[z_0](t) \\ &\quad = (K(s)G^{-1}(s) + K_1(s)\theta_2^T(\cdot)\frac{a(s)}{\Lambda(s)}(s+a_0))[z_1](t) \\ &\quad\quad + K_1(s)[s[\theta_{20}z] - zs[\theta_{20}] + a_0\theta_{20}z](t) + K_1(s)[\theta_3 r](t). \end{aligned} \tag{A.24}$$

Since the impulse response $k_1(t)$ of $K_1(s)$ satisfies

$$\int_0^\infty |k_1(t)|dt = \frac{n^*}{a},$$

there exists a constant $a^0 > 0$ such that for any constant $a > a^0$ in (A.21), the operator

$$T_0(s,t) \triangleq (1 + K_1(s)(a_0 - \theta_1^T(t)\frac{a(s)}{\Lambda(s)}(s + a_0)))^{-1} \tag{A.25}$$

is stable and proper. Then we let $a > a^0$ in both $K(s)$ and $K_1(s)$ so that (A.24) implies that

$$z_0(t) = T_1(s,\cdot)[z_1](t) + b_0(t) \tag{A.26}$$

for some stable and proper operator $T_1(s,t)$ and a bounded signal $b_0(t)$: $b_0(t) = T_0(s,\cdot)K_1(s)[\theta_3 r](t)$.

Denoting the reference model denominator polynomial by

$$P_m(s) = s^{n^*} + d_{n^*-1}s^{n^*-1} + \cdots + d_2 s^2 + d_1 s + d_0$$

we express $\xi(t)$ in (A.11) as

$$\begin{aligned}\xi(t) = & \frac{s^{n^*-1} + d_{n^*-1}s^{n^*-2} + \cdots + d_2 s + d_1}{P_m(s)}[\dot{\theta}^T W_m(s)[\omega]](t) \\ & + \frac{s^{n^*-2} + d_{n^*-1}s^{n^*-3} + \cdots + d_2}{P_m(s)}[\dot{\theta}^T s W_m(s)[\omega]](t) + \cdots \\ & + \frac{s + d_{n^*-1}}{P_m(s)}[\dot{\theta}^T s^{n^*-2} W_m(s)[\omega]](t) \\ & + \frac{1}{P_m(s)}[\dot{\theta}^T s^{n^*-1} W_m(s)[\omega]](t).\end{aligned} \tag{A.27}$$

Filtering both sides of the first equality of (A.10) through $\frac{1}{s+a_0}$ and using (A.20) again, we obtain

$$z_1(t) = \frac{1}{s + a_0}[y_m](t) + \frac{1}{s + a_0}[\epsilon - \rho\xi](t). \tag{A.28}$$

With the l^1 vector norm $\|\cdot\|_1$, the inequality

$$|\epsilon(t)| \leq \frac{|\epsilon(t)|}{m(t)}(1 + \|\zeta(t)\|_1 + |\xi(t)|) \tag{A.29}$$

and (A.26) - (A.28) imply that

$$|z_1(t)| \leq x_0(t) + T_2(s,\cdot)[x_1 T_3(s,\cdot)[|z_1|]](t) \tag{A.30}$$

for some $x_0(t) \in L^\infty$, $x_1(t) \in L^\infty \cap L^2$ and $x_1(t) \geq 0$, some operators $T_2(s,t)$ stable and strictly proper, and $T_3(s,t)$ stable and proper with a non-negative impulse response.

Let $z_2(t) = T_3(s,\cdot)[|z_1|](t)$. Operating $T_3(s,t)$ on both sides of (A.30) and observing the non-negativeness of the impulse response of $T_3(s,t)$, we see that

$$z_2(t) \leq b_1 + b_2 \int_0^t e^{-\alpha(t-\tau)} x_1(\tau) z_2(\tau) d\tau \tag{A.31}$$

for some positive constants α, b_1, b_2. If $\bar{z}_2(t)$ is defined to satisfy

$$\dot{\bar{z}}_2(t) = -\alpha \bar{z}_2(t) + b_2 x_1(t) \bar{z}_2(t) + \alpha b_1, \ \ \bar{z}_2(0) = b_1, \tag{A.32}$$

then it follows that

$$\bar{z}_2(t) = b_1 + b_1 b_2 \int_0^t e^{-\alpha(t-\tau)+b_2 \int_\tau^t x_1(\sigma) d\sigma} x_1(\tau) d\tau \tag{A.33}$$

and $\bar{z}_2(t) \geq z_2(t)$. Since $x_1(t) \in L^2$, using Schwarz inequality, we get

$$\int_\tau^t x_1(\sigma) d\sigma \leq \sqrt{\int_\tau^t d\sigma} \sqrt{\int_\tau^t x_1^2(\sigma) d\sigma} \leq b_3 \sqrt{t-\tau} \tag{A.34}$$

for some constant $b_3 > 0$. Hence (A.31) - (A.34) yield

$$z_2(t) \leq b_1 + b_1 b_2 \int_0^t e^{-\alpha(t-\tau)+b_2 b_3 \sqrt{t-\tau}} x_1(\tau) d\tau. \tag{A.35}$$

Because $x_1(t) \in L^\infty$ implies that $\int_0^t e^{-\alpha(t-\tau)+b_3\sqrt{t-\tau}} x_1(\tau) d\tau \in L^\infty$, it follows from (A.35) that $z_2(t) \in L^\infty$. So $z_1(t) \in L^\infty$ and $z_0(t)$ in (A.26) is bounded. This, in turn, implies that $\xi(t)$ in (A.27), $\zeta(t)$ in (A.11), $\epsilon(t)$ in (A.13), $y(t)$ in (A.10), and $u(t)$ in (A.23) are all bounded.

Then, from Lemma A.1, we see that $\epsilon(t) \in L^2 \cap L^\infty$, $\dot{\epsilon}(t) \in L^\infty$, so that

$$\int_0^t \epsilon^2(\tau) |\dot{\epsilon}(\tau)| d\tau \leq \sup_{t \geq 0} |\dot{\epsilon}(t)| \int_0^\infty \epsilon^2(\tau) d\tau < \infty$$

for any $t \geq 0$. This implies that $\lim_{t\to\infty} \int_0^t \epsilon^2(\tau)|\dot{\epsilon}(\tau)| d\tau$ exists and is finite, and, therefore, $\lim_{t\to\infty} \int_0^t \epsilon^2(\tau)\dot{\epsilon}(\tau) d\tau$ exists and is finite. To exploit this fact, we construct the identity

$$\epsilon^2(t) = |\epsilon^3(t)|^{\frac{2}{3}} = |3 \int_0^t \epsilon^2(\tau)\dot{\epsilon}(\tau) d\tau + \epsilon^3(0)|^{\frac{2}{3}},$$

which then implies that $\lim_{t\to\infty} \epsilon^2(t)$ exists and is finite. As we already know that $\epsilon(t) \in L^2$, this proves that $\lim_{t\to\infty} \epsilon(t) = 0$. Therefore, from (A.15) and (A.27), we have $\lim_{t\to\infty} \dot{\theta}(t) = 0$ and $\lim_{t\to\infty} \xi(t) = 0$. Finally, from (A.10), we have $\lim_{t\to\infty} e(t) = 0$.

Discrete-time case. For $\Lambda(D) = D^{n-1}$, $P_m(D) = D^{n^*}$, we have

$$\zeta(t) = \omega(t - n^*) \tag{A.36}$$

$$\omega_1(t) = (D^{-n+1}, \ldots, D^{-1})^T[u](t) \tag{A.37}$$

$$\omega_2(t) = (D^{-n+1}, \ldots, D^{-1})^T[y](t). \tag{A.38}$$

We then introduce a new regressor

$$\bar{\omega}(t) = (\omega_1^T(t), \omega_1^T(t), y(t))^T \tag{A.39}$$

and use (A.5) and (A.1) to obtain

$$\theta_1^{*T}\omega_1(t) + \theta_2^{*T}\omega_2(t) + \theta_{20}^* y(t) = u(t) - \theta_3^* y(t + n^*).$$

Combining (A.37) - (A.39), we see that $\bar{\omega}(t)$ satisfies

$$\begin{aligned} \bar{\omega}(t+1) &= A^*\bar{\omega}(t) + b^* y(t + n^*) \\ &= A^*\bar{\omega}(t) + b^* k_p \frac{Z(D)}{P(D)}[u](t + n^*) \end{aligned} \tag{A.40}$$

for some constant $A^* \in R^{(2n-1)\times(2n-1)}$ and $b^* \in R^{2n-1}$, such that

$$det(DI - A^*) = D^{n+n^*-1} Z(D). \tag{A.41}$$

Using (A.10), (A.36) and the first equality of (A.40), we have

$$\begin{aligned} \bar{\omega}(t+1) = {} & A^*\bar{\omega}(t) + b^*(y_m(t + n^*) \\ & + \epsilon(t + n^*) - \rho(t + n^*)(\theta(t + n^*) - \theta(t))^T \omega(t)). \end{aligned} \tag{A.42}$$

By (A.41) and assumption (M1), the eigenvalues of A^* are inside the unit circle, so that there exists a nonsingular $Q^* \in R^{(2n-1)\times(2n-1)}$ such that $\|Q^* A^* Q^{*-1}\|_2 < 1$, where $\|\cdot\|_2$ is the Euclidean (l^2) vector or induced matrix norm. We let the vector norm $\|\cdot\|$ in R^{2n-1} be $\|x\| = \|Q^* x\|_2$.

Introducing the auxiliary scalar signal

$$x(t) = \left|\frac{\epsilon(t + n^*)}{m(t + n^*)}\right| + \|\theta(t + n^*) - \theta(t)\|_2$$

we see from (A.42) and Lemma A.2 that $x(t) \in l^2$ and that for some constants $a_0 \in (0, 1)$, $c_1 > 0$, $c_2 > 0$, we have

$$\|\bar{\omega}(t+1)\| \le (a_0 + c_1 x(t))\|\bar{\omega}(t)\| + c_2. \tag{A.43}$$

Since $x(t) \in l^2$, there is a constant $c_3 > 0$ such that

$$\sum_{t=t_0}^{t_0+j} x(t) \le \sqrt{j+1}\sqrt{\sum_{t=t_0}^{t_0+j} x^2(t)} \le c_3\sqrt{j+1}$$

for any $j \ge 1$. Using this property, we express

$$\prod_{t=t_0}^{t_0+j}(a_0 + c_1 x(t)) \le (a_0 + \frac{c_1}{j+1}\sum_{t=t_0}^{t_0+j} x(t))^{j+1} \le (a_0 + \frac{c_1 c_3}{\sqrt{j+1}})^{j+1}$$
$$\le a_0^{j+1}(1 + \frac{c_1 c_3}{a_0\sqrt{j+1}})^{j+1} \le a_0^{j+1} e^{\frac{c_1 c_3}{a_0}\sqrt{j+1}},$$

from which it follows that

$$\lim_{k\to\infty}\sum_{j=1}^{k}\prod_{t=t_0}^{t_0+j}(a_0 + c_5 x(t)) < \infty. \tag{A.44}$$

Using (A.43) and (A.44), we conclude that $\|\bar{\omega}(t)\|$ is bounded; that is, all closed-loop signals are bounded. Then we have $\epsilon(t) \in l^2$, $\theta(t+1) - \theta(t) \in l^2$ so that $\lim_{t\to\infty}\epsilon(t) = 0$, $\lim_{t\to\infty}(\theta(t+1) - \theta(t)) = 0$, and $\lim_{t\to\infty}\xi(t) = 0$. Finally, from (A.10), we have $\lim_{t\to\infty} e(t) = 0$. ∇

Appendix B

Signal Boundedness in Continuous Time

This appendix contains the proofs of the closed-loop signal boundedness with continuous-time adaptive inverse controllers in Chapter 6. The proofs are given for the case when the dead-zone or hysteresis slopes are equal: $m = m_r = m_l$, $m = m_t = m_b$, so that adaptive inverse control schemes are designed with the parametrizations in Section 6.4.1 with equal slope estimates $\widehat{m} = \widehat{m_l} = \widehat{m_l}$, $\widehat{m} = \widehat{m_t} = \widehat{m_b}$. It also outlines of the proofs of Theorem 4.2 and Theorem 4.5 for continuous-time fixed inverse controllers in Sections 4.2.2 and 4.3.3.

Proof of Theorem 6.1 for $G(s)$ Known

As proved in Lemma 6.1, the adaptive law (6.21), with parameter projection modification or with switching σ-modification with parameter projection, ensures that $\theta_N(t)$, $\frac{\epsilon_N(t)}{m_N(t)}$, $\dot{\theta}_N(t)$ are bounded and that $\frac{\epsilon_N^2(t)}{m_N^2(t)}$, $||\dot{\theta}_N(t)||_2^2$ satisfy the smallness in the mean conditions (6.43), (6.45):

$$\int_{t_1}^{t_2} \frac{\epsilon_N^2(t)}{m_N^2(t)} dt \leq a_1 + b_1 \int_{t_1}^{t_2} \frac{d^2(t)}{m_N^2(t)} dt \tag{B.1}$$

$$\int_{t_1}^{t_2} ||\dot{\theta}_N(t)||_2^2 dt \leq a_2 + b_2 \int_{t_1}^{t_2} \frac{d^2(t)}{m_N^2(t)} dt \tag{B.2}$$

for some constants $a_i, b_i > 0$, $i = 1, 2$, and all $t_2 > t_1 \geq 0$. The parameter projection ensures that $\widehat{m}(t) \geq m_0 > 0$ for some $m_0 > 0$.

To prove the closed-loop signal boundedness, we use the dead-zone, backlash, or hysteresis characteristic $u(t) = N(v(t))$ in Chapter 2, and their inverses $v(t) = \widehat{NI}(u_d(t)))$ in Chapter 3, to express

$$u(t) = \frac{m}{\widehat{m}(t)} u_d(t) + d_1(t) \tag{B.3}$$

$$u(t) = mv(t) + d_2(t) \tag{B.4}$$

for some bounded $d_1(t)$, $d_2(t)$. As in Appendix A, we define the fictitious signals $z_0(t)$, $z_1(t)$ and the fictitious filters $K_1(s)$, $K(s)$ as

$$z_0(t) = \frac{1}{s+a_0}[u](t),\ z_1(t) = \frac{1}{s+a_0}[y](t) \tag{B.5}$$

$$sK_1(s) = 1 - K(s),\ K(s) = \frac{a^{n^*}}{(s+a)^{n^*}}, \tag{B.6}$$

where $a_0 > 0$ is arbitrary and $a > 0$ is to be chosen. Then we use (B.5), (B.6), and $u(t) = G(s)^{-1}[y](t)$ to obtain

$$z_0(t) + a_0 K_1(s)[z_0](t) - K_1(s)[u](t) = K(s)G^{-1}(s)[z_1](t). \tag{B.7}$$

Using (B.3) and (B.6) in (6.6) (or (6.15) for $\theta_{20}(t) = 0$), we have

$$\begin{aligned} u(t) = {} & \frac{m}{\widehat{m}(t)}\theta_1^{*T}\frac{a(s)}{\Lambda(s)}\frac{\widehat{m}(\cdot)}{m}(s+a_0)[z_0](t) \\ & + \frac{m}{\widehat{m}(t)}\theta_2^{*T}\frac{a(s)}{\Lambda(s)}(s+a_0)[z_1](t) + \frac{m}{\widehat{m}(t)}\theta_{20}^*(s+a_0)[z_1](t) \\ & + \frac{m}{\widehat{m}(t)}\theta_3^* r(t) + (1 - \frac{m}{\widehat{m}(t)}\theta_1^{*T}\frac{a(s)}{\Lambda(s)}\frac{\widehat{m}(\cdot)}{m})[d_1](t). \end{aligned} \tag{B.8}$$

Substituting (B.8) in (B.6), we see that $z_1(t)$ and $z_0(t)$ are related by

$$\begin{aligned} & (1 + K_1(s)(a_0 - \frac{m}{\widehat{m}(\cdot)}\theta_1^{*T}\frac{a(s)}{\Lambda(s)}\frac{\widehat{m}(\cdot)}{m}(s+a_0)))[z_0](t) \\ & = (K(s)G^{-1}(s) + K_1(s)\frac{m}{\widehat{m}(\cdot)}\theta_2^{*T}\frac{a(s)}{\Lambda(s)}(s+a_0) \\ & \quad + K_1(s)\frac{m}{\widehat{m}(\cdot)}\theta_{20}^*(s+a_0))[z_1](t) + K_1(s)[\frac{m}{\widehat{m}}\theta_3^* r](t) \\ & \quad + K_1(s)(1 - \frac{m}{\widehat{m}(\cdot)}\theta_1^{*T}\frac{a(s)}{\Lambda(s)}\frac{\widehat{m}(\cdot)}{m})[d_1](t). \end{aligned} \tag{B.9}$$

As defined in Appendix A, a single-input, single-output linear operator $T(s,t)$ with input $x(t)$ is stable and proper if

$$|T(s,\cdot)[x](t)| \leq \beta_1 \int_0^t e^{-\alpha(t-\tau)}|x(\tau)|d\tau + \beta_2|x(t)|$$

for some constants $\beta_1 \geq 0$, $\beta_2 > 0$, and $\alpha > 0$, and any $t \geq 0$, and $T(s,t)$ is stable and strictly proper if it is stable and proper with $\beta_2 = 0$. With these definitions, the facts that $\widehat{m}(t), \dot{\widehat{m}}(t) \in L^\infty$, and that $K_1(s)$ is stable and strictly proper, and the expressions

$$\frac{m}{\widehat{m}(t)}\theta_1^{*T}\frac{a(s)}{\Lambda(s)}\frac{\widehat{m}(\cdot)}{m}(s+a_0)[z_0](t)$$
$$= \frac{m}{\widehat{m}(t)}\theta_1^{*T}\frac{a(s)}{\Lambda(s)}[s[\frac{\widehat{m}}{m}z_0] - s[\frac{\widehat{m}}{m}]z_0 + a_0\frac{\widehat{m}}{m}z_0](t) \tag{B.10}$$

$$K_1(s)\frac{m}{\widehat{m}(\cdot)}\theta_{20}^*(s+a_0)[z_1](t)$$
$$= K_1(s)[s[\frac{m}{\widehat{m}}\theta_{20}^*z] - s[\frac{m}{\widehat{m}}\theta_{20}^*]z + a_0\frac{m}{\widehat{m}}\theta_{20}^*z](t) \tag{B.11}$$

imply that both $\frac{m}{\widehat{m}(t)}\theta_1^{*T}\frac{a(s)}{\Lambda(s)}\frac{\widehat{m}(t)}{m}(s+a_0)$ and $K_1(s)\frac{m}{\widehat{m}(\cdot)}\theta_{20}^*(s+a_0)$ are stable and proper operators.

From the definition of $K_1(s)$ in (B.6), the impulse response function $k_1(t)$ of $K_1(s)$ satisfies

$$\int_0^\infty |k_1(t)|dt = \frac{n^*}{a}.$$

Hence there exists $a^0 > 0$ such that for any finite $a > a^0$, the operator

$$T_0(s,t) = (1 + K_1(s)(a_0 - \frac{m}{\widehat{m}(t)}\theta_1^{*T}\frac{a(s)}{\Lambda(s)}\frac{\widehat{m}(t)}{m}(s+a_0)))^{-1}$$

is stable and proper. For a fixed $a > a^0$, (B.9) - (B.11) imply that

$$z_0(t) = T_1(s,\cdot)[z_1](t) + d_3(t), \tag{B.12}$$

where $T_1(s,t)$ is a stable and proper operator, and $d_3(t)$ is a bounded signal dependent on $r(t)$ and $d_1(t)$, which are both bounded.

Letting (A_f, B_f, C_f) be a minimal realization of $W(s) = \frac{k_p}{P_m(s)}(1 - \theta_1^{*T}\frac{a(s)}{\Lambda(s)})$ in (6.20), define $W_c(s) = C_w(sI - A_w)^{-1}$, $W_b(s) = (sI - A_w)^{-1}B_w$, and using the swapping lemma [71], [96], we express $\xi_N(t)$ as

$$\xi_N(t) = W_c(s)[W_b(s)(s+a_0)\frac{1}{s+a_0}[\omega_N^T]\dot{\theta}_N](t), \tag{B.13}$$

where $\omega_N(t)$ stands for $\omega_d(t)$, $\omega_b(t)$, and $\omega_h(t)$ for dead-zone, backlash, and hysteresis, as in (6.109), (3.36), and (6.110). In particular, for the hysteresis case, we use (3.58) and rewrite $\omega_h(t)$ in Section 6.4.1 as

$$\omega_h(t) = (-v(t) + (\widehat{\chi_r}(t) + \widehat{\chi_l}(t))v(t), -\widehat{\chi_t}(t), -\widehat{\chi_b}(t),$$
$$-\widehat{\chi_r}(t)v(t), \widehat{\chi_r}(t), -\widehat{\chi_l}(t)v(t), \widehat{\chi_l}(t))^T, \tag{B.14}$$

whose components, except for the term $-v(t)$, are all bounded. It follows from (B.4) and (B.5) that

$$\frac{1}{s+a_0}[\omega_N](t) = \frac{1}{m}(z_0(t), 0, 0)^T + d_4(t) \tag{B.15}$$

where $d_4(t)$ is a bounded signal.

Filtering both sides of (6.23): $\epsilon_N(t) = e(t) + \xi_N(t)$, through $\frac{1}{s+a_0}$ and using (B.5), we obtain

$$z_1(t) = \frac{1}{s+a_0}[y_m](t) + \frac{1}{s+a_0}[\epsilon_N - \xi_N](t). \tag{B.16}$$

Finally, using $m_N(t)$ in (6.25) and the inequality

$$|\epsilon_N(t)| \le \frac{|\epsilon_N(t)|}{m_N(t)}\,(1 + \|\zeta_N(t)\|_1 + |\xi_N(t)|)$$

and (B.12), (B.14) - (B.16), we get

$$|z_1(t)| \le |x_0(t)| + T_2(s,\cdot)[x_1 T_3(s,\cdot)[|z_1|]](t) \tag{B.17}$$

for some $x_0(t) \in L^\infty$ and some $x_1(t)$ such that

$$0 \le x_1(t) \le \frac{|\epsilon_N(t)|}{m_N(t)} + \|\dot{\theta}_N(t)\|_2. \tag{B.18}$$

From (B.1), (B.2), and (B.18), it follows that

$$\int_{t_1}^{t_2} x_1^2(t)dt \le \bar{k}_1 + \int_{t_1}^{t_2} \frac{\bar{k}_2}{1 + \zeta_N^T(t)\zeta_N(t) + \xi_N^2(t)}dt \tag{B.19}$$

with some constants $\bar{k}_1 > 0$, $\bar{k}_2 > 0$, and any $t_2 \ge t_1 \ge 0$, and $T_2(s,t)$ is a stable and strictly proper operator, while $T_3(s,t)$ is a stable and proper operator with a non-negative impulse response.

If $\zeta_N(t)$, $\xi_N(t)$ are bounded, then from Lemma 6.1 and (6.23) we see that $\epsilon_N(t)$, $y(t)$ are bounded, from (B.5), (B.12) that $z_1(t)$, $z_0(t)$ are bounded, and from (B.8), and (B.3) that $u(t)$ and $u_d(t)$ are bounded. Thus all the closed-loop signals are bounded. On the other hand, if $\zeta_N^T(t)\zeta_N(t) + \xi_N^2(t)$ grows unbounded, then the smallness of $x_1(t)$ in the sense of (B.19) results in the boundedness of $z_1(t)$ given by (B.17). This, in turn, implies that $z_0(t)$ in (B.12) is bounded, so that $u(t)$ in (B.8), $u_d(t)$ in (B.3), $\omega_N(t) = \omega_b(t)$ in (3.36) ($\omega_N(t) = \omega_d(t)$ or $\omega_h(t)$ in Section 6.4.1), $\xi_N(t)$ in (6.24), and $y(t)$ in (6.23) are bounded. Again, all the closed-loop signals are bounded.

Proof of Theorem 6.1 for $G(s)$ Unknown

In Section 6.3, we have presented two adaptive schemes: one with the gradient adaptive law (6.78) - (6.79) and the other with the Lyapunov-type adaptive law (6.93) - (6.94).

The gradient adaptive scheme. We first consider the adaptive inverse control scheme with the adaptive law (6.78) - (6.79). For this scheme, Lemma 6.2 states that $\theta(t)$, $\rho(t)$, $\dot{\theta}(t)$, $\dot{\rho}(t)$, $\frac{\epsilon^2(t)}{\sqrt{1+\varsigma^T(t)\varsigma(t)+\xi^2(t)}}$ are bounded and that (6.84) - (6.85) hold.

Using (B.3), (B.5), (B.6), (B.14) for $\omega_h(t)$ or (3.35) for $\omega_b(t)$ or the expression in Section 6.4.1 for $\omega_d(t)$, along with the definition of $\omega_4(t)$ in Section 6.3, we express

$$\theta_4^T(t)\omega_4(t) = \theta_z^T(t)\frac{a(s)}{\Lambda(s)}(s+a_0)[z_0](t) + d_5(t) \tag{B.20}$$

for some bounded $\theta_z(t) \in R^{n-1}$ for the controller (6.73) and $\theta_z(t) \in R^n$ for the controller (6.74) and some bounded $d_5(t)$.

Using (B.3), (6.73) or (6.74), (B.20), we obtain

$$\begin{aligned} u(t) = & \frac{m}{\widehat{m}(t)}\theta_z^T(t)\frac{a(s)}{\Lambda(s)}(s+a_0)[z_0](t) \\ & + \frac{m}{\widehat{m}(t)}\theta_2^T(t)\frac{a(s)}{\Lambda(s)}(s+a_0)[z_1](t) + \frac{m}{\widehat{m}(t)}\theta_{20}(t)(s+a_0)[z_1](t) \\ & + \frac{m}{\widehat{m}(t)}\theta_3(t)r(t) + \frac{m}{\widehat{m}(t)}d_5(t) + d_1(t). \end{aligned} \tag{B.21}$$

From (B.7) and (B.21) it follows that

$$\begin{aligned} & (1 + K_1(s)(a_0 - \frac{m}{\widehat{m}(\cdot)}\theta_z^T(\cdot)\frac{a(s)}{\Lambda(s)}(s+a_0)))[z_0](t) \\ & = (K(s)G^{-1}(s) + K_1(s)\frac{m}{\widehat{m}(\cdot)}\theta_2^T(\cdot)\frac{a(s)}{\Lambda(s)}(s+a_0) \\ & \quad + K_1(s)\frac{m}{\widehat{m}(\cdot)}\theta_{20}(\cdot)(s+a_0))[z_1](t) + K_1(s)[\frac{m}{\widehat{m}}\theta_3 r](t) \\ & \quad + K_1(s)[d_2](t) + K_1(s)[\frac{m}{\widehat{m}}d_5](t). \end{aligned} \tag{B.22}$$

In view of the facts that

$$\begin{aligned} & K_1(s)\frac{m}{\widehat{m}(\cdot)}\theta_{20}(\cdot)(s+a_0)[z_1](t) \\ & \quad = K_1(s)[s[\frac{m}{\widehat{m}}\theta_{20}z] - s[\frac{m}{\widehat{m}}\theta_{20}]z + a_0\frac{m}{\widehat{m}}\theta_{20}z](t) \end{aligned}$$

and that $\widehat{m}(t)$, $\dot{\widehat{m}}(t)$, $\theta_{20}(t)$, $\dot{\theta}_{20}(t)$ are all bounded, we see that in (B.22) the operator $K_1(s)\frac{m}{\widehat{m}(\cdot)}\theta_{20}(\cdot)(s+a_0)$ is a stable and proper one.

The remainder of the proof is analogous to (B.12) - (B.19), followed by a contradiction argument similar to that used for the case when $G(s)$ is known.

The Lyapunov-type adaptive scheme. For the adaptive inverse control scheme with the adaptive law (6.93) - (6.94), Lemma 6.3 states that $\theta(t)$, $\rho(t)$, $\epsilon(t)$ are bounded and that (6.95) - (6.97) hold. Using $z_0(t)$, $K_1(s)$, $K(s)$ defined in (B.9), (B.10), we obtain

$$z_0(t) + a_0 K_1(s)[z_0](t) - K_1(s)[u](t) = K(s)G^{-1}(s)\frac{1}{s+a_0}[y](t). \tag{B.23}$$

Using (B.3), (6.73) or (6.74), (B.20), we obtain

$$\begin{aligned} u(t) = & \frac{m}{\widehat{m}(t)}\theta_z^T(t)\frac{a(s)}{\Lambda(s)}(s+a_0)[z_0](t) \\ & + \frac{m}{\widehat{m}(t)}\theta_2^T(t)\frac{a(s)}{\Lambda(s)}[y](t) + \frac{m}{\widehat{m}(t)}\theta_{20}(t)y(t) \\ & + \frac{m}{\widehat{m}(t)}\theta_3(t)r(t) + \frac{m}{\widehat{m}(t)}d_5(t) + d_1(t). \end{aligned} \tag{B.24}$$

From (B.23) and (B.24) it follows that

$$\begin{aligned} & (1 + K_1(s)(a_0 - \frac{m}{\widehat{m}(\cdot)}\theta_z^T(\cdot)\frac{a(s)}{\Lambda(s)}(s+a_0)))[z_0](t) \\ & = (K(s)G^{-1}(s)\frac{1}{s+a_0} + K_1(s)\frac{m}{\widehat{m}(\cdot)}\theta_2^T(\cdot)\frac{a(s)}{\Lambda(s)} \\ & + K_1(s)\frac{m}{\widehat{m}(\cdot)}\theta_{20}(\cdot))[y](t) + K_1(s)[\frac{m}{\widehat{m}}\theta_3 r](t) \\ & + K_1(s)[d_1](t) + K_1(s)[\frac{m}{\widehat{m}}d_5](t). \end{aligned} \tag{B.25}$$

Similar to the derivation of (B.11) - (B.12), we see from (B.25), with a sufficiently large $a > 0$ in $K(s)$, $K_1(s)$, that

$$z_0(t) = \bar{T}_1(s,\cdot)[y](t) + \bar{d}_3(t) \tag{B.26}$$

for some stable and proper operator $\bar{T}_1(s,t)$ and some $\bar{d}_3(t) \in L^\infty$.

Similar to (B.13), letting (A_l, B_l, C_l) be a minimal realization of $\frac{1}{L(s)} = C_l(sI - A_l)^{-1}B_l$ and defining $W_c(s) = C_l(sI - A_l)^{-1}B_l$ and $W_b(s) = (sI - A_l)^{-1}B_l$, we express $\xi(t)$ in (6.92) as

$$\xi(t) = W_c(s)[W_b(s)[\omega^T]\dot{\theta}](t). \tag{B.27}$$

Substituting (B.27), (6.73) or (6.74), (6.7) for $\omega_2(t)$, (B.26) in $\omega(t)$ of (6.77), we obtain

$$\omega(t) = T_\omega(s,\cdot)[y](t) + d_\omega(t) \tag{B.28}$$

for some stable and proper operator $T_\omega(s,t)$ and some $d_\omega(t) \in L^\infty$.

Rewriting (6.92), we get

$$y(t) = y_m(t) + \epsilon(t) - W_m(s)L(s)[\rho\xi - \alpha\,\epsilon(1+\zeta^T\zeta+\xi^2) + \alpha\,\epsilon](t). \tag{B.29}$$

Similar to the derivation of (B.17), the expressions (B.27) - (B.29) and Lemma 6.3 imply that

$$|y(t)| \le |x_0(t)| + \bar{T}_2(s,\cdot)[x_1\bar{T}_3(s,\cdot)[|y|]](t), \tag{B.30}$$

where $x_0(t) \in L^\infty$, and $x_1(t) \ge 0$ satisfies

$$\int_{t_1}^{t_2} x_1^2(t)dt \le \bar{k}_3 + \int_{t_1}^{t_2} \frac{\bar{k}_4}{1+\zeta^T(t)\zeta(t)+\xi^2(t)}dt \tag{B.31}$$

for some constants $\bar{k}_3 > 0$, $\bar{k}_4 > 0$, and any $t_2 \ge t_1 \ge 0$. The operator $\bar{T}_2(s,t)$ is stable and strictly proper, while the operator $\bar{T}_3(s,t)$ is stable and proper with a non-negative impulse response. The closed-loop signal boundedness then follows from (B.30) and (B.31).

Proof of Theorem 4.2

For the continuous-time exact inverse control scheme in Section 4.2.2, the inequality (B.17) still holds with $x_0(t) \in L^\infty$, but in this case $x_1(t) \in L^\infty \cap L^2$ because of the exact nonlinearity cancellation: $u(t) = u_d(t)$, $t \ge 0$.

Letting $z_2(t) = T_3(s,\cdot)[|z_1|](t)$ and operating $T_3(s,t)$, which has a non-negative impulse response, on both sides of (B.17), we obtain

$$z_2(t) \le b_1 + b_2 \int_0^t e^{-\alpha(t-\tau)}|x_1(\tau)|z_2(\tau)d\tau \tag{B.32}$$

for some α, b_1, $b_2 > 0$. As in (A.32) - (A.35), since $x_1(t) \in L^2$: $\int_{t_1}^{t_2} x_1^2(t)dt \le \gamma_0$, for some $\gamma_0 > 0$, and $\forall t_2 > t_1 \ge 0$, we have

$$\begin{aligned} z_2(t) &\le b_1 + b_1 b_2 \int_0^t e^{-\alpha(t-\tau)+b_2\int_\tau^t |x_1(\sigma)|d\sigma}|x_1(\tau)|d\tau \\ &\le b_1 + b_1 b_2 \int_0^t e^{-\alpha(t-\tau)+b_2\sqrt{\gamma_0}\sqrt{t-\tau}}|x_1(\tau)|d\tau. \end{aligned} \tag{B.33}$$

We see from (B.33) that $z_2(t)$ is bounded, and so is $z_1(t)$ in (B.17). From this result, as in Appendix A, it follows that all closed-loop signals are bounded and that $\lim_{t\to\infty} y(t) - y_m(t) = 0$.

Proof of Theorem 4.5

For the continuous-time detuned inverse control scheme in Section 4.3.3, we also have (B.17) with $x_0(t) \in L^\infty$ but with $x_1(t) \in L^\infty$ and

$$\int_{t_1}^{t_2} x_1^2(t)dt \leq \bar{a}_1 + \bar{b}_1 \mu_N^2 (t_2 - t_1) + \int_{t_1}^{t_2} \frac{\bar{c}_1}{m^2(t)} \tag{B.34}$$

for some constants $\bar{a}_1 > 0$, $\bar{b}_1 > 0$, $\bar{c}_1 > 0$, and any $t_2 \geq t_1 \geq 0$, where $m(t)$ is defined in (4.78). For $\mu_N > 0$ small, (B.17) and (B.34) also lead to a bounded $z_1(t)$ and in turn to the closed-loop signal boundedness. With a detuned inverse, the tracking error $y(t) - y_m(t)$ may remain large.

Appendix C

Signal Boundedness in Discrete Time

Here we prove the closed-loop signal boundedness with discrete-time adaptive inverse controllers of Chapter 7. Unlike the continuous-time proofs, our proofs in discrete time are valid for the general case when the dead-zone or hysteresis slopes are unequal: $m_r \neq m_l$ or $m_t \neq m_b$. We also give outlines of the proofs of Theorem 4.3 and Theorem 4.6 for discrete-time fixed inverse controllers in Sections 4.2.2 and 4.3.3.

Proof of Theorem 7.2

Let us first consider the more general scheme in Section 7.3 where both the input nonlinearity $N(\cdot)$ and the linear part $G(D)$ are unknown. Recall from Lemma 7.2 that the adaptive law (7.40) - (7.41) guarantees that $\theta(t)$, $\rho(t)$, $\frac{\epsilon(t)}{m(t)} \in l^\infty$ and that

$$\sum_{t=t_1}^{t_2} \frac{\epsilon^2(t)}{m^2(t)} \leq a_1 + b_1 \sum_{t=t_1}^{t_2} \frac{d_0^2}{m^2(t)} \tag{C.1}$$

$$\sum_{t=t_1}^{t_2} \|\theta(t+1) - \theta(t)\|_2^2 \leq a_2 + b_2 \sum_{t=t_1}^{t_2} \frac{d_0^2}{m^2(t)} \tag{C.2}$$

for some constants $a_i > 0, b_i > 0$, $i = 1, 2$, and all $t_2 > t_1 \geq 0$, where

$$m(t) = \sqrt{1 + \omega^T(t-n^*)\omega(t-n^*) + \xi^2(t)}$$

and d_0 is the upper bound of $d(t)$: $|d(t)| \leq d_0$, $t \geq 0$.

Furthermore, let us define

$$\bar{\omega}(t) = (\omega_u^T(t), \omega_y^T(t))^T, \tag{C.3}$$

where

$$\omega_u(t) = a_\lambda(D)[u](t) \tag{C.4}$$
$$a_\lambda(D) = (D^{-n+1}, \ldots, D^{-1})^T[u](t)$$
$$\omega_y(t) = \omega_2(t) = b_\lambda(D)[y](t) \tag{C.5}$$
$$b_\lambda(D) = (D^{-n+1}, \ldots, D^{-1}, 1)^T[y](t)$$

and express the relationship between $\omega(t)$ and $\bar{\omega}(t)$ as

$$\omega(t) = F_1(t)\bar{\omega}(t) + g_1(t) \tag{C.6}$$

$$\bar{\omega}(t) = F_2\omega(t) + g_2(t) \tag{C.7}$$

for some bounded sequences $F_1(t)$, $g_1(t)$, $g_2(t)$ and some constant matrix F_2, of appropriate dimensions. (The expressions for $F_1(t)$, F_2, $g_1(t)$ and $g_2(t)$ can be found in [119] for the dead-zone case and can be similarly obtained for the backlash and hysteresis cases.)

Using (7.5) and the relationship $y(t) = k_p \frac{Z(D)}{P(D)}[u](t)$, we obtain

$$\theta_1^{*T}\omega_u(t) + \theta_2^{*T}\omega_y(t) = u(t) - \theta_3^* y(t+n^*). \tag{C.8}$$

With the help of (C.3) and (C.8), we express

$$\begin{aligned}\bar{\omega}(t+1) &= A^*\bar{\omega}(t) + b^* y(t+n^*) \\ &= A^*\bar{\omega}(t) + b^* k_p \frac{Z(D)}{P(D)}[u](t+n^*)\end{aligned} \tag{C.9}$$

for some constant matrix $A^* \in R^{(2n-1)\times(2n-1)}$ and some constant vector $b^* \in R^{2n-1}$. Since the first component of $\bar{\omega}(t)$ is $u(t-n+1)$, it follows from the second equality of (C.9) that, for $c^* = (1, 0, \ldots, 0)^T \in R^{2n-1}$,

$$D^{-n+1}[u](t) = c^{*T}(DI - A^*)^{-1} b^* k_p \frac{Z(D)}{P(D)} D^{n^*}[u](t). \tag{C.10}$$

Since (C.10) holds for any $u(t)$, we conclude that

$$det(DI - A^*) = D^{n+n^*-1}Z(D). \tag{C.11}$$

Recalling the definition of the estimation error $\epsilon(t)$:

$$\epsilon(t) = e(t) + \rho(t)(\theta(t) - \theta(t-n^*))^T\omega(t-n^*) \tag{C.12}$$

and using the first equality of (C.9), we obtain

$$\begin{aligned}\bar{\omega}(t+1) = A^*\bar{\omega}(t) + b^*(&y_m(t+n^*) \\ &+\epsilon(t+n^*) - \rho(t+n^*)(\theta(t+n^*) - \theta(t))^T\omega(t)).\end{aligned} \tag{C.13}$$

From (C.11), the eigenvalues of the closed-loop system matrix A^* are all inside the unit circle of the complex plane. Therefore there exists a nonsingular constant matrix $Q^* \in R^{(2n-1)\times(2n-1)}$ such that $\|Q^* A^* Q^{*-1}\|_2 < 1$. Define the vector norm $\|\cdot\|$ in R^{2n-1} by $\|x\| = \|Q^* x\|_2$. From (C.6) and (C.7), there exist constants $c_i > 0$, $\bar{c}_i > 0$, $i = 1, 2$, such that for all $t \geq 0$,

$$\|\bar{\omega}(t)\| \leq c_1 \|\omega(t)\|_2 + c_2 \tag{C.14}$$

$$\|\omega(t)\|_2 \leq \bar{c}_1 \|\bar{\omega}(t)\| + \bar{c}_2. \tag{C.15}$$

Introducing the auxiliary signals

$$\bar{\epsilon}(t+n^*) = \frac{\epsilon(t+n^*)}{m(t+n^*)}$$

$$x(t) = |\bar{\epsilon}(t+n^*)| + \|\theta(t+n^*) - \theta(t)\|_2$$

and using (C.1) - (C.5), (C.14), we see that

$$\sum_{t=t_1}^{t_1+t_2} x^2(t) \leq c_3 + \sum_{t=t_1-n^*}^{t_1+t_2} \frac{c_4}{1 + \|\bar{\omega}(t)\|^2} \tag{C.16}$$

for some constants $c_3 > 0$, $c_4 > 0$. In view of (C.11), (C.13), and (C.15), there exist constants $a_0 \in (0, 1)$, $c_5 > 0$, $c_6 > 0$ such that

$$\|\bar{\omega}(t+1)\| \leq (a_0 + c_5 x(t)) \|\bar{\omega}(t)\| + c_6. \tag{C.17}$$

Substituting (C.12) in (C.13), recalling the tracking error expression (7.39),

$$e(t) = k_p(\theta(t-n^*) - \theta^*)^T \omega(t-n^*) + d(t), \tag{C.18}$$

and using (C.6), we obtain

$$\bar{\omega}(t+1) = (A^* + b^*(\theta(t) - \theta^*)^T k_p F_1(t)) \bar{\omega}(t) + g_3(t) \tag{C.19}$$

for some bounded $g_3(t) \in R^{2n-1}$. Since $\theta(t) - \theta^*$, $F_1(t)$ and $g_3(t)$ are bounded, (C.19) implies that $\bar{\omega}(t)$ grows at most exponentially; that is, there exist constants $c_7 > 0$, $c_8 > 0$ such that

$$\|\bar{\omega}(t+1)\| \leq c_7 \|\bar{\omega}(t)\| + c_8, \ \forall t \geq 0. \tag{C.20}$$

Now we show the boundedness of $\bar{\omega}(t)$ by contradiction. Assume that $\bar{\omega}(t)$ grows unbounded. Then, in view of (C.20), given any $\delta_0 > 0$ and $t_2 > 0$, we can find $\delta \in (0, \delta_0]$ and $t_1 > 0$ such that

$$\|\bar{\omega}(t)\| \geq \frac{1}{\delta_0}, \ t \in \{t_1 - n^*, \ldots, t_1 - 1\} \tag{C.21}$$

$$\|\bar{\omega}(t)\| = \frac{1}{\delta}, \; t = t_1 \tag{C.22}$$

$$\|\bar{\omega}(t)\| \geq \frac{1}{\delta}, \; t \in \{t_1 + 1, \ldots, t_1 + t_2 + 1\}. \tag{C.23}$$

Therefore, in view of (C.16), (C.21) - (C.23), for $j \in \{0, \ldots, t_2\}$, the state transition function $\phi(t_1, t_1 + j)$ of

$$\|\bar{\omega}(t+1)\| = (a_0 + c_5 x(t))\|\bar{\omega}(t)\|$$

has the stability property

$$\begin{aligned}
\phi(t_1, t_1 + j) &= \prod_{t=t_1}^{t_1+j} (a_0 + c_5 x(t)) \\
&\leq (a_0 + \frac{c_5}{j+1} \sum_{t=t_1}^{t_1+j} x(t))^{j+1} \\
&\leq (a_0 + c_5 \delta_0 \sqrt{c_3(n^*+1)} + c_5 \frac{\sqrt{c_4}}{\sqrt{j+1}})^{j+1} \\
&\leq (a_0 + c_5 \delta_0 \sqrt{c_3(n^*+1)})^{j+1} e^{\frac{c_5 \sqrt{c_4}}{a_0 + c_5 \delta_0 \sqrt{c_3(n^*+1)}} \sqrt{j+1}}.
\end{aligned} \tag{C.24}$$

For any δ_0 such that

$$a_0 + c_5 \delta_0 \sqrt{c_3(n^*+1)} < 1$$

the inequalities (C.17) and (C.24) imply that

$$\|\bar{\omega}(t_1 + j + 1)\| \leq \prod_{t=t_1}^{t_1+j} (a_0 + c_5 x(t))\|\bar{\omega}(t_1)\| + c_0 \tag{C.25}$$

for some constant $c_0 > 0$ and that

$$\prod_{t=t_1}^{t_1+j} (a_0 + c_5 x(t)) < \frac{1}{2} \tag{C.26}$$

for any $j \geq j_1$ and some $j_1 \geq 0$. Hence, for $t_2 \geq j_1$ and

$$\delta_0 \in (0, \min\{\frac{1 - a_0}{c_5 \sqrt{c_3(n^*+1)}}, \frac{1}{2c_0}\}),$$

(C.25) and (C.26) imply that

$$\|\bar{\omega}(t_1 + j + 1)\| < \frac{1}{\delta}, \quad j \in \{j_1, \ldots, t_2\}$$

which contradicts (C.21) - (C.23). Hence $\bar{\omega}(t)$ is bounded, and so is $\omega(t)$. Thus, all closed-loop signals are bounded.

Proof of Theorem 7.1

For the adaptive inverse control scheme of Section 7.2, the adaptive law (7.8) ensures that $\theta_N(t)$ and $\frac{\epsilon_N(t)}{m_N(t)}$ are bounded and that (7.21) and (7.22) hold (see Lemma 7.1).

Let $\bar{\theta}_4(t) = -\theta_1^* \otimes \theta_d(t)$, $-\theta_1^* \otimes \theta_b(t)$ for dead-zone or backlash, or $\bar{\theta}_4(t) = -\theta_1^* \otimes (\theta_h(t), -1)^T$ for hysteresis, and introduce

$$\theta(t) = (\theta_N^T(t), \theta_1^{*T}, \theta_3^*, \bar{\theta}_4^T(t)),\ \theta^* = (\theta_N^{*T}, \theta_1^{*T}, \theta_3^*, \theta_4^{*T})^T.$$

Then, with $\rho(t) = k_p$, for $\xi(t)$, $\epsilon(t)$ of the forms in Section 7.3 and $\xi_N(t)$, $\epsilon_N(t)$ in (7.10), (7.9), it follows that

$$\xi(t) = \xi_N(t),\ \epsilon(t) = \epsilon_N(t). \tag{C.27}$$

Using (C.27), (7.21), and (7.22) and denoting $x_0(t) = \frac{\epsilon^2(t)}{m_N^2(t)}$ or $x_0(t) = \|\theta(t+1) - \theta(t)\|^2$, we get

$$\sum_{t=t_1}^{t_2} x_0(t) \le \bar{a}_1 + \sum_{t=t_1}^{t_2} \frac{\bar{b}_1}{m_N^2(t)} \tag{C.28}$$

for some $\bar{a}_1 > 0, \bar{b}_1 > 0$, and any $t_2 > t_1 \ge 0$, which is similar to (C.1) and (C.2). From (C.28), $x(t)$ in (C.17) satisfies

$$\sum_{t=t_1}^{t_1+t_2} x^2(t) \le \bar{c}_3 + \sum_{t=t_1-n^*}^{t_1+t_2} \frac{\bar{c}_4}{m_N^2(t)}$$

for some constants $\bar{c}_3 > 0$, $\bar{c}_4 > 0$. From here on, a contradiction argument shows that $m_N^2(t) = 1 + \zeta_N^T(t)\zeta_N(t) + \xi_N^2(t)$ is bounded, which implies that $\epsilon_N(t)$ and $\xi_N(t)$ are bounded. Then it follows that $e(t)$ and $y(t)$ are bounded and so is $u(t)$. Thus all signals in the closed-loop system are bounded.

Proof of Theorem 4.3

For the discrete-time exact inverse of Section 4.2.2, the exact nonlinearity cancellation is achieved: $u(t) = u_d(t)$, so that $\theta(t) \in l^\infty$, $\frac{\epsilon(t)}{m(t)} \in l^2 \cap l^\infty$, and $\theta(t+1) - \theta(t) \in l^2$ (see Lemma 4.2). In this case, (C.17) holds with $x(t) \in l^2$: $\sum_{t=t_1}^{t_1+t_2} x^2(t) \le \gamma_0$ for any $t_2 > t_1 \ge 0$ and some $\gamma_0 > 0$, which is sufficient for proving the boundedness of $\bar{\omega}(t)$. Hence, we conclude that all closed-loop signals are bounded and that $\epsilon(t) \in l^2$ and $\xi(t) \in l^2$ from (5.56). Finally from (5.57), it follows that $e(t) = y(t) - y_m(t) \in l^2$: $\sum_{t=0}^{\infty} e^2(t) < \infty$, which implies that $\lim_{t\to\infty} e(t) = 0$, as stated in Theorem 4.3.

Proof of Theorem 4.6

To see the properties of the discrete-time detuned inverse as stated in Theorem 4.6, we note that the inequality (C.17) also holds but with $x(t)$ satisfying

$$\sum_{t=t_1}^{t_1+t_2} x^2(t) \leq \bar{c}_3 + \sum_{t=t_1-n^*}^{t_1+t_2} \frac{\bar{c}_4}{1+\|\bar{\omega}(t)\|^2} + \bar{c}_5\mu_N \tag{C.29}$$

for any $t_2 > t_1 \geq 0$ and some $\bar{c}_i > 0$, $i = 3, 4, 5$. For $\mu_N > 0$ small, from (C.17) and (C.28), it can be shown that $\bar{\omega}(t)$ is bounded and in turn that all closed-loop signals are bounded.

Appendix D

Signal Boundedness for Output Inverses

The proofs for output nonlinearities are also given for the general case when the dead-zone and hysteresis slopes are unequal: $m_r \neq m_l$ or $m_t \neq m_b$.

Proof of Theorem 9.2

For the adaptive inverse control scheme in Section 9.3 for both $N(\cdot)$ and $G(D)$ unknown, Lemma 9.2 states that the adaptive law (9.51) with (9.52) or (9.53) guarantees that $\theta(t)$, $\frac{\epsilon(t)}{m(t)} \in l^\infty$ and that

$$\sum_{t=t_1}^{t_2} \frac{\epsilon^2(t)}{m^2(t)} \leq a_1 + \sum_{t=t_1}^{t_2} \frac{\bar{b}_1}{m^2(t)} \tag{D.1}$$

$$\sum_{t=t_1}^{t_2} ||\theta(t) - \theta(t-1)||_2^2 \leq a_2 + \sum_{t=t_1}^{t_2} \frac{\bar{b}_2}{m^2(t)} \tag{D.2}$$

for some constants $a_i > 0, \bar{b}_i > 0$, $i = 1, 2$, and all $t_2 > t_1 \geq 0$, where

$$m(t) = \sqrt{1 + \omega^T(t-n^*)\omega(t-n^*) + \xi^2(t)}.$$

To show the closed-loop signal boundedness, we introduce

$$\bar{\omega}(t) = (\omega_u^T(t), \omega_z^T(t))^T, \tag{D.3}$$

where

$$\omega_u(t) = \omega_1(t) = a_\lambda(D)[u](t),\ a_\lambda(D) = (D^{-n+1}, \ldots, D^{-1})^T \tag{D.4}$$

$$\omega_z(t) = b_\lambda(D)[z](t),\ b_\lambda(D) = (D^{-n+1}, \ldots, D^{-1}, 1)^T[z](t). \tag{D.5}$$

We then use the relationship $z(t) = \frac{Z(D)}{P(D)}[u](t)$ and (8.37) to obtain

$$\theta_1^{*T}\omega_u(t) + \theta_2^{*T}\omega_z(t) = u(t) - z(t+n^*) \tag{D.6}$$

and (D.3) - (D.6) to express

$$\begin{aligned}\bar{\omega}(t+1) &= A^*\bar{\omega}(t) + b^*z(t+n^*) \\ &= A^*\bar{\omega}(t) + b^*\frac{Z(D)}{P(D)}[u](t+n^*)\end{aligned} \tag{D.7}$$

for some constant matrix $A^* \in R^{(2n-1)\times(2n-1)}$ and some constant vector $b^* \in R^{2n-1}$. Since the first component of $\bar{\omega}(t)$ is $u(t-n+1)$, it follows from the second equality of (D.7) that

$$det(DI - A^*) = D^{n+n^*-1}Z(D),$$

which means that all eigenvalues of the closed-loop system matrix A^* are strictly inside the unit circle of the complex plane because $Z(D)$ is a stable polynomial.

Using $\theta_N(t)$, $\omega_N(t)$, $\omega_m(t)$, $\theta(t)$, and $\omega(t)$ given by (8.35), (8.36), (8.44), (9.47), and (9.49), respectively, and defining

$$\hat{e}(t) = \theta_N^T(t-n^*)(\omega_N(t) - \omega_m(t)), \tag{D.8}$$

we express the adaptive controller (9.43) as

$$u(t) = \theta^T(t)\omega(t) - \hat{e}(t+n^*). \tag{D.9}$$

Substituting (D.9) in (9.49) and using (9.48) and (D.8), we obtain

$$\begin{aligned}\hat{e}(t+n^*) &= \epsilon(t+n^*) - (\theta(t+n^*-1) - \theta(t))^T\omega(t) \\ &= \phi^T(t)\omega(t) - d(t).\end{aligned} \tag{D.10}$$

Since the adaptive update law (9.51) with parameter projection ensures that, for some $m_0 > 0$, $\frac{1}{\widehat{m_r(t)}} \geq m_0$, $\frac{1}{\widehat{m_l(t)}} \geq m_0$ for $\widehat{DI}(\cdot)$, $\frac{1}{\widehat{m(t)}} \geq m_0$ for $\widehat{BI}(\cdot)$, and $\frac{1}{\widehat{m_t(t)}} \geq m_0$, $\frac{1}{\widehat{m_b(t)}} \geq m_0$ for $\widehat{HI}(\cdot)$, we can express $z(t+n^*)$ as

$$z(t+n^*) = b_1(t)\hat{e}(t+n^*) + \bar{b}_1(t), \tag{D.11}$$

where

$$b_1(t) = \begin{cases} \frac{\widehat{m_r(t)}}{m_r}\chi_r(t+n^*) + \frac{\widehat{m_l(t)}}{m_l}(1-\chi_r(t+n^*)) & \text{for } \widehat{DI}(\cdot) \\ \frac{\widehat{m(t)}}{m} & \text{for } \widehat{BI}(\cdot) \\ \frac{\widehat{m_t(t)}}{m_t}\chi_t(t+n^*) + \frac{\widehat{m_b(t)}}{m_b}\chi_b(t+n^*) & \text{for } \widehat{HI}(\cdot) \end{cases}$$

is bounded, and $\bar{b}_1(t)$ is also a bounded signal.

Consider $\omega(t)$ and $\bar{\omega}(t)$ defined by (9.48) and (D.3). From (9.35), (9.36), and (D.6), using the boundedness of $\theta(t)$ and $d(t)$, we can also establish the relationship

$$\omega(t) = F_1(t)\bar{\omega}(t) + g_1(t) \tag{D.12}$$

$$\bar{\omega}(t) = F_2\omega(t) + g_2(t) \tag{D.13}$$

for some bounded sequences $F_1(t)$, $g_1(t)$, $g_2(t)$ and some constant matrix F_2. Since the eigenvalues of A^* are inside the unit circle of the complex plane, there exists a vector norm $\|\cdot\|$ such that the induced norm of A^* is $\|A^*\| < 1$. With this norm, based on (D.1), (D.2), (D.12), (D.13), the first equality of (D.7), and the first equality of (D.10), we also have the same forms of inequalities for $\omega(t)$ and $\bar{\omega}(t)$ as these in (C.14) - (C.17), for

$$x(t) = |\bar{\epsilon}(t+n^*)| + \|\theta(t+n^*-1) - \theta(t)\|_2.$$

Substituting the second equality of (D.10), (D.11) in the first equality of (D.7) and using (D.12), we obtain

$$\bar{\omega}(t+1) = (A^* + b^* b_1(t)\phi^T(t)F_1(t))\bar{\omega}(t) + g_3(t) \tag{D.14}$$

for some bounded sequence $g_3(t) \in R^{2n-1}$.

Since (D.14) has the same form as (C.19), the same procedure as that in (C.20) - (C.26) leads to a contradiction to the unboundedness hypothesis of $\bar{\omega}(t)$. In other words, the signal $\bar{\omega}(t)$ is bounded, and so are all other signals in the closed-loop system.

Proof of Theorem 9.1

For the adaptive inverse control scheme in Section 9.2 for $N(\cdot)$ unknown and $G(D)$ known, Lemma 9.1 states the adaptive law (9.14) or (9.18) ensures that $\theta_N(t)$ and $\frac{\epsilon_N(t)}{m_N(t)}$ are bounded and that $\frac{\epsilon_N^2(t)}{m_N^2(t)}$, $\|\theta_N(t) - \theta_N(t-1)\|_2^2$ has the smallness in the mean properties similar to those in (D.1) and (D.2).

Let $\bar{\theta}_5(t) = \theta_2^* \otimes \theta_N(t)$ and introduce

$$\theta(t) = (\theta_N^T(t), \theta_1^{*T}, \bar{\theta}_5^T(t)),\ \theta^* = (\theta_N^{*T}, \theta_1^{*T}, \theta_5^{*T})^T.$$

Then, for $\omega_{NN}(t)$, $\epsilon_N(t)$ in (9.9), (9.13) and $\omega(t)$, $\epsilon(t)$ in (9.49), (9.50) with the above new $\theta(t)$ and θ^*, it follows that $\epsilon(t) = \epsilon_N(t)$. With this result and the definition (D.7) of $\bar{\omega}(t)$, it can be shown that the same forms of (C.28) and (C.17) hold, with $x(t)$ satisfying a condition similar to (C.29). Hence, a contradiction argument shows that

$$m_N(t) = \sqrt{1 + \omega_{NN}^T(t-n^*)\omega_{NN}(t-n^*) + \xi_N^2(t)}$$

is bounded, which implies that $\epsilon_N(t)$ and $\xi_N(t)$ are bounded and so is $e(t)$; that is, $y(t)$ is bounded and so is $u(t)$. Thus all signals in the closed-loop system are bounded.

Bibliography

[1] Allin, A. and G. F. Inbar, "FNS control schemes for the upper limb," *IEEE Trans. on Biomedical Engineering*, vol. 33, no. 9, pp. 818–827, 1986.

[2] Anderson, B. O. D., R. R. Bitmead, C. R. Johanson, P. V. Kokotović, R. L. Kosut, I. M. Y. Mareels, L. Praly and B. D. Riedle, *Stability of Adaptive Systems: Passivity and Averaging Analysis*, the MIT Press, Cambridge, MA, 1986.

[3] Astrom, K. J. and B. Wittenmark, *Adaptive Control*, 2nd ed., Addison-Wesley, Reading, MA, 1995.

[4] Bai, E.-W. and S. Sastry, "Discrete-time adaptive control utilizing prior information," *IEEE Trans. on Automatic Control*, vol. 31, no. 8, pp. 779–782, 1986.

[5] Bernotas, L. A., P. E. Crago and H. J. Chizeck, "Adaptive control of electrically stimulated muscle," *IEEE Trans. on Biomedical Engineering*, vol. 34, no. 2, pp. 140–147, 1987.

[6] Bitmead, R. R., M. Gevers and V. Wertz, *Adaptive Optimal Control*, Prentice-Hall, Englewood Cliffs, NJ, 1990.

[7] Borisson, U., "Self-tuning regulators for a class of multivariable systems," *Automatica*, vol. 15, no. 2, pp. 209–217, 1979.

[8] Boyd. S and S. S. Sastry, "Necessary and sufficient conditions for parameter convergence in adaptive control," *Automatica*, vol. 22, no. 6, pp. 629–639, 1986.

[9] Campion, G. and G. Bastin, "Indirect adaptive state feedback control of linearly parametrized nonlinear systems," *International Journal of Adaptive Control and Signal Processing*, vol. 4, pp. 345–358, 1990.

[10] Canudas de Wit, C. A., *Adaptive Control for Partially Known Systems: Theory and Applications*, Elsevier Science Publishers B. V., 1988.

[11] Chalam, V. V., *Adaptive Control Systems: Techniques and Applications*, Marcel Dekker, New York, 1987.

[12] Chua, L. O. and S. C. Bass, "A generalized hysteresis model," *IEEE Trans. on Circuit Theory*, vol. 19, no. 1, pp. 36–48, 1972.

[13] Clary, J. P. and G. F. Franklin, "Self-tuning control with a priori plant knowledge," *Proc. of the 23rd IEEE Conference on Decision and Control*, pp. 369–374, Las Vegas, NV, 1984.

[14] Datta, A. and P. A. Ioannou, "Performance improvement versus robust stability in model reference adaptive control," *IEEE Trans. on Automatic Control*, vol. 39, no. 12, pp. 2370–2388, 1994.

[15] De Mathelin, M. and M. Bodson, "Frequency domain conditions for parameter convergence in multivariable recursive identification," *Automatica*, vol. 26, no. 4, pp. 757–767, 1990.

[16] Dion, J. M., L. Dugard and J. Carrillo, "Interactor and multivariable adaptive model matching," *IEEE Trans. on Automatic Control*, vol. 33, no. 4, pp. 399–401, 1988.

[17] Djaferis, T., M. Das and H. Elliott, "Reduced order adaptive pole placement for multivariable systems," *IEEE Trans. on Automatic Control*, vol. 29, no. 7, pp. 637–638, 1984.

[18] Duarte, M. A. and K. S. Narendra, "A new approach to model reference adaptive control," *International Journal of Adaptive Control and Signal Processing*, vol. 3, no. 1, pp. 53–73, 1989.

[19] Egardt, B., *Stability of Adaptive Controllers*, Springer-Verlag, Berlin, 1979.

[20] Elliott, H., R. Cristi and M. Das, "Global stability of adaptive pole placement algorithms," *IEEE Trans. on Automatic Control*, vol. 30, no. 4, pp. 348–356, 1985.

[21] Elliott, H. and W. A. Wolovich, "A parameter adaptive control structure for linear multivariable systems," *IEEE Trans. on Automatic Control*, vol. 27, no. 5, pp. 340–352, 1982.

[22] Elliott, H., W. A. Wolovich and M. Das, "Arbitrary adaptive pole placement for linear multivariable systems," *IEEE Trans. on Automatic Control*, vol. 29, no. 3, pp. 221–229, 1984.

[23] Feuer, A. and A. S. Morse, "Adaptive control of single-input, single-output linear systems," *IEEE Trans. on Automatic Control*, vol. 23, no. 4, pp. 557–569, 1978.

[24] Filippov, A. F., "Differential equations with discontinuous right hand sides," *Am. Math. Soc. Transl.*, vol. 42, pp. 199–231, 1964.

[25] Fu, L.-C. and S. Sastry, "Slow drift instability in model reference adaptive systems—an averaging analysis," *International Journal of Control*, vol. 45, no. 2, pp. 503–527, 1987.

[26] Gavel, D. T. and D. D. Siljak, "Decentralized adaptive control: structural conditions for stability," *IEEE Trans. on Automatic Control*, vol. 34, no. 4, pp. 413–425, 1989.

[27] Giri, F., M. M'Saad, L. Dugard and J. M. Dion, "Pole placement direct adaptive control for time-varying ill-modeled plants," *IEEE Trans. on Automatic Control*, vol. 35, no. 6, pp. 723–726, 1990.

[28] Goodwin, G. C. and D. Q. Mayne, "A parameter estimation perspective of continuous time model reference adaptive control," *Automatica*, vol. 23, no. 1, pp. 57–70, 1987.

[29] Goodwin, G. C., P. J. Ramadge and P. E. Caines, "Discrete time multivariable adaptive control," *IEEE Trans. on Automatic Control*, vol. 25, no. 6, pp. 449–456, 1980.

[30] Goodwin, G. C. and K. S. Sin, *Adaptive Filtering Prediction and Control*, Prentice-Hall, Englewood Cliffs, NJ, 1984.

[31] Hatwell, M. S., B. J. Oderkerk, C. A. Sacher and G. F. Inbar, "The development of a model reference adaptive controller to control the knee joint of paraplegics," *IEEE Trans. on Automatic Control*, vol. 36, no. 6, pp. 683–691, 1991.

[32] Hsu, L and R. R. Costa, "Bursting phenomena in continuous-time adaptive systems with a σ-modification," *IEEE Trans. on Automatic Control*, vol. 32, no. 1, pp. 84–86, 1987.

[33] Ioannou, P. A., "Robust adaptive controller with zero residual tracking errors," *IEEE Trans. on Automatic Control*, 31, no. 8, pp. 773–776, 1986.

[34] Ioannou, P. A. and and A. Datta (1991), "Robust adaptive controller: a unified approach," *Proc. of the IEEE*, vol. 79, pp. 1736–1768.

[35] Ioannou, P. A. and P. V. Kokotović, *Adaptive Systems with Reduced Models*, Springer-Verlag, Berlin, 1983.

[36] Ioannou, P. A. and P. V. Kokotović, "Instability analysis and improvement of robustness of adaptive control," *Automatica*, vol. 20, no. 5, pp. 583–594, 1984.

[37] Ioannou, P. A. and P. V. Kokotović, "Robust redesign of adaptive control," *IEEE Trans. on Automatic Control*, vol. 29, no. 3, pp. 202–211, 1984.

[38] Ioannou, P. A. and and J. Sun, "Theory and design of robust direct and indirect adaptive control schemes," *International Journal of Control*, vol. 47, no. 3, pp. 775–813, 1988.

[39] Ioannou, P. A. and J. Sun, *Stable and Robust Adaptive Control*, Prentice-Hall, Englewood Cliffs, NJ, 1995.

[40] Ioannou, P. A. and G. Tao, "Frequency domain conditions for strictly positive functions," *IEEE Trans. on Automatic Control*, vol. 32, no. 1, pp. 53–54, 1987.

[41] Ioannou, P. A. and G. Tao, "Dominant richness and improvement of performance of robust adaptive control," *Automatica*, vol. 25, no. 2, pp. 287–291, 1989.

[42] Ioannou, P. A. and K. Tsakalis, "A robust direct adaptive controller," *IEEE Trans. on Automatic Control*, vol. 31, no. 11, pp. 1033–1043, 1986.

[43] Ioannou, P. A. and K. Tsakalis, "Robust discrete time adaptive control," in *Adaptive and Learning Systems: Theory and Applications*, K. S. Narendra, ed., Plenum Press, New York, 1986.

[44] Isidori, A., *Nonlinear Control Systems*, 3rd ed., Springer-Verlag, Berlin, 1995.

[45] Kanellakopoulos, I., "Passive adaptive control of nonlinear systems," *International Journal of Adaptive Control and Signal Processing*, vol. 7, no. 5, pp. 339–352, 1993.

[46] Kanellakopoulos, I., P. V. Kokotović, and R. Marino, "An extended direct scheme for robust adaptive nonlinear control," *Automatica*, vol. 27, no. 2, pp. 247–255, 1991.

[47] Kanellakopoulos, I., P. V. Kokotović, and A. S. Morse, "Systematic design of adaptive controllers for feedback linearizable systems," *IEEE Trans. on Automatic Control*, vol. 36, no. 11, pp. 1241–1253, 1991.

[48] Koivo, H. N., "A multivariable self-tuning controller," *Automatica*, vol. 16, no. 3, pp. 354–366, 1980.

[49] Kokotović, P. V., ed., *Foundations of Adaptive Control*, Springer-Verlag, Berlin, 1991.

[50] Kokotović, P. V., H. K. Khalil and J. O'Reilly, *Singular Perturbations in Systems and Control: Analysis and Design*, Academic Press, New York, 1986.

[51] Kosut, R. L. and B. Friedlander, "Robust adaptive control: conditions for global stability," *IEEE Trans. on Automatic Control*, vol. 30, no. 7, pp. 610–624, 1985.

[52] Krasnoselskii, M. A. and A. V. Pokrovskii, *Systems with Hysteresis*, Springer-Verlag, Berlin, 1983.

[53] Kreisselmeier, G., "A robust indirect adaptive controller approach," *International Journal of Control*, vol. 43, no. 1, pp. 161–175, 1986.

[54] Kreisselmeier, G. and B. D. O. Anderson, "Robust model reference adaptive control," *IEEE Trans. on Automatic Control*, 31, no. 2, pp. 127–133, 1986.

[55] Kreisselmeier, G. and K. S. Narendra, "Stable model reference adaptive control in the presence of bounded disturbances," *IEEE Trans. on Automatic Control*, vol. 27, no. 6, pp. 1169–1175, 1982.

[56] Kreisselmeier, G. and G. Rietze-Augst, "Richness and excitation on an interval—with application to continuous-time adaptive control," *IEEE Trans. on Automatic Control*, vol. 35, no. 2, pp. 165–171, 1990.

[57] Krstić, M., I. Kanellakopoulos, P. V. Kokotović, "Adaptive nonlinear control without overparametrization," *Systems & Control Letters*, vol. 19, pp. 177–185, 1992.

[58] Krstić, M., I. Kanellakopoulos, P. V. Kokotović, "Nonlinear design of adaptive controllers for linear systems," *IEEE Trans. on Automatic Control*, vol. 39, no. 4, pp. 738–752, 1994.

[59] Krstić, M., I. Kanellakopoulos, P. V. Kokotović, *Nonlinear and Adaptive Control Design*, John Wiley & Sons, New York, 1995.

[60] Krstić, M. and P. V. Kokotović, "Adaptive nonlinear design with controller-identifier separation and swapping," *IEEE Trans. on Automatic Control*, vol. 40, no. 3, pp. 426–440, 1995.

[61] Landau, Y. D., *Adaptive Control: The Model Reference Approaches*, Marcel Dekker, New York, 1979.

[62] Ljung, L. and T. Soderstrom, *Theory and Practive of Recursive Identification*, the MIT Press, Cambridge, MA, 1983.

[63] Lozano, R. and X.-H. Zhao, "Adaptive ploe placement without excitation probing signals," *IEEE Trans. on Automatic Control*, vol. 39, no. 1, pp. 47–58, 1994.

[64] Mareels, I. M. Y. and R. R. Bitmead, "Nonlinear dynamics in adaptive control: chaotic and periodic stabilization," *Automatica*, vol. 22, no. 6, pp. 641–655, 1986.

[65] Marino, R. and P. Tomei, "Global adaptive output feedback control of nonlinear systems—part I: linear parametrization, " *IEEE Trans. on Automatic Control*, vol. 38, no. 1, pp. 17–32, 1993.

[66] Mayergoyz, I. D., *Mathematical Models of Hysteresis*, Springer-Verlag, Berlin, 1991.

[67] Merritt, H. E., *Hydraulic Control Systems*, John Wiley & Sons, Inc., New York, 1967.

[68] Middleton, R. H., G. C. Goodwin, D. J. Hill and D. Q. Mayne, "Design issues in adaptive control," *IEEE Trans. on Automatic Control*, vol. 33, no. 1, pp. 50–58, 1988.

[69] Miller, D. E. and E. J. Davison, "An adaptive controller which provides an arbitrary good transient and steady-state response," *IEEE Trans. on Automatic Control*, vol. 36, no. 1, pp. 68–81, 1991.

[70] Monopoli, R. V., "Model reference adaptive control with an augmented error signal," *IEEE Trans. on Automatic Control*, vol. 19, no. 5, pp. 474–484, 1974.

[71] Morse, A. S., "Global stability of parameter adaptive control systems," *IEEE Trans. on Automatic Control*, vol. 25, no. 6, pp. 433–439, 1980.

[72] Morse, A. S., "A 4(n + 1)-dimensional model reference adaptive stabilizer for any relative degree one or two, minimum phase system of dimension n or less," *Automatica*, vol. 23, no. 1, pp. 123–125, 1987.

[73] Morse, A. S., "Towards a unified theory of parameter adaptive control: tunability," *IEEE Trans. on Automatic Control*, vol. 35, no. 9, pp. 1002–1012, 1990.

[74] Morse, A. S., "Towards a unified theory of parameter adaptive control—part II: certainty equivalence and input tuning," *IEEE Trans. on Automatic Control*, vol. 37, no. 1, pp. 15–29, 1992.

[75] Morse, A. S., D. Q. Mayne and G. C. Goodwin, "Applications of hysteresis switching in parameter adaptive control," *IEEE Trans. on Automatic Control*, vol. 37, no. 9, pp. 1343–1354, 1992.

[76] Mosca, E., *Optimal, Predictive, and Adaptive Control*, Prentice-Hall, Englewood Cliffs, NJ, 1995.

[77] Naik, S. M., P. R. Kumar and B. E. Ydstie, "Robust continuous-time adaptive control by parameter projection," in *Foundations of Adaptive Control*, P. V. Kokotović, ed., pp. 153–199, Springer-Verlag, Berlin, 1991.

[78] Nam, K. and A. Arapostathis, "A model reference adaptive control scheme for pure-feedback nonlinear systems," *IEEE Trans. on Automatic Control*, vol. 33, no. 9, pp. 803–811, 1988.

[79] Narendra, K. S. and A. M. Annaswamy, "A new adaptive law for robust adaptation without persistent excitation," *IEEE Trans. on Automatic Control*, 32, no. 2, pp. 134–145, 1987.

[80] Narendra, K. S. and A. M. Annaswamy, *Stable Adaptive Systems*, Prentice-Hall, Englewood Cliffs, NJ, 1989.

[81] Narendra, K. S., Y. H. Lin and L. S. Valavani, "Stable adaptive controller design—part II: proof of stability," *IEEE Trans. on Automatic Control*, vol. 25, no. 3, pp. 440–448, 1980.

[82] Narendra, K. S. and L. S. Valavani, "Stable adaptive controller design—direct control," *IEEE Trans. on Automatic Control*, vol. 23, no. 4, pp. 570–583, 1978.

[83] Netushil, A., *Theory of Automatic Control*, Mir Publishers, Moscow, 1973.

[84] Ortega, R. and Y. Tang, "Robustness of adaptive controllers: a survey," *Automatica*, vol. 25, no. 5, pp. 651–677, 1989.

[85] Parks, P. C., "Lyapunov redesign of model reference adaptive control systems," *IEEE Trans. on Automatic Control*, vol. 11, no. 3, pp. 262–267, 1966.

[86] Peterson, B. B. and K. S. Narendra, "Bounded error adaptive control," *IEEE Trans. on Automatic Control*, vol. 27, no. 6, pp. 1161–1168, 1982.

[87] Physik Instrumente, *Products for Micropositioning*, Catalogue 108–12, Ed. E, 1990.

[88] Polycarpou, M. and P. A. Ioannou, "On the existence and uniqueness of solutions in adaptive control systems," *IEEE Trans. on Automatic Control*, vol. 38, no. 3, pp. 474–479, 1993.

[89] Pomet, J.-B. and L. Praly, "Adaptive nonlinear regulation: estimation from the Lyapunov equation, " *IEEE Trans. on Automatic Control*, vol. 37, no. 6, pp. 729–740, 1992.

[90] Praly, L., "Robust model reference adaptive controllers—part I: stability analysis," *Proc. of the 23rd IEEE Conference on Decision and Control*, pp. 1009–1014, Las Vegas, NV, 1984.

[91] Recker, D., P. V. Kokotović, D. Rhode and J. Winkelman, "Adaptive nonlinear control of systems containing a dead-zone," *Proc. of the 30th IEEE Conference on Decision and Control*, pp. 2111–2115, Brighton, England, 1991.

[92] Recker, D., *Adaptive Control of Systems Containing Piecewise Linear Nonlinearities*, Ph.D., Thesis, University of Illinois, Urbana, IL, 1993.

[93] Recker, D. and P. V. Kokotović, "Indirect adaptive nonlinear control of discrete-time systems containing a dead-zone," *Proc. of the 32nd IEEE Conference on Decision and Control*, pp. 2647–2653, San Antonio, TX, 1993.

[94] Riedle, B. D., B. Cyr and P. V. Kokotović, "Disturbance instabilities in an adaptive system," *IEEE Trans. on Automatic Control*, vol. 29, no. 9, pp. 822–824, 1984.

[95] Rohrs, C. E., L. Valavani, M. Athans and G. Stein, "Robustness of adaptive control algorithms in the presence of unmodeled dynamics," *Proc. of the 21st IEEE Conference on Decision and Control*, pp. 3–11, Orlando, FL, 1982.

[96] Sastry, S. and M. Bodson, *Adaptive Control: Stability, Convergence, and Robustness*, Prentice-Hall, Englewood Cliffs, NJ, 1989.

[97] Sastry, S. S. and A. Isidori, "Adaptive control of linearizable systems," *IEEE Trans. on Automatic Control*, vol. 34, no. 11, pp. 1123–1131, 1989.

[98] Singh, R. P. and K. S. Narendra, "Priori information in the design of multivariable adaptive controllers," *IEEE Trans. on Automatic Control*, vol. 29, no. 12, pp. 1108–1111, 1984.

[99] Sun, J., "A modified model reference adaptive control scheme for improved transient performance," *IEEE Trans. on Automatic Control*, vol. 38, no. 8, pp. 1255–1259, 1993.

[100] Tao, G. "Model reference adaptive control of multivariable plants with unknown interactor matrix," *Proc. of the 29th IEEE Conference on Decision and Control*, pp. 2780–2785, Honolulu, HI, 1990.

[101] Tao, G. "Robust adaptive control with reduced knowledge of unmodeled dynamics," *Proc. of the 29th IEEE Conference on Decision and Control*, pp. 3214–3219, Honolulu, HI, 1990.

[102] Tao, G. "New normalizing signals for robust adaptive control," *Proc. of the 1991 American Control Conference*, pp. 138–143, Boston, MA.

[103] Tao, G., "Model reference adaptive control of multivariable plants with delays," *International Journal of Control*, vol. 55, no. 2, pp. 393–414, 1992.

[104] Tao, G., "Adaptive control systems with $L^{1+\alpha}$ tracking," *Proc. of the 1994 American Control Conference*, pp. 1267–1268, Baltimore, MD.

[105] Tao, G., "Adaptive control of partially known systems," *IEEE Trans. on Automatic Control*, vol. 40, no. 10, pp. 1813–1818, 1995.

[106] Tao, G., "Adaptive control of systems with nonsmooth input and output nonlinearities," *Proc. of the 34th IEEE Conference on Decision and Control*, New Orleans, LA, 1995.

[107] Tao, G. and P. A. Ioannou, "Robust model reference adaptive control for multivariable plants," *International Journal of Adaptive Control and Signal Processing*, vol. 2, no. 3, pp. 217–248, 1988.

[108] Tao, G. and P. A. Ioannou, "Robust stability and performance improvement of multivariable adaptive control systems," *International Journal of Control*, vol. 50, no. 5, pp. 1825–1855, 1989.

[109] Tao, G. and P. A. Ioannou, "A model reference adaptive controller for multivariable plants with zero residual tracking error," *Proc. of the 28th IEEE Conference on Decision and Control*, pp. 1597–1600, Tampa, FL, 1989.

[110] Tao. G. and P. A. Ioannou, "Persistency of excitation and overparametrization in model reference adaptive control," *IEEE Trans. on Automatic Control*, vol. 35, no. 2, pp. 254–256, 1990.

[111] Tao, G. and P. A. Ioannou, "Robust adaptive control: a modified scheme," *International Journal of Control*, vol. 54, no. 1, pp. 241–256, 1991.

[112] Tao, G. and P. A. Ioannou, "Robust adaptive control of plants with unknown order and high frequency gain," *International Journal of Control*, vol. 53, no. 3, pp. 559–578, 1991.

[113] Tao, G. and P. A. Ioannou, "Stability and robustness of model reference adaptive control schemes," in *Advances in Robust Control Systems Techniques and Applications*, Academic Press, C. T. Leondes, ed., vol. 53, 1992.

[114] Tao, G. and P. A. Ioannou, "Model reference adaptive control of plants with unknown relative degree," *IEEE Trans. on Automatic Control*, vol. 38, no. 6, pp. 976–982, 1993.

[115] Tao, G. and P. V. Kokotović, "Adaptive control of systems with backlash," *Automatica*, vol. 29, no. 2, pp. 323–335, 1993.

[116] Tao, G. and P. V. Kokotović, "Adaptive control of plants with unknown dead-zones," *IEEE Trans. on Automatic Control*, vol. 39, no. 1, pp. 59–68, 1994.

[117] Tao, G. and P. V. Kokotović, "Adaptive control of systems with unknown output hystereses," *Proc. of the 1994 American Control Conference*, pp. 870–874, Baltimore, MD.

[118] Tao, G. and P. V. Kokotović, "Discrete-time adaptive control of systems with unknown nonsmooth input nonlinearities," *Proc. of the 33rd IEEE Conference on Decision and Control*, pp. 1171–1176, Lake Buena Vista, FL, 1994.

[119] Tao, G. and P. V. Kokotović, "Discrete-time adaptive control of systems with unknown dead-zones," *International Journal of Control*, vol. 61, no. 1, pp. 1–17, 1995.

[120] Tao, G. and P. V. Kokotović, "Adaptive control of plants with unknown hystereses," *IEEE Trans. on Automatic Control*, vol. 40, no. 2, pp. 200–212, 1995.

[121] Tao, G. and P. V. Kokotović, "Adaptive control of plants with unknown output backlash," *IEEE Trans. on Automatic Control*, vol. 40, no. 2, pp. 326–330, 1995.

[122] Tao, G. and P. V. Kokotović, "Adaptive control of plants with unknown output dead-zones," *Automatica*, vol. 31, no. 2, pp. 287–291, 1995.

[123] Tao, G. and P. V. Kokotović, "Continuous-time adaptive control of systems with unknown backlash," *IEEE Trans. on Automatic Control*, vol. 40, no. 6, pp. 1083–1087, 1995.

[124] Tao, G. and P. V. Kokotović, "Adaptive control of systems with unknown nonsmooth nonlinearities," *International Journal of Adaptive Control and Signal Processing*, vol. 10, 1996.

[125] Thaler, G. J. and M. P. Pastel, *Analysis and Design of Nonlinear Feedback Control Systems*, McGraw-Hill, New York, 1962.

[126] Tsakalis, K., "Robustness of model reference adaptive controllers: an input-output approach," *IEEE Trans. on Automatic Control*, vol. 37, no. 5, pp. 556–565, 1992.

[127] Tsakalis, K. and P. A. Ioannou, *Linear Time Varying Systems: Control and Application*, Prentice-Hall, Englewood Cliffs, NJ, 1993.

[128] Tsiligiannis, C. A. and S. A. Svoronos, "Multivariable self-tuning control via right interactor matrix," *IEEE Trans. on Automatic Control*, vol. 31, no. 4, pp. 987–989, 1986.

[129] Truxal, J. G., *Automatic Feedback Control System Synthesis*, McGraw-Hill, New York, 1955.

[130] Truxal, J. G., *Control Engineers' Handbook*, McGraw-Hill, New York, 1958.

[131] Tsypkin, Y. Z., *Adaptation and Learning in Automatic Systems*, Academic Press, New York, 1971.

[132] Utkin, V. I., *Sliding Modes in Control Optimization*, Springer-Verlag, Berlin, 1991.

[133] Vaughan, N. D. and J. B. Gamble, "The modeling and simulation of a proportional solenoid valve," *the ASME Winter Annual Meeting*, Dallas, TX, November 1990.

[134] Visintin, A., "Mathematical models of hysteresis," in *Topics in Nonsmooth Mechanics*, J. J. Moreau, P. D. Panagiotopoulos and G. Strang, eds., pp. 295–326, Birkhauser Verlag, Berlin, 1988.

[135] Wen, C., "A robust adaptive controller with minimal modification for discrete time-varying systems," *IEEE Trans. on Automatic Control*, vol. 39, no. 5, pp. 987–991, 1994.

[136] Wen, J. and M. Balas, "Robust adaptive control in Hilbert spaces," *Journal of Matematical Analysis and Applications*, vol. 143, pp. 1–26, 1989.

[137] Ydstie, B. E., "Transient performance and robustness of direct adaptive control," *IEEE Trans. on Automatic Control*, vol. 37, no. 8, pp. 1091–1105, 1992.

Index